Modelling Puzzles in First Order Logic

Adrian Groza

Modelling Puzzles in First Order Logic

 Springer

Adrian Groza ⓘ
Technical University of Cluj-Napoca
Cluj-Napoca, Romania

ISBN 978-3-030-62546-7   ISBN 978-3-030-62547-4   (eBook)
https://doi.org/10.1007/978-3-030-62547-4

This Springer imprint is published by the registered company Springer Nature Switzerland AG
The registered company address is: Gewerbestrasse 11, 6330 Cham, Switzerland

For Ştefan

Preface

This book provides a collection of logical puzzles modelled and solved in equational First-Order Logic (FOL).

The book can be seen both as an instrument to teach First Order Logic, and also for recreational exercises. Yet, there is more to puzzles than recreation.

The goal is to motivate students, and to increase their knowledge engineering skills by discussing a variety of puzzles and their formalisation in FOL. One aim is to convince undergraduate students that formal reasoning and its technicalities (e.g. resolution, conjunctive normal form, Skolem functions, interpretation models, and open-world assumption) are not *that* scary. There is no need for the students to avoid logic and to migrate to the tribe of machine learning. Solving puzzles in FOL can be an effective cognitive incentive to attract students to the knowledge representation and reasoning tribe.

Instructors in computer science are aware of the cognitive value of modelling puzzles and developing problem-solving skills. Logical puzzles are an efficient pedagogical instrument to engage students in learning artificial intelligence. Various authors have argued on the benefits of puzzle-based learning from various perspectives: mind Danesi (2018a), pedagogical Meyer et al. (2014), Szeredi (2003), or anthropological Danesi (2018b). My objective throughout is to show how some of the puzzles in the mathematical recreation domain can be modelled in FOL.

In most cases, it is difficult to trace the origin of a puzzle. Some puzzles formalised here are taken or adapted from books, while others are taken from what we would call today mathematical brain teasers sites: www.brainzilla.com, www. mathisfun.com, www.braingle.com, www.popularmechanics.com, www.geogebra. org. The puzzle books that I used are: "The Lady or the Tiger? and Other Logic Puzzles" by Raymond Smullyan Smullyan (2009), "The Moscow Puzzles: 359 Mathematical Recreations" by Boris Kordemsky Kordemsky (1992), "536 Puzzles and Curious Problems" by Henry Dudeney Dudeney (2016), "The Colossal Book of Short Puzzles and Problems" by Martin Gardner Gardner (2005), "Ahmes' Legacy Puzzles and the Mathematical Mind" by Marcel Danesi Danesi (2018a), "Math and Logic Puzzles for PC Enthusiasts" by J Clessa Clessa (1996), "The Riddle of Scheherazade and Other Amazing Puzzles Ancient and Modern" by

Raymond Smullyan Smullyan (1996), and "Introduction to Mathematical Logic" by Micha Walicki Walicki (2016). Instead of explaining the solution to the human agent, I focused on explaining how puzzles can be modelled in FOL.

To check each formalisation, I used two tools: Prover9 theorem prover and Mace4 model finder Mc-Cune et al. (2009). Some puzzles are easier to solve by the software agent (i.e. Prover9/Mace4), while others are easier for the human agent. One open question is: Which puzzle is easier to solve by the human agent and which by a software agent? For each puzzle, I present: (i) its formalisation in FOL, (ii) the models and proofs found by Mace4 or Prover9, and (iii) a drawing of the puzzle. Each drawing benefits from some great LaTeX packages like TikzPeople of Nils Fleischhacker Fleischhacker (2017), or the LogicPuzzle style of Josef Kleber Kleber (2013).

Puzzles are designed to test the ingenuity of humans Sutclie (2017). Thus, various books on solving puzzles provide hints and answers to the human agent. They aim to help either the learner to figure out the solution or to guide how to teach puzzle-based learning Meyer et al. (2014). Different from these books, my objective is to help the human agent to model the puzzles in FOL. Then, the software agent is responsible to find the solution. The solution found by the Prover9 or Mace4 is then explained to the human agent. The reasoning steps of the resolution mechanism represent the hints for a human agent on how to find the solution himself. One advantage is that the software agent is able to compute for the given puzzle not only for one but all solutions. After a modelling solution is found, most of the formalisations looks simple. However, before a modelling solution is clarified, a complete and correct formalisation is not always easy to design. That's because a small modelling error leads to wrong solutions or unsatisfiability. Yoda would say: "Hard to debug a FOL error is!".

The book starts with an introductory chapter for the tools of choice: Prover9 and Mace4. The more pressed reader, or those who are already familiar with Prover9 and Mace4, can skip this chapter and jump directly to Chap. 2. Then, there are 12 chapters, each containing 12 puzzles of increasing difficulty. By increasing the difficulty, the aim is to provide students with sufficient motivation to master the logical modelling patterns of FOL but also to avoid the drudgery normally associated with drill exercises. Chapter 2 starts with micro-arithmetic puzzles that aim to accustom the reader with the syntax of Prover9 and Mace4. Chapter 3 formalises puzzles on number properties. Chapter 4 introduces some more practical quizzes like route finding, ambiguous dates, or Golomb ruler. Chapter 5 provides solutions for the 12 puzzles from the "world of ladies and tigers" created by Smullyan Smullyan (2009). Chapter 6 provides formalisations of the various popular Einstein or zebra puzzles. Chapter 7 takes the reader on a trip to the island of truth, an island also discovered by Smullyan Smullyan (2011). Here, the knights are telling the truth while the knaves are telling lies. Chapter 8 collects puzzles related to interwoven family relationships, love, and marriage. Chapter 9 models grid puzzles starting with Latin or magic squares, and continuing with star battle puzzles or minesweeper. Chapter 10 takes the reader on a trip to Japan, by solving Sudoku-like puzzles: Killer Sudoku, Futoshiki, Kakurasu, Takuzo, Kakuro, or Kendoku.

Chapter 11 continues the trip in Russia, by modelling the Moscow puzzles created by Kordemsky Kordemsky (1992). Chapter 12 introduces the square-world of polyominoes, including tetrominoes, pentominoes Golomb (1996), or the broken-chess puzzle of Dudeney Dudeney (2002). Chapter 13 formalises some self-referencing brain teasers.

This book can be seen as an extension of the puzzle collection at the Thousands of Problems for Theorem Provers (TPTP) library Sutclie (2017). TPTP collection includes, among other problems, 138 puzzles. Here, I provide 144 newly formalised puzzles, with more focus on grid-based puzzles, Sudoku-like puzzles, polyominoes, or combinatorial puzzles of Kordemsky Kordemsky (1992) or Dudeney Dudeney (2016). This book can be an instrument to teach logic at high school and undergraduate levels. The puzzles here can be used as a warm-up and funny activity to start a lecture on first order logic and its friends.

Cluj-Napoca, Romania   Adrian Groza

Acknowledgments I would like to thank the book's reviewers: their specific suggestions have certainly been very helpful. I want to specifically thank to the editor of this book, Helen Desmond for her recommendations on how to make this book better. I would like to mention my collegues at the Intelligent Systemg Group at Technical University of Cluj-Napoca, Anca Mărginean and Radu Răzvan Slăvescu: Radu was the first to introduce logical puzzles in our discussions. I thank Loredana for brushing up the English in the book. I also mention my students whom I have been teasing with puzzles for the last two years.

References

Clessa, J. (1996). *Math and Logic Puzzles for PC Enthusiasts*. Courier Corporation.

Danesi, M. (2018a). *Ahmes' Legacy Puzzles and the Mathematical Mind*. Springer.

Danesi, M. (2018b). *An Anthropology of Puzzles: The Role of Puzzles in the Origins and Evolution of Mind and Culture*. Bloomsbury Publishing. typesetting logic puzzles.

Dudeney, H. E. (2002). *The Canterbury Puzzles*. Courier Corporation.

Dudeney, H. E. (2016). 536 *Puzzles and curious problems*. Courier Dover Publications.

Fleischhacker, N. (2017). *The tikzpeople package*.

Gardner, M. (2005). *The Colossal Book of Short Puzzles and Problems*. W.W. Norton.

Golomb, S.W. (1996). *Polyominoes: puzzles, patterns, problems, and packings*, volume 16. Princeton University Press.

Kleber, J. (2013). *A style for typesetting logic puzzles*.

Kordemsky, B. A. (1992). *The Moscow puzzles: 359 mathematical recreations*. Courier Corporation.

McCune, W. et al. (2009). Prover9 manual. URL: http://www.cs.unm.edu/mccune/prover9/manual/2009-11A.

Meyer, E. F., Falkner, N., Sooriamurthi, R., and Michalewicz, Z. (2014). *Guide to teaching puzzlebased learning*. Springer.

Smullyan, R. M. (1996). *The Riddle of Scheherazade and Other Amazing Puzzles Ancient and Modern*. New York: Knopf.

Smullyan, R. M. (2009). *The lady or the tiger?: and other logic puzzles*. Courier Corporation.

Smullyan, R. M. (2011). *What is the name of this book? The riddle of Dracula and other logical puzzles*. Dover Publications.

Sutcli_e, G. (2017). *The TPTP problem library and associated infrastructure: from CNF to TH0, TPTP v6. 4.0*. Journal of Automated Reasoning, pages 1–20.

Szeredi, P. (2003). *Teaching constraints through logic puzzles*. In *International Workshop on Constraint Solving and Constraint Logic Programming*, pages 196–222. Springer.

Walicki, M. (2016). *Introduction to Mathematical Logic: Extended Edition*. World Scientific Publishing Company.

Contents

Chapter 1
Getting Started with Prover9 and Mace4

Abstract In which, we introduce the tools of choice: Prover9 and Mace4. Prover9 searches for proofs; Mace4, for finite models of a given theory. The sentences they operate on could be written in propositional, first-order, or equational logic. We introduce the basic features of Prover9 and Mace4 to equip reader to run and understand the implementations in the following chapters.

> *Dubito → Cogito, Cogito → Sum*
>
> René Descartes

Prover9 searches for proofs; Mace4, for finite models of a given theory. The sentences they operate on could be written in propositional, first-order, or equational logic.

1.1 Installing Prover9 and Mace4

On Linux, you can simply type:

```
sudo apt-get update -y
sudo apt-get install prover9
```

The `update` command gets the latest package information, while the `apt-get` command installs Prover9, Mace4, and related utility tools. You can check the installation by typing in a terminal `prover9` or `mace4`.

Alternatively, you can download the current command-line version of the tool (LADR-2009-11A), which is available at https://www.cs.unm.edu/~mccune/mace4/download/LADR-2009-11A.tar.gz. Unpack it, change directory to LADR-2009-11A, then type `make all` and follow the instructions on the screen. Do not forget to add the `bin` folder to your `PATH` so that you can start Prover9 by simply typing `prover9` or `mace4` in a terminal, regardless your current directory.

© The Author(s), under exclusive license to Springer Nature Switzerland AG 2021
A. Groza, *Modelling Puzzles in First Order Logic*,
https://doi.org/10.1007/978-3-030-62547-4_1

1

1.2   Finding Proofs

First Example: "Dubito, ergo cogito, ergo sum"

Let the following statement: "I doubt, therefore I think, therefore I am" or in Latin: "dubito, ergo cogito, ergo sum". Let us assume the following hold:

1. If someone doubts, then he must think.

2. If someone thinks, then he must exist.

3. The reader of this book (i.e. you) is doubting.

Now try to prove that you exist.

   The assumptions in this tiny knowledge base could be represented formally using different types of logics. Right now, we employ the simplest one, propositional Logic, which comprises variable names for sentences plus a set of logical operators like $\wedge, \vee, \neg, \rightarrow, \leftrightarrow$. In our example, we'll use the following set of propositions:

D : The guy reading this book doubts (i.e. dubito—D)
C : The guy reading this book thinks (i.e. cogito—C)
S : The guy reading this book exists (i.e. sum—S)

   The Prover9 input file containing the implementation is presented in Listing 1.1.

Listing 1.1 Knowledge on "dubito, ergo cogito, ergo sum"

```
1    assign(max_seconds,5).
2    set(binary_resolution).
3    set(print_gen).
4
5    formulas(assumptions).
6      D.
7      D -> C.
8      C -> S.
9    end_of_list.
10
11   formulas(goals).
12     S.
13   end_of_list.
```

In order to run Prover9 with `cogito.in` as input file, type

prover9 -f cogito.in

where `-f` option is a shortcut for `file`. To redirect the output to the file `cogito.out`, you can use

prover9 -f cogito.in > cogito.out

When examining the `cogito.out` file, the text `Exiting with 1 proof` indicates that a proof has been found.

Prover9's Input File Explained

The input files for Prover9 comprise some distinct parts (McCune et al. 2009). In our example, there are three parts.

The first part contains some flags. For example, `assign(max_seconds,5)` limits the processing time at 5 s. The flag `set(binary_resolution)` allows the use of the binary resolution inference rule, while `clear (binary_resolution)` would do the opposite. The flag `set(print_gen)` instructs Prover9 to print all clauses generated while searching for the proof. You can find the description of all available flags in the online manual (McCune et al. 2009).

The second part between `formulas(assumptions)` and the corresponding `end_of_list` contains the actual knowledge base, i.e. the sentences which are assumed to be true. Sentence D means, as already mentioned, that the guy reading this book is doubting, while $D \rightarrow C$ says that if you are doubting, then you are thinking.

The third part between `formulas(goals)` and the corresponding `end_of_list` states the goal the prover must demonstrate, namely S in our case.

Note that every line ends with a period. Comment lines start with the "%" symbol which goes to the end of the line. You can use "-" for negation, "&" for logical conjunction, "|" for logical disjunction, "->" for implication, and "<->" for equivalence. By default, Prover9 uses names starting with u, v, w, x, y, z to represent variables in clauses; thus, for the time being, please avoid using sentence names starting with them.

Prover9's Output File Explained

The output file (e.g. `cogito.out`) starts with some information on the running process, a copy of the input, and a list of the formulas that are not in clausal form. These clauses are translated by Prover9 into the Clausal Normal Form:

```
4 D.     [assumption].
5 -D | C.   [clausify(1)].
6 -C | S.   [clausify(2)].
7 -S.    [deny(3)].
```

Prover9 uses the equivalence between $P \rightarrow Q$ and $\neg P \vee Q$, both in line 5 and line 6. You should also notice in line 7 that the goal has been added in the negated form $\neg S$. This is called *reductio ad absurdum*: if one accepts both the axioms and the denied conclusion, then a contradiction will be deduced. Prover9 actually searches for such a contradiction by repeatedly applying inference rules over the existing clauses till the empty clause is obtained. The SEARCH section lists all clauses inferred during search for the proof. The PROOF section shows only those clauses which are actually used in the proof. For our example, the PROOF section is

```
1 D -> C # label(non_clause).    [assumption].
2 C -> S # label(non_clause).    [assumption].
3 S # label(non_clause) # label(goal).    [goal].
4 D.    [assumption].
5 -D | C.    [clausify(1)].
6 -C | S.    [clausify(2)].
7 -S.    [deny(3)].
8 C.    [resolve(5,a,4,a)].
9 -C.    [resolve(7,a,6,b)].
10 $F.    [resolve(9,a,8,a)].
```

Note that each sentence that is given (lines 1–4) or inferred (lines 5–10) is numbered.

Lines 1–4 are the original sentences. Prover9 has correctly identified: (i) the first two sentences as non-clausal (since they contain an implication), (ii) the goal S (from the goals list in the input file), and (iii) the fact D (named here [assumption]).

Lines 5–10 are inferred, as follows: Lines 5 and 6 are the clausal forms of 1 and 2, while 7 is the goal denial. Line 8 shows how the binary resolution is applied over clauses 5 and 4. Propositional resolution inference rules says that if we have $P \vee Q$ and $\neg Q \vee R$, we can infer $P \vee R$. The first literal in clause 5 ($-D$, hence the index a) and the first literal in clause 4 (i.e. D) are "resolved" and the inferred clause is C. Similarly, line 9 shows how the binary resolution is applied over clauses 7 and 6. The first literal in clause 7 ($-S$, hence the index a) and the second literal in clause 6 (D, hence the index b) are "resolved" and the inferred clause is $\neg C$. Line 10 shows the inferred contradiction based on clause 8 (i.e. C) and clause 9 (i.e. $\neg C$). The contradiction is signalled by the F symbol.

You can obtain a graphical representation of the proof with

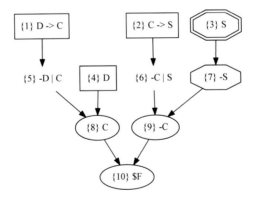

Fig. 1.1 Proving by resolution in propositional logic

```
prover9 -f cogito.in | prooftrans xml renumber |gvizify|
                dot -Tpdf > cogito.pdf
```

Here, I used the `prooftrans` tool that extracts proofs from Prover9 output files and transforms them in various formats. In our example, the proofs are converted to XML and the clauses are renumbered so that some numbers are skipped because they correspond to sentences (either given or inferred) that are not part of the proof. Both the `prooftrans` command and the `gvizify` command are installed together with Prover9. Given the XML proof, `gvizify` prepares the proof to be graphically deployed by the `dot` command (which you need to install yourself). The resulted `cogito.pdf` appears in Fig. 1.1. The goal to be proved is depicted with a double hexagon (i.e. clause {3}). The goal is negated, resulting in the clause {7}. The given clauses {1}, {2}, and {4} are depicted with rectangular boxes. The inferred clauses {5}–{10} are depicted with circled boxes.

Proofs in First-Order Logic

Our example can be formalised in FOL with

$$d(I) \tag{1.1}$$

$$\forall x \ d(x) \rightarrow c(x) \tag{1.2}$$

$$\forall x \ c(x) \rightarrow s(x) \tag{1.3}$$

with the goal to prove $s(I)$. The implementation in Prover9 appears in Listing 1.2. This `cogitoFOL.in` file can be run as usual: `prover9 -f cogitoFOL.in`.

Listing 1.2 Knowledge on "dubito, ergo cogito, ergo sum" in FOL

```
1   formulas(assumptions).
2     D(I).
3     all x (D(x) -> C(x)).
4     all x (C(x) -> S(x)).
5   end_of_list.
6
7   formulas(goals).
8     S(I).
9   end_of_list.
```

1.3 Finding Models

Models in Propositional Logic

Henceforth, we will use for our experiments a knowledge base named KB, comprising one or more clauses connected by a logical AND. We will focus on testing whether KB has at least one model (a set of assignments of truth values to its propositional variables which make it true) and whether a given goal sentence g could be proven based on the KB content.

We say a proposition is *satisfiable* if it has at least one model; otherwise, it is *unsatisfiable*. Let the sentence "At the movie I will eat popcorn or candy". The sentence can be formalised with the inclusive disjunction *popcorn* ∨ *candy*. The sentence is satisfiable in three cases:

popcorn	*candy*	*popcorn* ∨ *candy*
0	0	0
0	1	1
1	0	1
1	0	1

To find at least one or all satisfiable models, the Mace4 model finder can be used.

Mace4's Input File Explained

The syntax of Mace4 (McCune 2003) is similar with the Prover9's syntax. Some flags are specific to Mace4, while others to Prover9. The implementation in Listing 1.3, starts with the flag section. The first flag (i.e. *max_models*) is set to −1, stating that we are interested in all possible models of the given formula. The second flag (i.e. *domain_size*) is set to 2, meaning that each variable in the formula can take only two values: 0 and 1. The formula in lines 4–6 is checked by Mace4 for satisfiability. Different from Prover9, when searching for models, we do not specify the goal to be proved (a.k.a. the conjecture).

Listing 1.3 Finding models with Mace4 in propositional logic

```
1   assign(max_models,−1).
2   assign(domain_size,2).
3
4   formulas(assumptions).
5     popcorn | candy.
6   end_of_list.
```

In order to run Mace4 with the file or.in in Listing 1.3, you can type in a terminal:

```
mace4 -f or.in
```

Mace4 outputs information on how the given clauses were processed and the computed models. In order to keep only the models, the `interpformat` command can be used:

```
mace4 -f or.in | interpformat standard
```

The `interpformat` comes with the Prover9 distribution, and its `standard` option is used to display only the models.

Mace4's Output File Explained

For our example, Mace4 identifies three models. In the model `number=1`, the unary relation *candy* has value 0, while the unary relation *popcorn* has value 1. In the model `number=2`, the relation *candy* has value 1, while the relation *popcorn* has value 0. In the model `number=3`, the relation *candy* has value 1, while the relation *popcorn* has value 1.

```
interpretation( 2, [number = 1,seconds = 0], [
    relation(candy, [0]),
    relation(popcorn, [1])]).
interpretation( 2, [number = 2,seconds = 0], [
    relation(candy, [1]),
    relation(popcorn, [0])]).
interpretation( 2, [number = 3,seconds = 0], [
    relation(candy, [1]),
    relation(popcorn, [1])]).
```

Note that the "|" operator is the inclusive or. If you want to model the exclusive or between *candy* and *popcorn*, one option is to use the equivalence:

$$A \; xor \; B \equiv \neg(A \rightarrow B) \vee \neg(B \rightarrow A) \qquad (1.4)$$

Given the implementation in Listing 1.4 (i.e. file xor.in), Mace4 computes only two models:

Listing 1.4 Modelling XOR in Mace4

```
1  assign(max_models,−1).
2  assign(domain_size,2).
3
4  formulas(assumptions).
5    −(popcorn −> candy) | −(candy −> popcorn).
6  end_of_list.
```

```
interpretation( 2, [number = 1,seconds = 0], [
    relation(candy, [0]),
    relation(popcorn, [1])]).
interpretation( 2, [number = 2,seconds = 0], [
    relation(candy, [1]),
    relation(popcorn, [0])]).
```

Using the Arithmetic Module

Here is a puzzle for you: *Identify the distinct numbers satisfying the equation TWO + TWO = FOUR.* Three observations follow:

1. The letters represent digits from 0 to 9. Hence, the domain size needs to be extended to handle these values

2. You need to state that each letter is distinct from the others

3. You need the operations + and *.

All these three requirements are covered by Mace4 (see Listing 1.5). In line 2, we specify that each variable in the domain can take values from the interval [0 ... 9]. The *list(distinct)* block is used to specify lists of variables that are distinct. Here, we have two lists. In the first one, we state that all variables occurring in the equation TWO + TWO = FOUR are distinct. In the second list, we state that the variables *T* and *F* should be distinct from zero since they occur on the first position of number TWO, respectively FOUR. By importing the arithmetic module (line 3), we have access to the operators $+$, $-$, $*$, div, mod. These operations are used in line 11 to formalise the given equation.

Listing 1.5 Using the arithmetic module

```
1   assign(max_models,-1).
2   assign(domain_size,10).
3   set(arithmetic).
4
5   list(distinct).
6     [T,W,O,F,U,R].
7     [T,F,0].
8   end_of_list.
9
10  formulas(assumptions).
11    T * 200 + W * 20 + 2 * O = 1000 * F + 100 * O + 10 * U + R.
12  end_of_list.
```

By calling Mace4 with the implementation in Listing 1.5 (i.e. file two.in), we obtain seven models. Each model represents a solution for the equation.

```
mace4 -f two.in | interpformat standard
```

Models	F	O	R	T	U	W
1	1	4	8	7	6	3
2	1	5	0	7	3	6
3	1	6	2	8	7	3
4	1	6	2	8	9	4
5	1	7	4	8	3	6
6	1	8	6	9	5	2
7	1	8	6	9	7	3

These are the basic features of Prover9 and Mace4 to equip reader to run and understand the implementations in the following chapters. Other features of the tools will be gradually introduced only when an example requires more advanced features. Recall that the focus is on solving logical puzzles not on mastering a tool of choice, hence let's start solving some puzzles.

Source Code Available

The source code of each puzzle from this book is available at http://users.utcluj.ro/~agroza/puzzles/maloga

References

McCune, W. (2003). Mace4 reference manual and guide. arXiv:cs/0310055.

McCune, W., et al. (2009). Prover9 manual. http://www.cs.unm.edu/mccune/prover9/manual/2009-11A.

Chapter 2
Micro-arithmetic Puzzles

Abstract In which, we introduce micro-arithmetic puzzles that aim to accustom the reader with the syntax of Prover9 and Mace4. First, the reader learns how to solve logic equations in $n \times n$ grids. Second, puzzles on arithmetic reasoning are modelled in Equational First-Order Logic.

$$\forall x \; refuses(x, do(arithmetic)) \rightarrow doomed(x, talk(nonsense))$$

John McCarthy

The chapter starts with four logic equations. You have to figure out the variables values by solving the logic equations using $n \times n$ grid. Each variable represents unique integers ranging from 1 to the number of variables n. A 4×4 grid with two logic equations is depicted below:

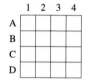 $$D \leq 2 \rightarrow A + C \leq 4$$
$$A + B \leq 6 \rightarrow C < 1$$

In a logic equation, arithmetical operations interleave with logical operations. In the example above, arithmetic inequations interleave with the logical implication. The unique solution ($A = 3$, $B = 4$, $C = 1$, $D = 3$) is depicted in the right part of the picture. Even if there are only two formulas, we can find the values for all four variables, since there are two additional constraints: the variables are distinct and within the [1...4] interval. Logical equations presented here are taken from Brainzilla[1] and Math is Fun.[2] Solving such logic equations is a good exercise to get used to the syntax and to warm-up your brain.

The chapter continues with six puzzles on arithmetic reasoning taken from Dudeney's collection of "536 Puzzles and curious problems" (Dudeney 2016). Some

---

[1] www.brainzilla.com.

[2] www.mathsisfun.com.

of these puzzles are used to introduce the concept of isomorphic models. These puzzles are not challenging for the human mind, but they are useful to practise formalisation of a complete task into FOL. Recall that the focus is not on human reasoning, but on modelling, for the logical agent to solve it.

Puzzle 1. Logic equation

In this 4×4 logic equation, you have to find unique integer values for the variables A, B, C, D—ranging from 1 to 4—to make the following statements true: $2 \times C = B$ and $4 \times C = D$. (puzzle taken from Brainzilla—www.brainzilla.com)

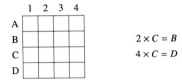

Listing 2.1 Importing the arithmetic module to solve equations

```
1   set(arithmetic).
2   %set(trace).
3   assign(max_models,−1).
4   assign(domain_size,5).
5
6   list(distinct).
7     [0,A,B,C,D].
8   end_of_list.
9
10  formulas(assumptions).
11    2 ∗ C = B.
12    4 ∗ C = D.
13  end_of_list.
```

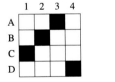

Solution

The human agent prefers to propagate constraints instead of backtracking. For instance, the human agent might start with the equation $4 * C = D$ to quickly deduce that $C = 1$ is the only value that keeps D in domain [1...4].

Differently, the software agent (i.e. Mace4) interleaves backtracking with reasoning. To solve the equations, we need to import the arithmetic module (line 1 Listing 2.1). With $assign(max\_models, -1)$, we state that we are interested in all solutions. With domain size set to 5 (line 4), each variable can take values from the interval [0...4]. Note that the puzzle asks for distinct solutions in the domain [1...4]. This is formalised in line 7 where the list $[0, A, B, C, D]$ contains only distinct elements.

Mace4 finds a single solution: $A = 3, B = 2, C = 1, D = 4$. With $set(trace)$, the interested user can see how backtracking and constraint propagation search for solution. First, the system assigns an order for processing the variables: (e.g. B, C, D, A). Second, it updates a set of current constraints; the initial set of constraints C_0 is

$$2 * C = B, B + D > 4 * C, 4 * C = D,$$
$$0 \neq A, 0 \neq B, 0 \neq C, 0 \neq D, A \neq B, A \neq C, A \neq D, B \neq C, B \neq D, C \neq D \quad (2.1)$$

Third, it starts to assign values, then propagate constraints, and backtracks when needed:

Selection				Active constraints	Result
B=	C=	D=	A=	C_0	assigning value
B=1	C=	D=	A=	$C_1 = C_0 \cup \{C \neq 1, D \neq 1, A \neq 1\}$	assigning value
B=1	C=2	D=	A=	$C_2 = C_1 \cup \{D \neq 2, A \neq 2\}$	fail
B=1	C=3	D=	A=	$C_2 = C_1 \cup \{D \neq 3, A \neq 3\}$	fail
B=1	C=4	D=	A=	$C_2 = C_1 \cup \{D \neq 4, A \neq 4\}$	fail
B=2	C=	D=	A=	$C_1 = C_0 \cup \{C \neq 2, D \neq 2, A \neq 2\}$	assigning value
B=2	C=1	D=	A=	$C_2 = C_1 \cup \{D \neq 1, A \neq 1\}$	assigning value
B=2	C=1	D=3	A=	$C_2 = C_1 \cup \{A \neq 3, A = 4\}$	fail
B=2	C=1	D=4	A=	$C_2 = C_1 \cup \{A \neq 4, A = 3\}$	finding model
B=3	C=	D=	A=	$C_1 = C_0 \cup \{D \neq 1, A \neq 1\}$	fail
B=4	C=	D=	A=	$C_1 = C_0 \cup \{D \neq 4, A \neq 2\}$	assigning values
B=4	C=2	D=	A=	$C_1 = C_0 \cup \{D \neq 5, A \neq 5\}$	fail
...	...	...	...	...	...

As we asked for all models, the search continues until all the solutions are found for the given domain.

Puzzle 2. Logic equation

In this 4×4 logic equation, you have to find unique integer values for the variables
A, B, C, D—ranging from 1 to 4—to make all the following statements true: (puzzle
taken from Brainzilla—www.brainzilla.com)

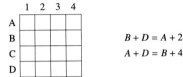

$$B + D = A + 2$$
$$A + D = B + 4$$

Listing 2.2 Formalising the 4×4 logical equation

```
1    set(arithmetic).
2    assign(max_models,−1).
3    assign(domain_size,5).
4
5    list(distinct).
6      [0,A,B,C,D].
7    end_of_list.
8
9    formulas(assumptions).
10     B + D = A + 2.
11     A + D = B + 4.
12   end_of_list.
```

Solution

A human agent may subtract the two equations to get $A - B = B - A + 2$. That is
$B + 1 = A$. It means that $B \neq 4$, $A \neq 1$, and they are consecutive. Replacing A with
$B + 1$ in the given equations you get $D = 3$. As A and B are consecutive, the only
possible solution is $B = 1$ and $A = 2$. This leaves C with value 4.

The formalisation needs to specify: (1) the domain size of each variable (line 3); (2) the
values of variables are distinct and different from zero (line 6); and (3) the two equations (lines 10–11).

Mace4 also finds this unique model: $A = 2$, $B = 1$, $C = 4$, $D = 3$. Note in the illustration below that each column contains only one value (i.e. each variable is distinct). The
value for C is also found, even if the variable does not appear in the equations.

$$B + D = A + 2$$
$$A + D = B + 4$$

Puzzle 3. Logic equation 5×5

In this 5×5 logic equation, you have to find unique integer values for the variables A, B, C, D, E—ranging from 1 to 5—to make all statements true: (puzzle taken from Brainzilla—www.brainzilla.com)

$$C = A + E$$
$$E = B + 2$$
$$B * E + 3 * E \neq B \rightarrow A * A + D > E$$

Listing 2.3 Interleaving arithmetic with logical operators

```
1    set(arithmetic).
2    assign(max_models,-1).
3    assign(domain_size,6).
4
5    list(distinct).
6       [0,A,B,C,D,E].
7    end_of_list.
8
9    formulas(assumptions).
10      C = A + E.
11      E = B + 2.
12      B * E + 3 * E != B -> (A * A + D > E).
13   end_of_list.
```

Solution

Note here that the last constraint is a logical implication. As we will see throughout the book, Mace4 interleaves arithmetic with logical reasoning. To have the domain of the variables from 1 to 5, we set the domain size to 6 and add 0 to the list of distinct variables. With the arithmetic constraints in lines 10–11 and the logical constraint in line 12, Mace4 finds the unique solution $A = 2, B = 1, C = 5, D = 4, E = 3$ (Fig. 2.1).

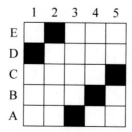

Fig. 2.1 The unique solution found by Mace4 for the 5×5 logical equation

Puzzle 4. Logic equation 9×9

Find distinct values for the variables ranging from 1 to 9 to make all statements true.
(puzzle taken from Brainzilla—www.brainzilla.com)

$$4 * F = E \qquad\qquad (2.2)$$
$$I = G * H \qquad\qquad (2.3)$$
$$C + E \neq 12 \to G = D + E + 2 \qquad\qquad (2.4)$$
$$B = A + H \qquad\qquad (2.5)$$
$$A + G \neq C \qquad\qquad (2.6)$$

Listing 2.4 Five equations and one logical implication for nine variables

```
1   set(arithmetic).
2   assign(max_models,-1).
3   assign(domain_size,10).
4
5   list(distinct).
6     [0,A,B,C,D,E,F,G,H,I].
7   end_of_list.
8
9   formulas(assumptions).
10    4 * F = E.
11    I = G * H.
12    C + E != 12 -> G = D + E + 2.
13    B = A + H.
14    A + G != C.
15  end_of_list.
```

Solution

The puzzle contains four equations and one implication for identifying nine variables. The human agent would try to avoid backtracking. For instance, one can start with Eq. (2.3). As G and H are distinct, there are only four cases where $I \leq 9$:

$$I = G \times H : \qquad 6 = 2 \times 3, \qquad 6 = 3 \times 2, \qquad 8 = 2 \times 4, \qquad 8 = 4 \times 2$$

Hence, $I \in \{6, 8\}$ and $G, H \in \{2, 3, 4\}$. In equation $4 \times F = E$ (2.2), if $F > 2$, then there is no solution. Hence $F \in \{1, 2\} \wedge E \in \{4, 8\}$.
Differently, for the software agent, we have to specify: (i) the domain of the variables which is from [0...9] (line 3); (ii) the fact that values are distinct among them and also 0 is not in the domain (line 6); and (iii) the given constraints (lines 10–14). Mace4 has no problem to find the model:

$$A = 7, \; B = 9, \; C = 8, \; D = 5, \; E = 4, \; F = 1, \; G = 3, \; H = 2, \; I = 6$$

We learn also from Mace4 that this is the only solution.

Puzzle 5. Three squares

Find three distinct squares such that one is the arithmetic mean of the other two. (puzzle taken from www.mathisfun.com)

Listing 2.5 Avoiding isomorphic models by introducing an order relation among variables

```
1   set(arithmetic).
2   assign(max_models, -1).
3
4   list(distinct).
5      [a,b,c].
6   end_of_list.
7
8   formulas(assumptions).
9      2 * b = a + c.
10     a = d * d.
11     b = e * e.
12     c = f * f.
13
14     % a < c.   % remove isomorphic models
15  end_of_list.
```

Solution

Let the three squares a, b, c. We assume distinct values (line 5). Let b the arithmetic mean of a and c (line 9). The values are constrained to be squares in lines 10–12.
As we don't know how large the solution would be, we let Mace4 iterate over various domain sizes. We did this by not specifying the domain size. In this case, Mace4 starts searching with $domain\_size = 2$ and incrementally increases the domain size. It finds two models, both for the $domain\_size = 50$:

Model	a	b	c	d	e	f
1	1	25	49	1	5	7
2	49	25	1	7	5	1

Note that these two models represent the same solution from the human agent perspective; $b = 25$ and $a, c = \{1, 49\}$. To avoid such *isomorphic models*, one can set an order relation between a and c (the commented line 14). Based on $a < c$ and $2 * b = a + c$ the system deduces $a < b < c$. It also follows that $d < e < f$. With this additional constraint, Mace4 outputs a single solution: $a = 1$, $b = 25$, and $c = 49$.
Since we asked for all the models (line 2 in Listing 2.5, Mace4 continues to search for models by increasing the domain size (e.g. 51, 52, ...)

Puzzle 6. Pocket money

"When I got to the station this morning," said Harold Tompkins, at his club, "I found I was short of cash. I spent just one-half of what I had on my railway ticket, and then bought a nickel's worth of candy. When I got to the terminus I spent half of what I had left and ten cents for a newspaper. Then I spent half of the remainder on a bus ticket and gave fifteen cents to that old beggar outside the club. Consequently, I arrive here with this single nickel. How much did I start out with?" (puzzle 17 from Dudeney 2016)

Listing 2.6 Reasoning backwards

```
1   set(arithmetic).
2
3   formulas(assumptions).
4     t3 = 5.                    % after the third transaction
5     t2 = 2 * (t3 + 15).        % after the second transaction
6     t1 = 2 * (t2 + 10).        % after the first transaction
7     t0 = 2 * (t1 + 5).         % initial sum
8   end_of_list.
```

Solution

We do not know initially the domain size of the variables, so we let Mace4 figure by itself this size. Let t_i the amount at time i. After performing three transactions $t_0 \rightsquigarrow t_1 \rightsquigarrow t_2$, Harold has a nickel (i.e. five cents). That is, now, at time $t_3 = 5$. Before the last transaction, we have: $t_2 = 2 * (t_3 + 15)$. Before that, the amount of money should have been: $t_1 = 2 * (t_2 + 10)$. The initial value should have been: $t_0 = 2 * (t_1 + 5)$. Mace4 finds the unique solution: $t_3 = 5$, $t_1 = 40$, $t_2 = 100$, $t_0 = 210$:

$$210 \rightsquigarrow 100 \rightsquigarrow 40 \rightsquigarrow 5$$

This solution was found for the domain size 211, which is the smallest domain (i.e. [0...200]) to which all the values belong. Since, we didn't ask for all models, Mace4 stops after finding the first model.

Puzzle 7. Dividing the legacy

A man left 100 dollars to be divided between his two sons Alfred and Benjamin. If one-third of Alfred's legacy was taken from one-fourth of Benjamin's, the remainder would be 11 dollars. What was the amount of each legacy? (puzzle 15 from Dudeney 2016)

Listing 2.7 Dividing the legacy

```
1  set(arithmetic).
2  assign(domain_size,101).
3  assign(max_models,−1).
4
5  formulas(dividing_legacy).
6    (3 * b) + ((−4) * a) = 11 * 12.
7    a + b = 100.
8  end_of_list.
```

Solution

As we know the maximum value (i.e. 100 dollars), we can set the domain size to 101. Let b be Benjamin's legacy and a be Alfreds'. Without using the div operator, the equation can be formalised as in Listing 2.7. Mace4 finds a unique solution:

Model	a	b
1	24	76

Puzzle 8. Boys and girls

Nine boys and three girls agreed to share equally their pocket money. Every boy gave an equal sum to every girl, and every girl gave another equal sum to every boy. Every child then possessed exactly the same amount. What was the smallest possible amount that each child then possessed? (puzzle 26 from Dudeney 2016)

Listing 2.8 Nine boys and three girls exchange money

```
1  assign(max_models,1).
2  set(arithmetic).
3
4  formulas(boys_and_girls).
5    b != g & b != 0 & g != 0.      % we assume different init values
6
7    sb > 0.                         % sum given by the boys
8    sg > 0.                         % sum given by the girls
9
10   b > 3 * sb.                     % boys have more money than given
11   g > 9 * sg.                     % girls have more money than given
12
13   bafter = b + −3 * sb + 3 * sg.  % boys sum after transaction
14   gafter = g + −9 * sg + 9 * sb.  % girls sum after transaction
15   bafter = gafter.
16 end_of_list.
```

Solution

By default, Mace4 starts with domain size 2 and increases it until a solution is found. Hence, the first solution will correspond to the smallest domain. We ask only for one model (line 1 in Listing 2.8) and we do not specify the domain size.

We assume positive and different initial values for boys (b) and girls (g): $b \neq g \wedge b \neq 0 \wedge g \neq 0$. We can also assume that the sums given by boys (sb) and girls (sg) are both larger than 0: $sb > 0 \wedge sg > 0$. Moreover, the boys and girls should have more money than the given sum: $b > 3 \times sb$, respectively, $g > 9 * sg$. Each boy gives the value sb to each one of the three girls ($3 \times sb$ in total) and receives sg from each one of the girls ($3 \times sg$ in total). The final sum of each boy ($bafter$) is

$$bafter = b - 3 * sb + 3 * sg \tag{2.7}$$

Each girl gives the value sg to each one of the nine boys ($9 \times sg$ in total) and receives sb from each one of the boys ($9 \times sg$ in total). The final sum of each girl ($gafter$) is

$$gafter = g - 9 * sg + 9 * sb \tag{2.8}$$

Mace4 finds the first solution for the domain size 20: $b = 7$, $bafter = 10$, $g = 19$, $gafter = 10$, $sb = 1$, $sg = 2$. Here, each boy has at the beginning 7 dollars and gives 1 dollar to each girl. Each girl has 19 dollars and gives 2 dollars to every boy. In the end, the boys and girls will have 10 dollars each.

Puzzle 9. Robinson's age

"How old are you, Robinson?" asked Colonel Crackham one morning. "Well, I forget exactly," was the reply; "but my brother is two years older than me; my sister is four years older than him; my mother was twenty when I was born; and I was told yesterday that the average age of the four of us is thirty-nine years." What was Robinson's age? (puzzle 45 from Dudeney 2016)

Listing 2.9 Finding Robinson's age

```
1   set(arithmetic).
2
3   list(distinct).
4     [brother,sister,robinson,mother,0].
5   end_of_list.
6
7   formulas(assumptions).
8     brother = robinson + 2.
9     sister  = brother  + 4.
10    mother  = robinson + 20.
11    mother  = brother  + 18.
12    mother  = sister   + 14.
13
14    robinson + brother + sister + mother = 39 * 4.
15  end_of_list.
```

> **Solution**
>
> From the puzzle, we deduce that the members of the family have different ages (line 4 in Listing 2.9). The puzzle is straightforward formalised with six equations (lines 8–14). Mace4 finds a single model:
>
> | Model | robinson | brother | sister | mother |
> |-------|----------|---------|--------|--------|
> | 1 | 32 | 34 | 38 | 52 |

> **Puzzle 10. Five cards**
>
> I have five cards bearing the figures 1, 3, 5, 7, and 9. How can I arrange them in a row so that the number formed by the first pair multiplied by the number formed by the last pair, with the central number substracted, will produce a number composed of repetitions of one figure? Thus, in the example I have shown, 31 multiplied by 79 and 5 subtracted will produce 2444, which would have been all right if that 2 had happened to be another 4. Of course, there must be two solutions, for the pairs are clearly interchangeable (puzzle 103 from Dudeney 2016).

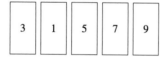

Listing 2.10 Five cards

```
1   assign(domain_size,10).
2   assign(max_models,−1).
3   set(arithmetic).
4
5   list(distinct).
6     [0,c1,2,c2,4,c3,6,c4,8,c5].
7   end_of_list.
8
9   formulas(5cards).
10    (c1*10 + c2) * (c5*10 + c4) + −c3 = d1*1000 + d2*100 + d3*10 + d4.
11    d1 = d2 & d2 = d3 & d3 = d4.
12  end_of_list.
```

Solution

The list of distinct values in line 5 specifies that the five cards c_i have values from the set $\{1, 3, 5, 7, 9\}$. The number formed with the first two cards c_1 and c_2 multiplied with the number composed by the last cards c_4 and c_5, if decreased by the middle card c_3 gives a number with four equal digits d_i:

$$(c_1 * 10 + c_2) * (c_5 * 10 + c_4) - c_3 = d_1 * 1000 + d_2 * 100 + d_3 * 10 + d_4 \quad (2.9)$$

where $d_1 = d_2 = d_3 = d_4$. Mace4 finds two models: 39,157 or 57,139. In either case, the product of the two pairs 39 and 57, minus 1, results in 2,222.

Puzzle 11. A square family

A man had nine children, as follows:

1. all born at regular intervals, and
2. the sum of the squares of their ages was equal to the square of his own.

What was the age of each? Every age was an exact number of years. (puzzle 41 from Dudeney 2016)

Listing 2.11 A square family

```
1   assign(max_models,1).
2   set(arithmetic).
3
4   formulas(square_family).
5     a1 < a2 & a2 < a3 & a3 < a4 & a4 < a5 & a5 < a6 & a6 < a7 &
6     a7 < a8 & a8 < a9 & a9 < father.
7
8     a1 * a1 + a2 * a2 + a3 * a3 + a4 * a4 + a5 * a5 +
9     a6 * a6 + a7 * a7 + a8 * a8 + a9 * a9 = father * father.
10
11    a2 + -a1 = a3 + -a2.
12    a3 + -a2 = a4 + -a3.
13    a4 + -a3 = a5 + -a4.
14    a5 + -a4 = a6 + -a5.
15    a6 + -a5 = a7 + -a6.
16    a7 + -a6 = a8 + -a7.
17    a8 + -a7 = a9 + -a8.
18  end_of_list.
```

Let a_i, $i \in [1...9]$ the ages of the nine children. To avoid repetitive models, we set an order relation between children (lines 5–6 in Listing 2.11), where a_1 was the first child). Of course, the father is the oldest one. The first constraint (lines 11–17) is

$$\forall i, \; i \in [1...8] \; a_{i+1} - a_i = a_{i+2} - a_{i+1} \qquad (2.10)$$

Another modelling would be $\forall i \in [1...8] \; a_{i+1} - a_i = k$, which is solved by Mace4 slower compared to Eq. (2.10). The second constraint is handled by the equation:

$$\sum_{i=1}^{9} a_i^2 = father^2 \qquad (2.11)$$

This equation is modelled in lines 8–9 in Listing 2.11. The unique solution is found for domain size 49. The ages of the nine children were, respectively 2, 5, 8, 11, 14, 17, 20, 23, 26, and the age of the father was 48.

Puzzle 12. Family ages

A man and his wife had three children, John, Ben, and Mary, and the difference between their parents' ages was the same as between John and Ben and between Ben and Mary. The ages of John and Ben, multiplied together, equalled the age of the father, and the ages of Ben and Mary multiplied together equalled the age of the mother. The combined ages of the family amounted to 90 years. What was the age of each person? (puzzle 37 from Dudeney 2016)

Listing 2.12 A family with three children

```
1   assign(domain_size,91).
2   assign(max_models,−1).
3   set(arithmetic).
4
5   formulas(assumptions).
6     Father != 0.  Mother != 0.  John != 0.  Ben != 0.  Mary != 0.
7
8     abs(Father + −Mother) = abs(John + −Ben).
9     abs(Father + −Mother) = abs(Ben + −Mary).
10    John * Ben = Father.
11    Mary * Ben = Mother.
12    Father + Mother + John + Ben + Mary = 90.
13  end_of_list.
```

> **Solution**
>
> We know that the combined ages are 90, so we can limit the domain size to this value. However, this is a large approximation as we do not expect, for instance, that the father is 89 and the mother is 1. We assume that all family members have ages different from 0 (line 6 in Listing 2.12). We also don't know who is older: father or mother. We model this with the *abs* function (lines 8–9). The unique solution is found for domain size 37. *Mather* and *Father* are both 36 and the three children are 6.

Reference

Dudeney, H. E. (2016). *536 puzzles and curious problems*. New York: Courier Dover Publications.

Chapter 3
Strange Numbers

Abstract In which, we formalise puzzles related to unusual numbers. The reader practises here constraint formalisation, in the style of constraints programming. Mace4 is quick at computing all the solutions, not only a single one, as most puzzles require from the human agent. Actually, in most cases, the human agent will be surprised by the huge number of interpretation models of a FOL theory. We will see throughout the book that, in many cases, the main task of the programmer will be to gradually reduce the number of models until a solution is found.

$$\exists x \, \exists y \, (x \neq y \land (\forall z \, (z = x \lor z = y)) \land Master(x) \land Apprentice(y))$$

Yoda

This chapter formalises puzzles related to unusual numbers, puzzles taken from Dudeney (2016), Kordemsky (1992), Clessa (1996), or from the "Math is fun" website.[1] Again, Mace4 is responsible to find solutions. The reader practises here constraint formalisation, in the style of constraints programming. Since the constraints are straightforward, the puzzles here are easier to be solved by the software agent (i.e. Mace4) than by the human agent. Moreover, Mace4 is quick at computing all the solutions, not only a single one, as most puzzles require from the human agent. Actually, in most cases, the human agent will be surprised by the huge number of interpretation models of a FOL theory. The following example tries to surprise you on this line.

Here is a puzzle for you: Assume the text "Romeo and Julieta are in love" is automatically translated by a machine learning-based translator (this is a tool from the tribe of AI enthusiasts that speak statistics instead of logics) into "Romeo is in love" and "Julieta is in love". These two sentences are formalised in FOL with:

$$A_1: \quad \exists x, \, love(romeo, x) \qquad A_2: \quad \exists x, \, love(julieta, x)$$

How many interpretation models of this first-order logic theory are there?

---

[1] www.mathisfun.com.

A. Groza, *Modelling Puzzles in First Order Logic*,
https://doi.org/10.1007/978-3-030-62547-4_3

The human agent has a single interpretation model: there are two individuals, Romeo (r) and Julieta (j), that love each other (see Fig. 3.1). Here, r stands for *romeo* and j for *julieta*.

Instead, the logical agent has a huge number of interpretation models, even for tiny theories like this. Mace4 is able to compute interpretation models of first-order logic theories with finite domains. Here, we close the domain to four individuals (i.e. we assume that there are only four objects in the domain). Axiom A_1 says that *romeo* loves an individual x, while axiom A_2 says that *julieta* loves an individual x, that is not necessarily the same, since each variable has its own existential quantifier. Given the implementation in Listing 3.1, Mace4 solves the puzzle for you: there are 278,528 models.

Fig. 3.1 The unique interpretation model of the human agent

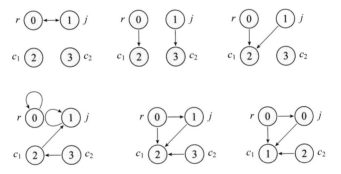

Fig. 3.2 Sample of interpretation models for the software agent

Listing 3.1 Finding models with domain closed to 4 individuals

```
1   assign(max_models, -1). assign(domain_size, 4).
2   formulas(assumptions).
3       exists x love(romeo,x).
4       exists x love(julieta,x).
5   end_of_list.
```

Since this number of models might be unexpected (recall that the domain was restricted to four individuals only), let us analyse the generated models. A sample of these models is illustrated in Fig. 3.2. Here, c_1 and c_2 are the Skolem constants generated for the existential quantifiers in A_1 and A_2, respectively. Since the domain is closed to four individuals, we work only with the set of integers $\{0, 1, 2, 3\}$. The first model (first row, left) is consistent with the human interpretation: *romeo* and *julieta* do love each other. Note also that all four individuals are distinct: $r \rightarrow 0$, $j \rightarrow 1$,

$c_1 \to 2$, $c_2 \to 3$. In the second model (first row, centre), *romeo* loves an individual c_1, while *julieta* loves a distinct individual c_2. In the third model (first row, right), both *romeo* and *julieta* love the same individual c_1. Moreover, no one said that the *love* relation is not reflexive. One such model is the fourth one (second row, left), where both *romeo* and *julieta* love each other. The variety of the models is also increased by different possible love relations involving c_1 and c_2. For instance, in the fourth model, c_1 loves *julieta* and c_2 loves c_1. Similarly, no one said that someone can love only one person at the same time. Therefore, the fifth model (second row, centre) is possible. Here, *romeo* loves both *julieta* and c_1. The largest influence is given by the fact that the logical agent can interpret that some individuals are not distinct. In the sixth model (second row, right), *romeo* and *julieta* are interpreted by Mace4 as the same individual (i.e. 0), which is referred by two distinct names ($r \to 0$, $j \to 0$).

The above analysis explains how the logical agent has indeed 278,528 interpretation models for the simple sentence *Romeo and Julieta are in love*. Moreover, all these models are equally plausible for the software agent. Given this gap (278,528 models vs. one model of the human agent), the natural question is *How the two agents would understand each other?*

We will see throughout the book, that in many cases, the main task of the programmer will be to gradually reduce the number of models until a solution is found. Until then, let us find some models for a dozen of unusual numbers.

Puzzle 13. Adding their cubes

The numbers 407 and 370 have this peculiarity, that they exactly equal the sum of the cubes of their digits. Thus, the cube of 4 is 64, the cube of 0 is 0, and the cube of 7 is 343. Add together 64, 0, and 343, and you get 407. Again, the cube of 3 (i.e. 27), added to the cube of 7 (i.e. 343), is 370. Can you find a number not containing a 0 that will work in the same way? Of course, we bar the absurd case of 1. (puzzle 143 from Dudeney (2016))

Listing 3.2 Formalising *adding their cubes* puzzle

```
1    assign(domain_size,1000).
2    assign(max_models,−1).
3    set(arithmetic).
4
5    formulas(addingcubes).
6        X = D2 * 100 + D1 * 10 + D0.
7        X = D2 * D2 * D2 + D1 * D1 * D1 + D0 * D0 * D0.
8        D0 != 0 & D1 != 0 & D2 != 0.
9        D0 < 10 & D1 < 10 & D2 < 10.
10   end_of_list.
```

Solution

We assume numbers of three digits, hence, we limit the domain size to 1000. The number X is formalised with

$$X = 100 * D_2 + 10 * D_1 + D_0 \qquad (3.1)$$

The values D_2, D_1, D_0 represent digits, hence there are less then 10 (line 9 in Listing 3.2). We learn that the number X does not contain any 0 (line 8). The constraint of the puzzle says also that $X = D_2^3 + D_1^3 + D_0^3$ (line 7). Mace4 finds two solutions:

Model	D_2	D_1	D_0	X
1	3	7	1	371
2	1	5	3	153

Puzzle 14. Multiplication

How many solutions are for: A B C D E F * 3 = B C D E F A? (puzzle from Math is fun - www.mathisfun.com)

Listing 3.3 Modelling the equation A B C D E F * 3 = B C D E F A

```
1    set(arithmetic).
2    assign(domain_size,10).
3    assign(max_models, −1).
4
5    list(distinct).
6    [A, B, C, D, E, F].
7    end_of_list.
8
9    formulas(assumptions).
10   A != 0 & B != 0.
11   C1 < 3 & C2 < 4 & C3 < 4 & C4 < 4 & C5 < 4.
12
13   3 * F       = A + C1 * 10.
14   3 * E + C1 = F + C2 * 10.
15   3 * D + C2 = E + C3 * 10.
16   3 * C + C3 = D + C4 * 10.
17   3 * B + C4 = C + C5 * 10.
18   3 * A + C5 = B.
19   end_of_list.
```

The six letters A, B, C, D, E, and F take values from 0 to 9. Hence, the domain size is set to 10. We need to specify that each digit is distinct (line 6). The first digits of a number cannot be 0: $A \neq 0$, $B \neq 0$.

One option is to model the problem with transport. For instance, $3 * F = A + C_1 * 10$, where C_1 is the transport. By multiplying any figure by 3, the largest value is 27. Because we can also have transport from previous multiplication, variables C_i cannot be larger than 3. Only for C_1, there is no previous transport, hence C_1 is maximum 2. Hence,

$$C_1 < 3 \wedge C_2 < 4 \wedge C_3 < 4 \wedge C_4 < 4 \wedge C_5 < 4. \tag{3.2}$$

Note that the constraints in Eq. 3.2 help Mace4 during search: the system will not consider values from the entire domain (i.e. [0..9] for these variables). However, Mace4 is able to find the solution without these optional constraints (lines 10–11 in Listing 3.3). Mace4 outputs two models:

Model	A	B	C	D	E	F	C_1	C_2	C_3	C_4	C_5
1	1	4	2	8	5	7	2	1	2	0	1
2	2	8	5	7	1	4	1	0	2	1	2

The two solutions are: 142,857 * 3 = 428,571 and 285,714 * 3 = 857,142.

The driver of the taxi was wanting in civility, so Mr. Wilkins asked him for his number. "You want my number, do you?" said the driver. "Well, work it out for yourself. If you divide my number by 2, 3, 4, 5, or 6 you will find there is always 1 over; but if you divide it by 11 there is no remainder. What's more, there is no other driver with a lower number who can say the same." What was the fellow's number? (puzzle 223 from Dudeney (2016))

Listing 3.4 The arithmetical cabby

```
1   set(arithmetic).
2
3   list(distinct).
4     [r2,r3,r4,r5,r6,r11].
5   end_of_list.
6
7   formulas(cabby).
8     number != 0.
9     number = 2  * r2  + 1.
10    number = 3  * r3  + 1.
11    number = 4  * r4  + 1.
12    number = 5  * r5  + 1.
13    number = 6  * r6  + 1.
14    number = 11 * r11.
15  end_of_list.
```

Solution

We are searching for the smallest *number* satisfying the constraints. Let r_2, r_3, r_4, r_5, r_6, and r_{11} the remainders. Hence, we don't specify the domain size, and we ask only for one model. In this way, Mace4 will find the first value for the variable *number*. As we have divisions with different values, also the remainders r_2, r_3, r_4, r_5, r_6, and r_{11} are distinct (line 4 in Listing 3.4). Recall that this hint only helps Mace4 during search, but the system is able to find the solutions without this constraint. The first solution is found for the domain size 122:

Model	*number*	r_2	r_3	r_4	r_5	r_6	r_{11}
1	121	60	40	30	24	20	11

The next number having the same properties would be 1,331.

Puzzle 16. The ten barrels

A merchant had ten barrels of sugar, which he placed in the form of a pyramid, as shown below in the left picture. Every barrel bore a different number, except one, which was not marked. It will be seen that he had accidentally arranged them so that the numbers in the three sides added up alike, that is, to 16. Can you rearrange them so that the three sides shall sum up to the smallest number possible? Of course the central barrel (which happens to be 7 in the illustration) does not come into the count. (puzzle 462 from Dudeney (2016))

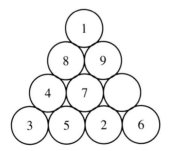
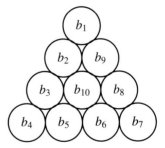

Listing 3.5 Ten barrels in a pyramid

```
1    assign(start_size,14).
2    assign(max_models,1).
3    set(arithmetic).
4
5    list(distinct).
6      [b1, b2, b3, b4,b5, b6, b7, b8, b9, b10].
7    end_of_list.
8
9    formulas(tenbarrels).
10     p > 12.
11     3 * p = 45 + b1 + b4 + b7 + −b10.
12
13     b1 + b2 + b3 + b4 = p.
14     b4 + b5 + b6 + b7 = p.
15     b7 + b8 + b9 + b1 = p.
16
17     b1 < 10.   b2 < 10.   b3 < 10.   b4 < 10.   b5 < 10.
18     b6 < 10.   b7 < 10.   b8 < 10.   b9 < 10.   b10 < 10.
19
20     % p = 13.
21     % b10 != 9.
22   end_of_list.
```

Solution

Let b_i the notation for each barrel with $i \in [1..10]$ (right part in the illustration). Let p the sum on each size. In the left illustration, $p = 16$ on each side. We aim to find models with $p < 16$. The main constraint of the puzzle is:

$$b_1 + b_2 + b_3 + b_4 = p \tag{3.3}$$
$$b_4 + b_5 + b_6 + b_7 = p \tag{3.4}$$
$$b_7 + b_8 + b_9 + b_1 = p \tag{3.5}$$

Here, b_1, b_4, and b_7 are the vertices of the triangle, hence they appear twice, and b_{10} is the barrel in the centre. If we add Eqs. (3.3), (3.4), and (3.5), we have

$$\sum_{i=1}^{9} b_i + b_1 + b_4 + b_7 = 3 \times p \tag{3.6}$$

$$45 + b_1 + b_4 + b_7 - b_{10} = 3 \times p \tag{3.7}$$

Here, 45 is the sum of the first 10 numbers. The value of p is minimum for $b_1 + b_4 + b_7 = 3$ and $b_{10} = 9$. Minimum $p = 39/3 = 13$. Hence, we start searching directly with the domain size 14 (line 1 in Listing 3.5). Note that this helps Mace4 to search less, but the system is able to figure for itself that there is no solution for $p < 13$. The first model is found for the domain size 14. The computed sum is $p = 13$. Two such models are:

Model	b_1	b_2	b_3	b_4	b_5	b_6	b_7	b_8	b_9	b_{10}	p
1	2	8	3	0	5	7	1	6	4	9	13
2	0	3	8	2	4	6	1	5	7	9	13

These two models are also depicted in Fig. 3.3.
For this value $p = 13$, Mace4 finds 96 models. You can simply add the constraint
$p = 13$ in Listing 3.5. Note that, in all these 96 models, b_10 has the maximum possible
value, i.e. $b_{10} = 9$. One can test it by adding the constraint $b_{10} \neq 9$ and by observing
that there are no models for domain size 14.

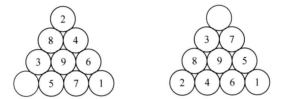

Fig. 3.3 Two of the 96 solutions found by Mace4 for the minimum sum on each side $p = 13$

Puzzle 17. Find the triangle

The sides and height of a triangle are four consecutive whole numbers. What is the
area of the triangle? (puzzle 230 from Dudeney (2016))

Listing 3.6 Find the triangle when sides and height are four consecutive numbers

```
1   assign(max_models,1).
2   set(arithmetic).
3
4   list(distinct).
5     [0, a, b, c, h].
6   end_of_list.
7
8   formulas(assumptions).
9     a + b > c & a + c > b & b + c > a.
10    b = a + 1 & c = b + 1.
11    a = h + 1 | h = c + 1.
12
13    2 * S = a * h | 2 * S = b * h | 2 * S = c * h.
14
15    2 * p = a + b + c.
16    S * S = p * (p + -a) * (p + -b) * (p + -c).
17  end_of_list.
```

Solution

Let a, b, c the sides of the triangle ($a < b < c$) and h its height. We can formalise a first constraint from the triangle inequality:

$$(a + b > c) \wedge (a + c > b) \wedge (b + c > a) \tag{3.8}$$

Values a, b, c are consecutive: $b = a + 1$ and $c = b + 1$. The height h can be smaller than a (i.e. $h + 1 = a$) or larger than c (i.e. $c + 1 = h$). There is no explicit stating on to which side the height h belongs. Hence, there are three possible formulas for the area S:

$$(2 * S = a * h) \vee (2 * S = b * h) \vee (2 * S = c * h) \tag{3.9}$$

The Heron formula is also helpful here:

$$S * S = p * (p - a) * (p - b) * (p - c) \tag{3.10}$$

where the semi-perimeter p is given by $2 * p = a + b + c$ (line 15 in Listing 3.6). Mace4 finds a solution for the domain size 85:

Model	a	b	c	h	p	S
1	13	14	15	12	21	84

Mace4 fails to find another case in which the height complies with the given conditions.

Puzzle 18. A digital difficulty

Arrange the ten digits, 1 2 3 4 5 6 7 8 9 0, in such order that they shall form a number that may be divided by every number from 2 to 18 without in any case a remainder. As an example, if I arrange them thus, 1,274,953,680, this number can be divided by 2, 3, 4, 5, and so on up to 16, without any remainder, but it breaks down at 17. (puzzle 118 from Dudeney (2016))

Listing 3.7 Divisibility of a large number

```
1   assign(domain_size,10).
2   assign(max_models,-1).
3   set(arithmetic).
4
5   list(distinct).
6      [c0,c1,c2,c3,c4,c5,c6,c7,c8,c9].
7      [c0,0].                    % the first digit cannot be zero
8   end_of_list.
```

```
10 │ formulas(assumptions).
11 │    c9 = 0.                                        % divisibility  with  10
12 │
13 │    % divisibility  with  11
14 │    abs(c0 + c2 + c4 + c6 + c8 + −c1 + −c3 + −c5 + −c7 + −c9) = 11.
15 │
16 │    % we use  (A + B) % C = (A % C + B % C) % C
17 │
18 │    (c0 >= 2) &
19 │       ((((c9 + c8*10 + c7*100 + c6*1000 + c5*10000 + c4*100000+c3*1000000 +
20 │             c2*10000000 + c1*100000000+1000000000) mod 1113840) +
21 │    ((1000000000 mod 1113840)*(c0 + − 1)) ) mod 1113840 = 0).
22 │ end_of_list.
```

Solution

The challenge here is to help Mace4 to find the solution quickly. We model the number N with 10 digits $c_0, c_1, c_2, c_3, c_4, c_5, c_6, c_7, c_8, c_9$ as a sum:

$$N = c_9 + c_8 * 10 + c_7 * 10^2 + c_6 * 10^3 + c_5 * 10^4 + c_4 * 10^5 + c_3 * 10^6 + c_2 * 10^7 + c_1 * 10^8 + c_0 * 10^9 \tag{3.11}$$

Here, the digits c_i are distinct (line 6). In this way, we keep the domain size to the small value 10. Moreover, the first digit of the number cannot be 0 ($c_0 \neq 0$ in line 7). The number is divisible by 10 if the last digit is 0: $c_9 = 0$. Hence, it is also divisible by 2 and 5. The number is divisible by 9 if the sum of its digits is also divisible by 9. In our case, the sum of numbers up to 9 is 9*10/2=45. Hence, the N is divisible by 9 and without the need to state explicit constraints. Given that the last digit is 0, N is also divisible by 18. As the number is already divisible by 3 and 5, then it is also divisible by 15. The number is divisible by 11 if the difference between digits on the even positions and digits on the odd positions gives a number that is divisible by 11:

$$|c_0 - c_1 + c_2 - c_3 + c_4 - c_5 + c_6 - c_7 + c_8 - c_9| = k \tag{3.12}$$

where $k \bmod 11 = 0$. Note in Listing 3.7 that k is directly instantiated with 11 (line 14). To be divisible by 7, 13, and 17, the number should be divisible by $7 \times 13 \times 17 = 1,547$. The number should be divisible by 16, therefore by 8 and 4.

$$(N \bmod 1547 = 0) \wedge (N \bmod 16 = 0) \tag{3.13}$$

To use only one statement, one can multiply $16 \times 1,547 = 23,312$.

$$A \bmod C = 0 \Leftrightarrow A - [A/C] * C = 0 \tag{3.14}$$

There might be a problem on systems where the numbers are limited to 32 bits. Because the number can be larger than 32 bits (e.g. $9,876,543,210 > 2^{32}$), one might want to split the sum N in two values:

$$A = c_9 + c_8 * 10 + c_7 * 10^2 + c_6 * 10^3 + c_5 * 10^4 + c_4 * 10^5 + c_3 * 10^6 + c_2 * 10^7 + c_1 * 10^8$$

$$B = c_0 * 10^9$$

A contains c_9 to c_1 and B for c_0. Now we can use the modulo property:

$$(A + B) \bmod C = ((A \bmod C) + (B \bmod C)) \bmod C \qquad (3.15)$$

Mace4 computes four solutions:

Model	c_0	c_1	c_2	c_3	c_4	c_5	c_6	c_7	c_8	c_9
1	2	4	3	8	1	9	5	7	6	0
2	3	7	8	5	9	4	2	1	6	0
3	4	7	5	3	8	6	9	1	2	0
4	4	8	7	6	3	9	1	5	2	0

Puzzle 19. The archery match

On a target on which the scoring was 40 for the bullseye, and 39, 24, 23, 17, and 16, respectively, for the rings from the centre outwards, as shown in the illustration, three players had a match with six arrows each. The result was: Miss Dora Talbot, 120 points; Reggie Watson, 110 points; Mrs. Finch, 100 points. Every arrow scored, and the bullseye was only once hit. Can you, from these facts, determine the exact six hits made by each competitor? (puzzle 467 from Dudeney (2016))

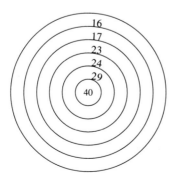

Listing 3.8 Modelling the archery match

```
1    set(arithmetic).
2    assign(domain_size,6).
3    assign(max_models,−1).
4
5    formulas(assumptions).
6      p10 + p20 + p30 = 1.               % center is hit once
7      p10 + p11 + p12 + p13 + p14 + p15 = 6.   % there are six attempts
8      p20 + p21 + p22 + p23 + p24 + p25 = 6.
9      p30 + p31 + p32 + p33 + p34 + p35 = 6.
10
11     p10*40 + p11*39 + p12*24 + p13*23 + p14*17 + p15*16 = 120. % Dora    120p
12     p20*40 + p21*39 + p22*24 + p23*23 + p24*17 + p25*16 = 110. % Reggie 110p
13     p30*40 + p31*39 + p32*24 + p33*23 + p34*17 + p35*16 = 100. % Finch  100p
14   end_of_list.
```

Solution

First, we try to reduce the domain size as much as possible. For this, let the following associations for each target: (0,40), (1,39), (2,24), (3,23), (4,17), and (5,16). There are also three players: Dora Talbot (1), Reggie Watson (2), and Mrs Finch (3). With the above notations, a domain size of 6 (i.e. [0..5]) suffices to model both the three players and the six targets.

Let p_{xy} represent the number of hits on target y by player x. We know that the centre is hit only once: $p_{10} + p_{20} + p_{30} = 1$. Also, we learn that each player had six attempts:

$$p_{10} + p_{11} + p_{12} + p_{13} + p_{14} + p_{15} = 6 \qquad (3.16)$$

$$p_{20} + p_{21} + p_{22} + p_{23} + p_{24} + p_{25} = 6 \qquad (3.17)$$

$$p_{30} + p_{31} + p_{32} + p_{33} + p_{34} + p_{35} = 6 \qquad (3.18)$$

We need to state the number of points won by each player: 120 points for the first player, 110 for the second, and 100 for the third player:

$$p_{10} * 40 + p_{11} * 39 + p_{12} * 24 + p_{13} * 23 + p_{14} * 17 + p_{15} * 16 = 120 \quad (3.19)$$

$$p_{20} * 40 + p_{21} * 39 + p_{22} * 24 + p_{23} * 23 + p_{24} * 17 + p_{25} * 16 = 110 \quad (3.20)$$

$$p_{30} * 40 + p_{31} * 39 + p_{32} * 24 + p_{33} * 23 + p_{34} * 17 + p_{35} * 16 = 100 \quad (3.21)$$

Mace4 finds a single solution:

p_{10}	p_{11}	p_{12}	p_{13}	p_{14}	p_{15}	p_{20}	p_{21}	p_{22}	p_{23}	p_{24}	p_{25}	p_{30}	p_{31}	p_{32}	p_{33}	p_{34}	p_{35}
1	0	0	0	0	5	0	0	0	2	0	4	0	0	0	0	4	2

The solution corresponds to the following interpretation (see Fig. 3.4)

Miss Dora Talbot (1) hits 40 points one time and 16 points five times: $40 * 1 + 16 * 5 = 120$
Reggie Watson (2) hits 23 points two times and 16 points four times: $23 * 2 + 16 * 4 = 110$
Mrs. Finch (3) hits 17 points four times and 16 points two times: $17 * 4 + 16 * 2 = 100$

Note that we used propositional variables p_{xy} instead of predicates $p(x, y)$. Here, the player x is in [1..3], while the hit value y in [0..5]. Since the domain size is set to 6 for all variables (including x), if one prefers the formalisation with predicates, then he/she needs to add constraints for x in [4..5] to avoid isomorphic models.

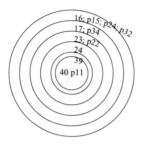

Fig. 3.4 A solution for the archery match

Puzzle 20. Target practice

Colonel Crackham paid a visit one afternoon by invitation to the Slocomb-on-Sea Tox-ophilite Club, where he picked up the following little poser. Three men in a competition had each six shots at a target and the result is shown in our illustration, where they all hit the target every time. The bullseye scores 50, the next ring 25, the next 20, the next 10, the next 5, the next 3, the next 2, and the outside ring scores only 1. It will be seen that the hits on the target are one bullseye, two 25's, three 20's, three 10's, three 1's, and two hits in every other ring. Now the three men tied with an equal score. Next morning the Colonel asked his family to show the exact scoring of each man. Will it take the reader many minutes to find the correct answer? (puzzle 468 from Dudeney (2016))

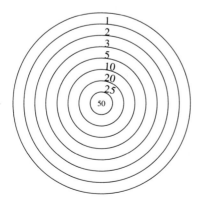

Listing 3.9 Target practice

```
1    assign(domain_size,8).
2    assign(max_models,-1).
3    set(arithmetic).
4
5    formulas(assumptions).
6      all x all y (p(x,y) < 4).
7      all x (x >= 3 -> all y (p(x,y) = 0)).
8
9      target(y) = p(0,y) + p(1,y) + p(2,y).
10     target(0) = 1.   target(1) = 2. target(2) = 3.  target(3) = 3.
11     target(4) = 2.   target(5) = 2. target(6) = 2.  target(7) = 3.
12
13     hits(x) = p(x,0) + p(x,1) + p(x,2) + p(x,3) +
14                p(x,4) + p(x,5) + p(x,6) + p(x,7).
15     hits(0) = 6 & hits(1) = 6 & hits(2) = 6.
16
17     p(0,0)*50 + p(0,1)*25 + p(0,2)*20 + p(0,3)*10 +
18     p(0,4)*5  + p(0,5)*3  + p(0,6)*2  + p(0,7)     =
19     p(1,0)*50 + p(1,1)*25 + p(1,2)*20 + p(1,3)*10 +
20     p(1,4)*5  + p(1,5)*3  + p(1,6)*2  + p(1,7).
21
22     p(0,0)*50 + p(0,1)*25 + p(0,2)*20 + p(0,3)*10 +
23     p(0,4)*5  + p(0,5)*3  + p(0,6)*2  + p(0,7)     =
24     p(2,0)*50 + p(2,1)*25 + p(2,2)*20 + p(2,3)*10 +
25     p(2,4)*5  + p(2,5)*3  + p(2,6)*2  + p(2,7).
26   end_of_list.
```

Solution

On the target, there are eight values [50,25,20,10,5,3,2,1], hence we need the domain size of at least 8 values. Let function $p(x, y)$ count the number of hits for the player x in point y. Hits on the target are: [(50,1),(25,2),(20,3),(10,3),(5,2),(3,2),(2,2),(1,3)]. Hence, for any player x, the number of hits in a circle y is less then four: $\forall x \, \forall y \, p(x, y) < 4$ (line 6 in Listing 3.9). There are only three players (i.e. 0, 1, 2), hence, in line 7, we do not care for larger values:

$$\forall x \, (x \geq 3 \rightarrow (\forall y \, p(x, y) = 0)) \tag{3.22}$$

Let the function $target(y)$ representing how many times the value y was hit. There are eight circles on the archery, hence the eight values for the variable y. To keep the domain size small, we let y in [0..7], where zone 0 corresponds to 50 points, zone 1–25 points, and so on. With eight zones in the domain [0..7] and three players in [0..2], we can set the domain size at 8. Without these translations, we had to set the domain size to 51, to accommodate the largest value 50.

The puzzle explicitly specifies the number of hits in each area:

y	0	1	2	3	4	5	6	7
$target(y)$	1	2	3	3	2	2	2	3

This information appears in lines 10–11. The function $target(y)$ equals the number of hits obtained by each player:

$$target(y) = p(0, y) + p(1, y) + p(2, y) \tag{3.23}$$

Let function $hits(x)$ counting the number of hits for the payer x. We know that each player hits the target six times:

$$(hits(0) = 6) \wedge (hits(1) = 6) \wedge (hits(0) = 6) \tag{3.24}$$

The function $hits(x)$ is defined as the sum of hits performed by player x in all eight zones on the archery:

$$hits(x) = \sum_{y=0}^{7} p(x, y)$$

Next, we state that the three men tied with an equal score. We state that the points obtained by player 0 are the same with those of player 1 (lines 17–20) and the points of player 0 are the same with player 2 (lines 22–25).
One solution computed by Mace4 is:

Player	50	25	20	10	5	3	2	1
0	1	0	0	1	1	1	1	1
1	0	1	2	0	0	1	1	1
2	0	1	1	2	1	0	0	1

This solution represents the first 21 values from the function $p(x, y)$ Since we can permute each player, there are 3!=6 isomorphic solutions.

Puzzle 21. The nine barrels

In how many different ways may these nine barrels be arranged in three tiers of three so that no barrel shall have a smaller number than its own below it or to the right of it? The first correct arrangement that will occur to you is 1 2 3 at the top, then 4 5 6 in the second row, and 7 8 9 at the bottom, and my sketch gives a second arrangement. How many are there altogether? (puzzle 138 from Dudeney (2016)) What about the case when there are $n \times n$ barrels. How many models are there?

Listing 3.10 Nine barrels: smaller values are not allowed below or on the right

```
1    set(arithmetic).
2    assign(domain_size,10).
3    assign(max_models,-1).
4
5    list(distinct).
6     [0,b1,b2,b3,b4,b5,b6,b7,b8,b9].
7    end_of_list.
8
9    formulas(nine_barrels).
10    r(x,y)  <-> x > y.
11    on(x,y) <-> x < y.
12
13    on(b1,b4) & on(b2,b5) & on(b3,b6) & on(b4,b7) & on(b5,b8) & on(b6,b9).
14    r(b2,b1)  & r(b3,b2)  & r(b5,b4)  & r(b6,b5)  & r(b8,b7)  & r(b9,b8).
15   end_of_list.
```

Solution

Let the nine barrels noted with b_i, $i \in [1..9]$ as in the illustration above. We define two predicates: $right(x, y)$ and $on(x, y)$. The only two conditions are: that the number is smaller than the one on its right (line 10) and below it (line 11):

$$right(x, y) \Leftrightarrow x > y \tag{3.25}$$

$$on(x, y) \Leftrightarrow x < y \tag{3.26}$$

We also need to give the relative positions of each barrel (lines 13–14). Mace4 finds 42 models. Six of them are:

Model	b_1	b_2	b_3	b_4	b_5	b_6	b_7	b_8	b_9
1	1	2	3	4	5	6	7	8	9
2	1	2	4	3	5	7	6	8	9
3	1	2	5	3	4	8	6	7	9
4	1	3	4	2	5	8	6	7	9
5	1	3	5	2	4	8	6	7	9
6	1	3	6	2	5	7	4	8	9

Figure 3.5 illustrates model 6 (left), model 3 (centre), and model 2 (right). Note that the first position is fixed to the minimum value ($b_1 = 1$) and the last position is fixed

to the maximum value ($b_9 = 9$). Also, the value 2 should be either below 1 or right to 1 ($b_4 = 2 \vee b_2 = 2$). In both cases, the value 3 has two positions. For the first case when $b_4 = 2$, the value 3 is beneath the 2 or right to 1. If the value 3 is to the right of 1 there are five arrangements with 4 under the 2, five with 5 under the 2, four with 6 under the 2, and two with 7 under 2. That is 21 arrangements. Similarly, for the second case when $b_2 = 2$, there are 21 isomorphic arrangements.

Fig. 3.5 Three solutions among 42 found by Mace4

Puzzle 22. Upside-down year

An upside-down year is a year which reads the same when the "year number" is turned upside down. That is when the digits which form the year appear the same when rotated 180 degrees. How many upside-down years have been from 0BC until now? When was the latest year that is the same upside down? Which is the next upside-down year? (puzzle 39 from Kordemsky (1992)).

Listing 3.11 Formalising an upside-down year

```
1   set(arithmetic).
2   assign(domain_size, 10).
3   assign(max_models, -1).
4
5   formulas(assumptions).
6       eq(x,y) <->  (x = 0 & y = 0) |
7                    (x = 1 & y = 1) |
8                    (x = 6 & y = 9) |
9                    (x = 8 & y = 8) |
10                   (x = 9 & y = 6).
11
12      1000 * a + 100 * b + 10 * c + d < 2020.
13
14      n = 1 | n = 2 | n = 3 | n = 4.
15
16      (n = 4) -> a!= 0 & d!= 0        & eq(a,d) & eq(b,c).
17      (n = 3) -> a = 0 & b!= 0 & d!= 0 & eq(b,d) & eq(c,c).
18      (n = 2) -> a = 0 & b = 0 & c!= 0 & d != 0 & eq(c,d).
19      (n = 1) -> a = 0 & b = 0 & c = 0 & eq(d,d).
20  end_of_list.
```

Solution

Possible upside-down digits are 0, 1, 8, and the pair (6,9). We define the predicate $eq(x, y)$ to signal the pairs that are upside down, i.e. (8,8), (1,1), (6,9), (9,6), (0,0):

$$eq(x, y) \leftrightarrow (x = 0 \wedge y = 0) \vee (x = 1 \wedge y = 1) \vee (x = 6 \wedge y = 9)$$
$$\vee (x = 8 \wedge y = 8) \vee (x = 9 \wedge y = 6)$$

The year should be less than 2020: $1000 \times a + 100 \times b + 10 \times c + d < 2020$.
One difficulty is that the year can have four digits, or three, or two, or one digit. We note with n the number of digits in a year. Hence, $n = 1 \vee n = 2 \vee n = 3 \vee n = 4$.
First, if the year has four digits (i.e. $\overline{abcd}$), then the first digit cannot be 0 ($a \neq 0$). Second, since the corresponding upside-down year has also four digits, its last digit cannot be 0 also ($d \neq 0$). Third, the upside-down year looks the same if pairs (a, d) and (b, c) satisfy the predicate eq. These three conditions are formalised with:

$$n = 4 \rightarrow a \neq 0 \wedge d \neq 0 \wedge eq(a, d) \wedge eq(b, c) \tag{3.27}$$

If the year has three digits (i.e. $\overline{bcd}$, $a = 0$), its first and last digits are not 0, while the pairs (b, d) and (c, c) should satisfy the eq predicate:

$$n = 3 \rightarrow a = 0 \wedge b \neq 0 \wedge d \neq 0 \wedge eq(b, d) \wedge eq(c, c) \tag{3.28}$$

For two digits (i.e. $\overline{cd}$, $a = 0$, $b = 0$), $c \neq 0$, $d \neq 0$, and their pair should look the same:

$$n = 2 \rightarrow a = 0 \wedge b = 0 \wedge c \neq 0 \wedge d \neq 0 \wedge eq(c, d) \tag{3.29}$$

If the year has only one digit ($a = 0$, $b = 0$, $c = 0$), that digit d should be different from 0 and it should also satisfy the predicate eq:

$$n = 1 \rightarrow a = 0 \wedge b = 0 \wedge c = 0 \wedge d \neq 0 \wedge eq(d, d) \tag{3.30}$$

There are 24 solutions in interval [0..2020], among which 3 solutions for $n = 1$, 4 solutions for $n = 2$, 12 solutions for $n = 3$, and 5 solutions for $n = 4$:

Models	a b c d n		Models	a b c d n
1	0 0 0 0 1		13	0 6 8 9 3
2	0 0 0 1 1		14	0 8 0 8 3
3	0 0 0 8 1		15	0 8 1 8 3
4	0 0 1 1 2		16	0 8 8 8 3
5	0 0 6 9 2		17	0 9 0 6 3
6	0 0 8 8 2		18	0 9 1 6 3
7	0 0 9 6 2		19	0 9 8 6 3
8	0 1 0 1 3		20	1 0 0 1 4
9	0 1 1 1 3		21	1 1 1 1 4
10	0 1 8 1 3		22	1 6 9 1 4
11	0 6 0 9 3		23	1 8 8 1 4
12	0 6 1 9 3		24	1 9 6 1 4

The latest upside-down year was 1961 (model 24). To find the next upside-down year, one needs to replace in line 12 of Listing 3.11 the maximum value of 2020 with 9999:

$$1000 \times a + 100 \times b + 10 \times c + d < 9999 \tag{3.31}$$
$$1000 \times a + 100 \times b + 10 \times c + d > 2020 \tag{3.32}$$

Mace4 quickly finds that the next upside-down year is 6009.

Puzzle 23. Rotate digits

In the grid shown, the digits 1–9 are arranged so that the first row added to the second row equals the bottom row (i.e. 583 + 146 = 729). Now if the grid is rotated clockwise at 90 degrees, you will see that the first two rows still add up to the last row (i.e. 715 + 248 = 963). Can you find another combination of the digits 1–9 which has the same property? (puzzle 60 from Clessa (1996))

5	8	3
1	4	6
7	2	9

7	1	5
2	4	8
9	6	3

Listing 3.12 Rotating nine distinct digits a_i

```
1   set(arithmetic).
2   assign(domain_size,10).
3   assign(max_models,−1).
4
5   list(distinct).
6     [0,a1,a2,a3,a4,a5,a6,a7,a8,a9].
7   end_of_list.
8
9   formulas(assumptions).
10    100∗a1 + 10∗a2 + a3 + 100∗a4 + 10∗a5 + a6 = 100∗a7 + 10∗a8 + a9.
11    100∗a7 + 10∗a4 + a1 + 100∗a8 + 10∗a5 + a2 = 100∗a9 + 10∗a6 + a3.
12  end_of_list.
```

Solution

First, we need to specify that the numbers are distinct and in the interval [1..9]. This is done in lines 2 and 6. Second, we need to add two equations:

$$100 * a_1 + 10 * a_2 + a_3 + 100 * a_4 + 10 * a_5 + a_6 = 100 * a_7 + 10 * a_8 + a_9 \quad (3.33)$$

$$100 * a_7 + 10 * a_4 + a_1 + 100 * a_8 + 10 * a_5 + a_2 = 100 * a_9 + 10 * a_6 + a_3 \quad (3.34)$$

Mace4 finds two models: model 1 was given as example and model 2 appears in Fig. 3.6.

Model	a_1	a_2	a_3	a_4	a_5	a_6	a_7	a_8	a_9
2	5	8	3	1	4	6	7	2	9
1	4	8	2	1	5	7	6	3	9

4	8	2
1	5	7
6	3	9

6	1	4
3	5	8
9	7	2

Fig. 3.6 The second solution for the rotated digits

Puzzle 24. An unusual number

Find a six-digit number which, when multiplied by an integer between 2 and 9 inclusive, gives the original six-digit number with its digits reversed. Thus, if the original number was 123,456, and the chosen integer is 8, then 123,456 x 8 should equal 654,321, which, of course, it doesn't. However, it is possible to find more than one solution to this problem, but I'll accept anyone that meets the required condition (puzzle 34 from Clessa (1996))

Listing 3.13 Formalising a unusual number

```
1   set(arithmetic).
2   assign(domain_size,10).
3   assign(max_models,−1).
4
5   list(distinct).
6     [0,1,n].
7     [a,0].
8     [f,0].
9   end_of_list.
10
11  formulas(assumptions).
12    (a*100000 + b*10000 + c*1000 + d*100 + e*10 +f) * n =
13      f*100000 + e*10000 + d*1000 + c*100 + b*10 + a.
14  end_of_list.
```

Solution

Let n the integer between 2 and 9 formalised by the domain size set to 10 and the distinct list in line 6 from Listing 3.13. Let the number composed of six digits $\overline{abcdef}$ (lines 12–13). Its first digit a and its last digit f should be different from 0 (lines 7–8). Mace4 finds two models, one for $n = 4$ and the other one for $n = 9$:

Model	a	b	c	d	e	f	n
1	2	1	9	9	7	8	4
2	1	0	9	9	8	9	9

The two solutions are: 219,978 * 4 = 879,912 and 109,989 * 9 = 989,901.

References

Clessa, J. (1996). *Math and logic puzzles for PC enthusiasts*. Courier Corporation.
Dudeney, H. E. (2016). *536 Puzzles and curious problems*. Courier Dover Publications.
Kordemsky, B. A. (1992). *The Moscow puzzles: 359 mathematical recreations*. Courier Corporation.

Chapter 4
Practical Puzzles

Abstract In which, we group 12 puzzles that have some practical touch. Take for instance the "Golomb ruler". The task is to construct a ruler so that no two pairs of marks measure the same length. We also introduce cryptarithmetic and map colouring puzzles, that are easy to be modelled in Mace4. Measuring and weighing puzzles are formalised here as a planning task for the theorem prover Prover9. Following this line, the eager student can further investigate the connection between theorem proving and planning.

$$\forall x \ \forall y \ can(x, do(y)) \rightarrow does(x, y)$$
$$\forall x \ \forall y \ \neg can(x, do(y)) \rightarrow teaches(x, y)$$

George Bernard Shaw

This chapter groups 12 puzzles that have some practical touch. Take, for instance, the "Golomb ruler". A Golomb ruler is a sequence of positive integers so that every difference between two integers in the sequence is distinct. The task is to construct a ruler so that no two pairs of marks measure the same length. The number of marks is called the *order* of the Golomb ruler. Currently, the highest order for Golomb ruler is 27, while ongoing search aims to discover a Golomb ruler of order 28. Golomb rulers are used in a wide variety of applications including radio astronomy, X-ray crystallography, information theory, or pulse phase modulation Polash et al. (2017). For instance, in radio astronomy achieving high-resolution imagery is not cost-effective with a single aperture. Instead, combining rather small apertures enables high resolution data in a cost-effective manner. Hence, the task is to find the geometric configuration and optimal number of apertures. Indeed, telescopes collect more accurate information if they are placed on the marks of a Golomb ruler Memarsadeghi (2016).

Since most of the books about recreational mathematics contain cryptarithmetic puzzles, such a puzzle is also included. In cryptarithmetic puzzles, letters or other symbols have replaced the digits and we are challenged to find the original numbers. Many such puzzles have been created by M. Pigeolet (you can see a sample of 92 cryptarithms retrieved from journal Sphinx, 1933–34 editions at http://cryptarithms.

A. Groza, *Modelling Puzzles in First Order Logic*,
https://doi.org/10.1007/978-3-030-62547-4_4

awardspace.us/collection.html), H. E. Dudeney in Dudeney (2016) (for instance, the initial version SEND + MORE = MONEY was also created H. E. Dudeney in the July 1924 issue of Strand Magazine), or M. Broke Brooke (1963). These puzzles are easy to be modelled in Mace4. Since most of them follow the same modelling pattern, one example suffices for a student to be able to formalise other cryptarithmetic puzzles. I adapted here the "SEND + MOST = MONEY" puzzle mainly to illustrate that Mace4 is not able to order models. Mace4 searches for unsorted finite models only. Moreover, the domain size is the same for all functions and relations in the input file. The eager student might be interested to move to the multi-sorted setting, where each sort may have a different domain size Reger et al. (2016). The same eager student may also explore the conceptual and technical instrumentation available to order the models according to some preference relation. Another eager student may use Mace4 as a support tool for creating their own cryptarithmetic puzzles.

Another example is the puzzle on "Colouring Dracula's land", that is an instance of the classical map colouring problem.

The "Ambiguous dates" puzzle has a remarkable tiny implementation. Given a date like 7/4/71, and not knowing which system is being used, the task is to compute how many dates in a year are ambiguous in this two-slash notation. The formalisation provided in this chapter uses four equations with three variables only.

The "One landlord and 100 servants" puzzle is another example in which the human agent has difficulties to find all the solutions. One reason is that the puzzle refers to four variables constrained by three equations. The eager student will be quick to notice that the puzzle is an instance of a diophantine problem. Benefiting from the fact that both the puzzle and Mace4 work on the integer's domain only, Mace4 has no problem to correctly find all the models.

Finding "How many routes are in a given map" is another practical puzzle. Some of the students might be surprised to learn that this task is solved with only one line of code. Actually, this would be the take-home idea of this chapter: some tasks—including practical ones—are terribly easy to be modelled in FOL, comparing with non declarative programming languages.

This chapter introduces the first two examples of using Prover9 by means of measuring and weighing puzzles. These types of puzzles have a long history, with one example given by the Nicccoló Tartaglia who asks for 25 ounces of balsam to be divided into three equal parts using vessels of 5, 11, and 13 ounces Darling (2004). In the seventeenth century, Claude Gaspard Bachet asked for the minimum number of weights needed to weigh any integer value from 1 to 40. The solution is 1, 3, 9, and 27 Darling (2004). The "Buying wine" puzzle in this chapter is an instance of two water jug puzzles. Given two jugs of different sizes, the task is to measure a specific value. In our case, the task is to prove that all values from 1 to 9 can be measured with a jug of 4 and one of 9. Prover9 is used to find a proof for each value in [1..9]. The puzzle is modelled as a planning problem, by defining the possible actions and the initial state. The goal state is the theorem that Prover9 has to demonstrate. The proof built by Prover9 represents the plan to measure the current value. Following this line, the eager student can further investigate the connection between theorem proving and planning.

You are sending the following crypto-arithmetic message asking money from your parents. Each letter is unique. How much money you will receive in the worst case and in the best case? (the initial version SEND + MORE = MONEY was published by H. E. Dudeney in the July 1924 issue of Strand Magazine)

$$
\begin{array}{ccccc}
S & E & N & D & + \\
M & O & S & T & \\
\hline
M & O & N & E & Y
\end{array}
$$

Listing 4.1 Formalising the SEND+MOST=MONEY puzzle

```
1   set(arithmetic).
2   assign(domain_size, 10).
3   assign(max_models,−1).
4
5   list(distinct).
6     [S, E, N, D, M, O, T, Y].
7   end_of_list.
8
9   formulas(assumptions).
10    S != 0.
11    M = 1.
12    S * 1000 + E * 100 + N * 10 + D + M * 1000 + O * 100 + S * 10 + T =
13    M * 10000 + O * 1000 + N * 100 + E * 10 + Y.
14  end_of_list.
```

Solution

To answer the question, we have to find all models (line 3), and to manually identify the maximum and the minimum values among the computed interpretations.
We model the problem as a single arithmetic equation. We help the search of Mace4 with two optional constraints in lines 10–11. There are 16 models satisfying the sum:

Models	D	E	M	N	O	S	T	Y
1	2	3	1	4	0	9	5	7
2	2	3	1	4	0	9	6	8
3	2	4	1	5	0	9	6	8
4	2	6	1	7	0	9	3	5
5	2	7	1	8	0	9	3	5
6	2	7	1	8	0	9	4	6
7	3	5	1	6	0	9	4	7
8	3	6	1	7	0	9	2	5
9	3	6	1	7	0	9	5	8
10	3	7	1	8	0	9	2	5
11	4	5	1	6	0	9	3	7
12	4	7	1	8	0	9	2	6
13	5	3	1	4	0	9	2	7
14	5	6	1	7	0	9	3	8
15	6	3	1	4	0	9	2	8
16	6	4	1	5	0	9	2	8

Three such models (1, 2, and 3 from the table above) are:

```
  9 3 4 2 +          9 4 5 6 +          9 5 6 4 +
  1 0 9 5            1 0 9 2            1 0 9 3
  ---------          ---------          ---------
1 0 4 3 7          1 0 5 4 8          1 0 6 5 7
```

The minimum value for MONEY is: 10,437 (model 1).
The maximum value for MONEY is: 10,876 (model 6).
Since the interpretations represent unsorted finite models only, Mace4 is not able to order itself the models. Mace4 searches for unsorted finite models only. Each model has one underlying finite set, (i.e. the domain) and structures (i.e. functions and relations over the domain), corresponding to the functions and relations in the input file. Note that the domain size is the same for all functions and relations. The eager student might be interested in moving to the multi-sorted setting where each sort may have a different domain size.

Puzzle 26. Ambiguous dates

In this country, a date such as July 4, 1971, is often written 7/4/71, but in other countries, the month is given second and the same date is written 4/7/71. If you do not know which system is being used, how many dates in a year are ambiguous in this two-slash notation? Gardner (2005)

Listing 4.2 Computing the ambiguous dates in the last month

```
1   assign(domain_size,13).         %ambigous date has day less than 13
2   assign(max_models,−1).
3   set(arithmetic).
4
5   formulas(ambigous_dates).
6       day   != 0.                 %day in [1..12]
7       month != 0.                 %month in [1..12]
8       day != month.               %if day = month there is no ambiguity
9       month = 12.
10  end_of_list.
```

Solution

If a date is larger than 12, there is no ambiguity. We are searching for models in which both the *day* and the *month* are smaller or equal to 12. Thereby, we need a domain size of 13. Of course, both the day and the month cannot be 0 (lines 6–7). Notice when the day equals the month, there is no ambiguity (line 8), for instance, 12/12.

Mace4 computes 132 models. Each month has 11 ambiguous dates, making 132 in all. To obtain all the 11 ambiguous dates in one specific month, one should add a constraint such as $month = 12$. For December ($month = 12$ in line 9), Mace4 outputs the following 11 ambiguous dates:

Model	1	2	3	4	5	6	7	8	9	10	11
day	1	2	3	4	5	6	7	8	9	10	11
month	12	12	12	12	12	12	12	12	12	12	12

Puzzle 27. One landlord and 100 servants

A certain head of a household had 100 servants. He ordered that they be given 100 modia of corn as follows. The men should receive 3 modia; the women, 2; and the children, half a modium. Thus how many men, women, and children were there? (taken from Danesi (2018)) Can you find all seven solutions?

Listing 4.3 A diophantine problem

```
1  set(arithmetic).
2  assign(max_models,-1).
3  assign(domain_size,100).
4
5  formulas(assumptions).
6    m + f + c = 100.
7    3 * m + 2 * f + c/2 = 100.
8    c = 2 * n.
9  end_of_list.
```

Solution

There are 100 man (m), women (f) and children (c): $m + f + c = 100$. The 100 modia is split with:

$$3 * m + 2 * f + c/2 = 100 \tag{4.1}$$

The eager student may notice that the puzzle is a diophantine problem since there are more variables than equations.

In this puzzle, a hidden constraint is that the number of children should be even, in order to have a solution. We also assume that there is at least one child, one woman and one man: $m \neq 0$, $f \neq 0$, and $c \neq 0$.

Mace4 finds six models satisfying the constraints:

Model	c	f	m	n
1	68	30	2	34
2	70	25	5	35
3	72	20	8	36
4	74	15	11	37
5	76	10	14	38
6	78	5	17	39

Without the assumption $m \neq 0$, $f \neq 0$, and $c \neq 0$, there would be a seventh model: 20 man, no women and 80 children.

Puzzle 28. No consecutive numbers in adjacent nodes

Place numbers 1 through 8 on nodes so that:

1. Each number appears exactly once;

2. No connected nodes have consecutive numbers

How many solutions exist?

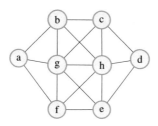

Listing 4.4 No consecutive digits in adjacent nodes

```
1   set(arithmetic).
2   assign(domain_size,9).
3   assign(max_models,−1).
4
5   list(distinct).
6    [0,a,b,c,d,e,f,g,h].
7   end_of_list.
8
9   formulas(assumptions).
10    abs(a + (−b)) != 1.    abs(a + (−g)) != 1.    abs(a + (−f)) != 1.
11    abs(b + (−c)) != 1.    abs(b + (−h)) != 1.    abs(b + (−g)) != 1.
12    abs(c + (−d)) != 1.    abs(c + (−h)) != 1.    abs(c + (−g)) != 1.
13    abs(d + (−h)) != 1.    abs(d + (−e)) != 1.
14    abs(e + (−f)) != 1.    abs(e + (−g)) != 1.    abs(e + (−h)) != 1.
15    abs(f + (−g)) != 1.    abs(f + (−h)) != 1.
16    abs(g + (−h)) != 1.
17   end_of_list.
```

Solution

Since each number between 1 and 8 appears only once, we use the list of distinct elements (line 6 in Listing 4.4). Connected nodes x, y do not contain consecutive numbers if their difference is larger than one. For instance, node a has three connections:

$$(|a - b| \neq 1) \wedge (|a - g| \neq 1) \wedge (|a - f| \neq 1) \tag{4.2}$$

Equation (4.2) is implemented in line 10. After formalising all the connections (lines 11–16), Mace4 finds 4 solutions (Fig. 4.1):

Model	a	b	c	d	e	f	g	h
1	2	5	3	7	4	6	8	1
2	7	4	6	2	5	3	1	8
3	2	6	4	7	3	5	8	1
4	7	3	5	2	6	4	1	8

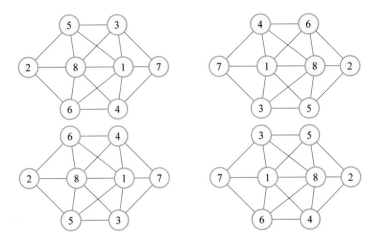

Fig. 4.1 The four solutions found by Mace4

Puzzle 29. Colouring Dracula's land in red

Can the following map of Romania be coloured with only three colours? What about four colours? Find a solution in which Transylvania, the birth place of Dracula, is coloured red. How many solutions exist?

Listing 4.5 Colouring Dracula's country with four colors

```
1   set(arithmetic).
2   assign(max_models,-1).
3   assign(domain_size,9). % 9 regions
4
5   formulas(remove_isomorphic).
6     Red=0. Yellow=1. Blue=2. Green=3.
7
8     T < Maramures.          Maramures < Bucovina.    Bucovina < Moldova.
9     Moldova < Dobrogea.     Dobrogea < Muntenia.     Muntenia < Oltenia.
10    Oltenia < Banat.        Banat < Crisana.
11  end_of_list.
12
13  formulas(assumptions).
14    n(x,y) -> color(x) != color(y).  % neighbors have different colors
15    n(x,y) <-> n(y,x).
16
17    n(T,Crisana).           n(T,Maramures).          n(T,Bucovina).
18    n(T,Moldova).           n(T,Muntenia).           n(T,Oltenia).
19    n(T,Banat).             n(Crisana,Maramures).    n(Crisana,Banat).
20    n(Oltenia,Banat).       n(Oltenia,Muntenia).     n(Maramures,Bucovina).
21    n(Bucovina,Moldova).    n(Moldova,Dobrogea).     n(Moldova,Muntenia).
22    n(Dobrogea,Muntenia).
23
24    -n(T,Dobrogea).         -n(Maramures,Banat).     -n(Maramures,Oltenia).
25    -n(Maramures,Muntenia). -n(Maramures,Moldova).   -n(Maramures,Dobrogea).
26    -n(Bucovina,Crisana).   -n(Bucovina,Banat).      -n(Bucovina,Oltenia).
27    -n(Bucovina,Muntenia).  -n(Bucovina,Dobrogea).   -n(Moldova,Oltenia).
28    -n(Moldova,Banat).      -n(Moldova,Crisana).     -n(Dobrogea,Crisana).
29    -n(Dobrogea,Banat).     -n(Dobrogea,Oltenia).    -n(Muntenia,Banat).
30    -n(Muntenia,Crisana).   -n(Oltenia,Crisana).
31
32    color(T) = Red.                  %Dracula's birth place should be red
33  end_of_list.
```

Solution

As there are nine regions, we need a domain size of nine values. We use the predicate $n(x, y)$ to state that the neighbours have different colours:

$$n(x, y) \rightarrow color(x) \neq color(y) \tag{4.3}$$

The predicate is symmetric $n(x, y) \leftrightarrow n(y, x)$. Let T the shortcut for Transylvania. We need to specify the neighbours (lines 18–25), and also that the regions which are not connected (lines 27–38). Last line states that Transylvania is red.

To remove the isomorphic models we introduce an order relation both on the colours (line 6) and on the regions (lines 8–10).

Assume in Listing 4.5, there are only three colours available. In this case, Mace4 cannot find any model. By introducing the fourth colour, Mace4 outputs 252 models (most of them isomorphic). One such solution is depicted below. The regions are assigned to the following values from the domain:

Banat	Crisana	Bucovina	T	Maramures	Moldova	Muntenia	Oltenia	Dobrogea
7	8	2	0	1	3	5	6	4

while the computed *color* function is:

$$function(color(\_), [0, 1, 2, 1, 0, 2, 1, 2, 3])$$

Note that the Transylvania has value 0 which corresponds to the first position in the vector of colours. This first position contains the value 0, which corresponds to the red colour (see line 6). For another example, let Banat which has been given the value 7. The seventh position in the vector *color* contains the value 2, which is the notation for the *blue* colour.

For the constraints in Listing 4.5, Mace4 finds 981,306 models.

Note that Mace4 works under the open world assumption. That's why we had to explicitly specify the regions that are not neighbours. It would be helpful to have a directive (e.g. $close(p)$), to close some predicate p, or a list where to enumerate closed predicates. The eager student can investigate the Careful Closed World Assumption (CCWA) Gelfond and Przymusinska (1986). CCWA is an extension of General Closed World Assumption that closes to world only for specific predicates. The rest of the predicates are permitted to vary in the process of closure (Fig. 4.2).

Fig. 4.2 Coloured map of Romania with Mace4

Puzzle 30. Golomb ruler

Define a ruler with $M = 4$ marks (e.g. a, b, c, d) so that the distances between any
two marks are different. That is, if you can measure an integer distance with this
ruler, there should be only one way of making this measurement with your ruler. Your
ruler should be able to measure all the integer distances up to length $L = 6$. (adapted
from Memarsadeghi (2016))

$$0 \quad 1 \quad 2 \quad 3 \quad 4 \quad 5 \quad 6$$

Marks: a   b   c   d

Listing 4.6 Modelling a Golomb ruler with Mace4

```
1    set(arithmetic).
2    assign(domain_size,7).
3    assign(max_models,−1).
4
5    formulas(assumptions).
6      a < b & b < c & c < d.                %the marks are distinct
7      a = 0.                                %ruler starts with 0
8
9      abs(a + (−b)) = 1 | abs(b + (−c)) = 1 | abs(c + (−d)) = 1 |
10     abs(a + (−c)) = 1 | abs(a + (−d)) = 1 | abs(b + (−d)) = 1.
11
12     abs(a + (−b)) = 2 | abs(b + (−c)) = 2 | abs(c + (−d)) = 2 |
13     abs(a + (−c)) = 2 | abs(a + (−d)) = 2 | abs(b + (−d)) = 2.
14
15     abs(a + (−b)) = 3 | abs(b + (−c)) = 3 | abs(c + (−d)) = 3 |
16     abs(a + (−c)) = 3 | abs(a + (−d)) = 3 | abs(b + (−d)) = 3.
17
18     abs(a + (−b)) = 4 | abs(b + (−c)) = 4 | abs(c + (−d)) = 4 |
19     abs(a + (−c)) = 4 | abs(a + (−d)) = 4 | abs(b + (−d)) = 4.
20
21     abs(a + (−b)) = 5 | abs(b + (−c)) = 5 | abs(c + (−d)) = 5 |
22     abs(a + (−c)) = 5 | abs(a + (−d)) = 5 | abs(b + (−d)) = 5.
23
24     abs(a + (−b)) = 6 | abs(b + (−c)) = 6 | abs(c + (−d)) = 6 |
25     abs(a + (−c)) = 6 | abs(a + (−d)) = 6 | abs(b + (−d)) = 6.
26   end_of_list.
```

Solution

Formally, a Golomb ruler consists of set of integer marks $\mathcal{M} = \{a_1, a_2, ..., a_m\}$ where $a_1 < a_2 < ... < a_m$ such that $\forall x > 0$ there is at most one solution to the equation $x = a_j - a_i$, $\forall a_i, a_j \in \mathcal{M}$.

In our case, let $\mathcal{M} = \{a, b, c, d\}$ the set containing the four marks on the ruler. These marks should be distinct. The ruler is able to measure integers up to 6, hence we set the domain size to 7. To remove some isomorphic models, we set an order relation between the values a, b, c, and d, e.g. $a < b < c < d$ (line 6 in Listing 4.6). We help Mace4 by stating that the ruler starts with zero (i.e. $a = 0$). The ruler should measure all the integers up to 6. To be able to measure the value 1 we need:

$$|a - b| = 1 \vee |b - c| = 1 \vee |c - d| = 1 \vee |a - c| = 1 \vee |a - d| = 1 \vee |b - d| = 1. \tag{4.4}$$

Similarly, to be able to measure the value 2, we need:

$$|a - b| = 2 \vee |b - c| = 2 \vee |c - d| = 2 \vee |a - c| = 2 \vee |a - d| = 2 \vee |b - d| = 2. \tag{4.5}$$

Similar equations are needed for all the values up to 6 (lines 15–25). Note that this is a particular implementation for $M = 4$ and $L = 6$. A different approach is required to formalise a general solution for any M and L.

Mace4 computes the following two Golomb rulers:

Model	a	b	c	d
1	0	1	4	6
2	0	2	5	6

These two models are depicted in Fig. 4.3, where we can see that all values up to 6 can be measured.

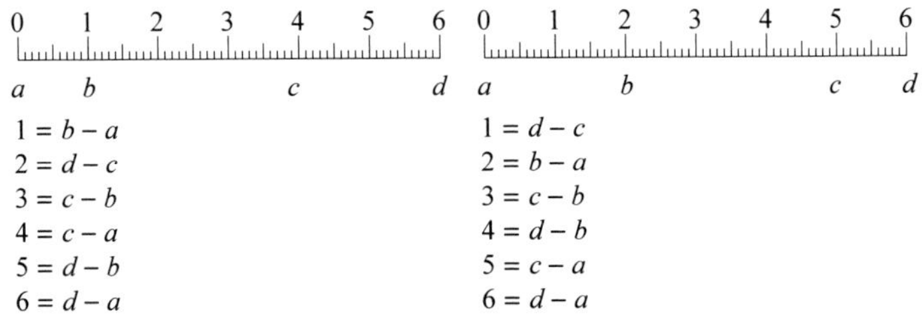

Fig. 4.3 Two Golomb rulers for $M = 4$ marks and length $L = 6$

Puzzle 31. Two cube calendar

In Grand Central Terminal in New York I saw in a store window an unusual desk calendar The day was indicated simply by arranging the two cubes so that their front faces
gave the date. The face of each cube bore a single digit, 0 through 9, and one could
arrange the cubes so that their front faces indicated any date from 01, 02, 03, . . . , to
31. What are the four digits that cannot be seen on the left cube and the three that cannot be seen on the right cube? It is a bit trickier than one might expect. (puzzle 1.4
from Gardner (2005))

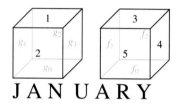

J A N U A R Y

Listing 4.7 Formalisng the two cube calendar in Mace4

```
1   assign(domain_size,9).      %6 can be used for 9
2   assign(max_models,−1).
3   set(arithmetic).
4
5   list(distinct).
6     [f0,f1,f2,3,4,5].
7     [g0,g1,g2,g3,1,2,3,4,5].
8   end_of_list.
9
10  formulas(assumptions).
11    (f0 = 0) & (f1 = 1) & (f2 = 2).
12    (g0 = 0) | (g1 = 0) | (g2 = 0) | (g3 = 0).
13  end_of_list.
```

Solution

First, we observe that the same face can be used for 6 and 9, depending on how the
cube is turned. Since we need only nine values instead of ten, the domain size is set to
9.

Let g_0, g_1, g_2, g_3 the hidden faces on the left cube. Let f_0, f_1, f_2 the hidden faces
on the right cube. For each cube, these values should be distinct (lines 6–7). To avoid
isomorphic models, we set an order relation for these values: $f_0 < f_1 < f_2$ and $g_0 <
g_1 < g_2 < g_3$.

Each cube must bear a 0, 1, and 2. This leaves only six faces for the remaining seven digits, but fortunately the same face can be used for 6 and 9, depending on how the cube is turned. The picture shows 3, 4, 5 on the right cube, and therefore its hidden faces must be 0, 1, and 2 (line 11). On the left cube one can see 1 and 2, and so its hidden faces must be 0, 6 or 9, 7, and 8.

Given the formalisation in Listing 4.7, Mace4 finds a single model (see Fig. 4.4):

Model	f_0	f_1	f_2	g_0	g_1	g_2	g_3
1	0	1	2	0	6	7	8

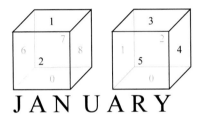

J A N U A R Y

Fig. 4.4 The solution for the two cube calendar

Puzzle 32. How many routes?

In our Mathematics Circle we diagrammed 16 blocks of our city. How many different routes can we draw from the bottom-left corner to the top-right corner moving only upward and to the right? Different routes may, of course, have portions that coincide. What answer should we give these students? (puzzle 50 from Kordemsky (1992))

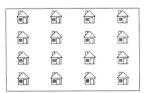

Listing 4.8 Modelling the puzzle as an equation with 8 variables: each solution represents a route

```
1   set(arithmetic).
2   assign(domain_size,2).
3   assign(max_models,-1).
4
5   formulas(assumptions).
6     m1 + m2 + m3 + m4 + m5 + m6 + m7 + m8 = 4.
7   end_of_list.
```

Solution

This problem is not easy for the human agent, but surprisingly easy for the software agent.

Observe that any route contains exactly 8 moves: 4 upward moves and 4 right moves. We code the upward move with 1 and the right move with 0. Hence the domain size is 2. As there are 4 right moves and 4 left moves, the sum of the moves m_i should be exactly four (see Listing 4.8)

$$m_1 + m_2 + m_3 + m_4 + m_5 + m_6 + m_7 + m_8 = 4 \tag{4.6}$$

Mace4 outputs 70 models for this equation. It seems that the problem is easy for Mace4.

A different solution would be to approach the problem as a planning task for Prover9 (see Listing 4.9). We are interested in finding all the proofs from the init state to the goal state. Let the init state $J(0, 0)$ and the goal state $J(4, 4)$. The "right" move increases the y coordinate (line 4), while the "up" move increases the x coordinate (line 5):

$$J(x, y) \wedge y < 4 \rightarrow J(x, y + 1) \tag{4.7}$$
$$J(x, y) \wedge x > 4 \rightarrow J(x + 1, y)W \tag{4.8}$$

e use the production mode and also the #answer directive to print the clauses that are activated. The proof prints the moves from the init state to the goal state:

"up" # "up" # "up" # "up" # "right" # "right" # "right" # "right"

The proof represents one route. To find all the routes one might use the directive assign(max_proofs, 3) combined with some other search flags from Prover9.

Listing 4.9 Modelling the puzzle as a planning task for Prover9: a proof represents a route

```
1   set(production).
2
3   formulas(usable).
4       J(x,y) & y < 4 -> J(x,y+1)    #answer("up").
5       J(x,y) & x < 4 -> J(x+1,y)    #answer("right").
6   end_of_list.
7
8   formulas(assumptions).
9       J(0,0)                        #answer("Init state: J(0,0)").
10  end_of_list.
11
12  formulas(goals).
13      J(3,4).
14  end_of_list.
```

Puzzle 33. Going to church

A man living in the house shown in the diagram wants to know what is the greatest number of different routes by which he can go to the church. The possible roads are indicated by the lines, and he always walks either to the N, to the E, or NE; that is, he goes so that every step brings him nearer to the church. Can you count the total number of different routes from which he may select? (puzzle 417 from Dudeney (2016))

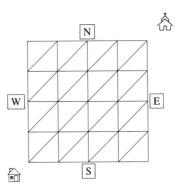

Listing 4.10 Finding a route as model finding task for Mace4

```
1    assign(domain_size, 8).
2    assign(max_models,−1).
3    set(arithmetic).
4
5    formulas(going_to_church).
6      all x (f(x) = 0 | f(x) = 1 | f(x) = 2 | f(x) = 3).
7      f(x) = 0 <-> g(x) = 0 & h(x) = 1.              % goes N
8      f(x) = 1 <-> g(x) = 1 & h(x) = 0.              % goes E
9      f(x) = 2 <-> g(x) = 1 & h(x) = 1.              % goes NE
10     f(x) = 3 <-> g(x) = 0 & h(x) = 0.              % stays
11
12       all x (f(x)  = 3 -> (all y (y > x -> f(y)=3))).
13       x < 4 -> f(x) != 3.
14
15     ((g(0) + g(1) + g(2) + g(3) = 4) &                %4 moves
16      (h(0) + h(1) + h(2) + h(3) = 4) &
17         (x>3 -> f(x)=3))  |
18     ((g(0) + g(1) + g(2) + g(3) + g(4) = 4) &         %5 moves
19      (h(0) + h(1) + h(2) + h(3) + h(4) = 4) &
20         (x>4 -> f(x)=3))  |
21     ((g(0) + g(1) + g(2) + g(3) + g(4) + g(5) = 4) &  %6 moves
22      (h(0) + h(1) + h(2) + h(3) + h(4) + h(5) = 4) &
23         (x>5 -> f(x)=3 ))  |
24     ((g(0) + g(1) + g(2) + g(3) + g(4) + g(5) + g(6) = 4) &   %7 moves
25      (h(0) + h(1) + h(2) + h(3) + h(4) + h(5) + h(6) = 4) &
26         f(7) = 3)  |
27     ((g(0) + g(1) + g(3) + g(4) + g(5) + g(6) + g(2) + g(7) = 4) & %8 moves
28      (h(0) + h(1) + h(2) + h(3) + h(4) + h(5) + h(6) + h(7)= 4)).
29    end_of_list.
```

┌─ **Solution** ───┐

We assume the house is in $(0, 0)$ and the church is in $(4, 4)$. There are three possible moves: $f(x) = 0$ for N (up), $f(x) = 1$ for E (right), and $f(x) = 2$ for NE. For instance, a sequence of actions to reach $(4, 4)$ would be $s_0 = [E, N, NE, NE, E, N, stay, stay]$ with the corresponding formalisation $f = [1, 0, 2, 2, 1, 0, stay, stay]$. Note that the shortest sequence has 4 actions, e.g. $s_1 = [NE, NE, NE, NE, stay, stay, stay, stay]$, while the longest sequence has 8 actions, e.g. $s_2 = [N, E, N, E, N, E, N, E]$. To accommodate this length difference, we introduce the move $stay$ with $f(x) = 3$. The move $stay$ occurs only when the agent reaches the destination in less than 8 moves. The agent is not allowed to stay until it reaches the destination (line 13 in Listing 4.10). That is, we don't allow staying in the first four moves:

$$x < 4 \rightarrow f(x) \neq 3. \tag{4.9}$$

and once a $stay$ operation has occurred ($f(x) = 3$), the agent should continuously $stay$ for all the remaining steps y:

$$\forall x \, (f(x) = 3 \rightarrow (\forall y \, (y > x \rightarrow f(y) = 3))) \tag{4.10}$$

For the sequence s_1, at each step the initial position $(0, 0)$ is increased with 1 for both coordinates:

$$(0, 0) + (1, 1) + (1, 1) + (1, 1) + (1, 1) = (4, 4) \tag{4.11}$$

For the sequence s_2, at each step only one coordinate is increased:

$$(0, 0) + (1, 0) + (0, 1) + (1, 0) + (0, 1) + (1, 0) + (0, 1) + (1, 0) + (0, 1) = (4, 4) \tag{4.12}$$

Our task here is to find all the tuples

$$(x_1, y_1) + (x_2, y_2) + ... + (x_n, y_n) = (4, 4) \tag{4.13}$$

└──┘

Listing 4.11 Finding a route as a planning task for Prover9

```
1   set(production).
2
3   formulas(demodulators).
4     n = 4.   %rows
5     m = 3.   %columns
6   end_of_list.
7
8   formulas(usable).
9     J(x, y) & y < n            -> J(x, y+1)   #answer("E").
10    J(x, y) & x < m            -> J(x+1, y)   #answer("N").
11    J(x, y) & x < n & y < m -> J(x+1, y+1) #answer("NE").
12  end_of_list.
13
14  formulas(assumptions).
15    J(0, 0)                 #answer("Init state: J(0,0)").
16  end_of_list.
17
18  formulas(goals).
19    J(n,m).
20  end_of_list.
```

Solution

To solve the above task, we formalise each tuple with two functions $(g(x), h(x))$. There are four possible tuples: $(0, 0)$, $(0, 1)$, $(1, 0)$ and $(1, 1)$:

1.	Going North:	$f(x) = 0 \leftrightarrow g(x) = 0 \wedge h(x) = 1$
2.	Going East:	$f(x) = 1 \leftrightarrow g(x) = 1 \wedge h(x) = 0$
3.	Going North-East:	$f(x) = 2 \leftrightarrow g(x) = 1 \wedge h(x) = 1$
4.	Stay	$f(x) = 2 \leftrightarrow g(x) = 0 \wedge h(x) = 0$

For instance, for a sequence of five moves, Mace4 needs to find all combinations of tuples such that $(x_1, y_1) + (x_2, y_2) + (x_3, y_3) + (x_4, y_4) + (x_5, y_5) = (4, 4)$. That is $x_1 + x_2 + x_3 + x_4 + x_5 = 4$ and $y_1 + y_2 + y_3 + y_4 + y_5 = 4$. In our notation that is $g(0) + g(1) + g(2) + g(3) + g(4) = 4$ and $h(0) + h(1) + h(2) + h(3) + h(4) = 4$. As the domain size is 8 (line 1 in Listing 4.10), we need to fix $g(5) = g(6) = g(7) = 1$ and $h(5) = h(6) = h(7) = 1$.

We need to compute all models for each sequence: of length 4, 5, 6, 7, and 8. For instance, for length 4 we have:

$g(0) + g(1) + g(2) + g(3) = 4 \wedge (h(0) + h(1) + h(2) + h(3) = 4 \wedge (x > 4) \rightarrow g(x) = 1 \wedge h(x) = 1$

The total number of different routes from $(0, 0)$ to $(4, 4)$ is 321. One such model is:

$$\text{function(f(\_), [2, 2, 1, 0, 2, 3, 3, 3])} \leftrightarrow [\nearrow, \nearrow, \rightarrow, \uparrow, \nearrow, stay, stay, stay]$$
$$\text{function(g(\_), [1, 1, 1, 0, 1, 0, 0, 0])}$$
$$\text{function(h(\_), [1, 1, 0, 1, 1, 0, 0, 0])}$$

A second solution is to approach the puzzle as a planning task for Prover9 (see Listing 4.11). For this we activate the production mode (line 1 in Listing 4.11). We are interested in finding all the proofs from the init state to the goal state. We can even assume a generic grid of $n \times m$. In Listing 4.11, $n = 4$ and $m = 3$. Let the init state $J(0, 0)$ and the goal state $J(n, m)$. The east move increases the column coordinate y (line 9), the north move increases the row coordinate x (line 10), while north-east increases both coordinates (line 11):

$$J(x, y) \wedge y < n \rightarrow J(x, y + 1) \tag{4.14}$$

$$J(x, y) \wedge x < m \rightarrow J(x + 1, y) \tag{4.15}$$

$$J(x, y) \wedge y < n \wedge x < m \rightarrow J(x + 1, y + 1) \tag{4.16}$$

The activated clauses are: *"NE" # "NE" # "NE" # "NE" # "Init state: J(0,0)"*. The proof represents one route. To find all the routes one might use the directive `assign(max_proofs,3)` combined with some other search flags from Prover9.

Puzzle 34. Buying wine

A wine seller has two wine jugs, o small one of 4 L capacity, and a larger one of 9 L. There is no measuring label mentioned on either of these two jugs i.e. he cannot know the exact amount filled in the jug. Can we measure all values from 1 to 9 using these unmarked jugs? (a generalisation of classical water jugs problem)

Solution

Let $j_1 = 4$ be the capacity of the small jug and $j_2 = 9$ the capacity of the large jug. We use a *state space search*, where each state is represented with $J(x, y)$: x is the current amount of wine in jug j_1 and y the amount in jug j_2. The initial state is $J(0, 0)$. We use production rules to change the states of the system. This reasoning mode is activated with $set(production)$ (line 1 in Listing 4.12). For the current state $J(x, y)$, the following eight actions are possible:

1. Fill-in the small jug: $J(x, y) \rightarrow J(j_1, y)$
2. Empty the small jug: $J(x, y) \rightarrow J(0, y)$
3. Fill-in the large jug: $J(x, y) \rightarrow J(x, j_2)$
4. Empty the large jug: $J(x, y) \rightarrow J(x, 0)$
5. Empty the small jug into the large jug, if capacity allows this: $J(x, y) \wedge x + y \leq j_2 \rightarrow J(0, y + x)$
6. If j_2 does not suffice to empty j_1 then move some amount from the small jug to the larger jug, until j_2 is full: $J(x, y) \wedge x + y > j_2 \rightarrow J(x - (j_2 - y), j_2)$
7. If capacity of j_2 allows, empty the large jug into the small jug: $J(x, y) \wedge x + y \leq j_1 \rightarrow J(x + y, 0)$
8. If capacity of j_1 does not suffice to empty j_2 then move some amount from j_2, until the j_1 is full: $J(x, y) \wedge x + y > j_1 \rightarrow J(j_1, y - (j_1 - x))$

Note in Listing 4.12, that rules are written in clausal form, in order to allow variables in the *#answer* directive. The *#answer* directive is useful to print the steps for reaching the goal. The puzzle is a kind of Golomb ruler with water jugs instead of marks on a ruler.

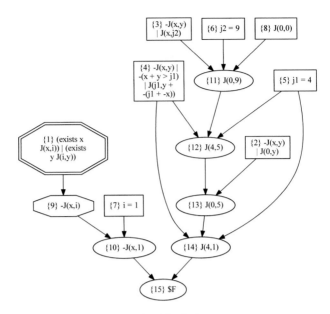

Fig. 4.5   A proof by resolution for measuring 1 L of wine

Listing 4.12 Measuring all numbers from 1 to 9 with two jugs of 4 and 9

```
1    set(production).
2
3    formulas(demodulators).
4     j1 = 4.      j2 = 9.                    %small jug and large jug
5     i  = 2.                                 %value to measure i=[1..9]
6    end_of_list.
7
8    formulas(usable).
9     −J(x,y) | J(j1,y) #answer("fill the small jug")  #answer(J($j1,y)).
10    −J(x,y) | J(0,y) #answer("empty the small jug")  #answer(J(0,y)).
11    −J(x,y) | J(x,j2) #answer("fill the big jug")      #answer(J(x,$j2)).
12    −J(x,y) | J(x,0) #answer("empty the big jug")      #answer(J(x,0)).
13    −J(x,y) | −(x+y <= j2) | J(0,y+x)
14       #answer("empty the small jug into the big jug") #answer(J(0,x+y)).
15    −J(x,y) | −(x+y > j2) | J(x+ −(j2+ −y),j2)
16       #answer("small into big, until full") #answer(J(x+ −(j2+ −y),$j2)).
17    −J(x,y) | −(x+y <= j1) | J(x+y,0)
18       #answer("empty the big jug into the small jug") #answer(J(x+y, 0)).
19    −J(x,y) | −(x+y >  j1) | J(j1,y + −(j1+ −x))
20       #answer("big into small, until full") #answer(J($j1,y+ −($j1+ −x))).
21    end_of_list.
22
23    formulas(assumptions).
24     J(0,0)                                  #answer("Init state: J(0,0)").
25    end_of_list.
26
27    formulas(goals).
28     exists x J(x,i) | exists y J(i,y).
29    end_of_list.
```

Solution

For each measured value, Prover9 computes the corresponding proof. For instance, how 1 L is measured is illustrated by the proof in Fig. 4.5. The theorem asks to find a state in which either the first jug or the second one contains only 1 L. For $\exists x\, J(x, 1) \vee \exists y\, J(1, y)$, the $Proof_1$ has 5 steps (see Fig. 4.5).

$Proof_1$	#answer(Action)	#answer(State)
1	"init state"	$J(0, 0)$
2	"fill the big jug"	$J(0, j_2)$
3	"big into small, until full"	$J(j_1, 9 - (j_1 - 0))$
4	"empty the small jug"	$J(0, 5)$
5	"big into small, until full"	$J(j_1, 5 - (j_1 - 0))$

For measuring two liters, we need $\exists x\, J(x, 2) \vee \exists y\, J(2, y)$. Prover9 finds the $Proof(2)$ in 11 steps:

$Proof_2$	#answer(Action)	#answer(State)
1	"init state"	$J(0, 0)$
2	"fill the big jug"	$J(0, j_2)$
3	"big into small, until full"	$J(j_1, 9 - (j_1 - 0))$
4	"empty the small jug"	$J(0, 5)$
5	"big into small, until full"	$J(j_1, 5 - (j_1 - 0))$
6	"empty the small jug"	$J(0, 1)$
7	"empty the big jug into the small jug"	$J(0 + 1, 0)$
8	"fill the big jug"	$J(1, j_2)$
9	"big into small, until full"	$J(j_1, 9 - (j_1 - 1))$
10	"empty the small jug"	$J(0, 6)$
11	"big into small, until full"	$J(j_1, 6 - (j_1 - 0))$

Note that the first five steps in $Proof_2$ are similar with $Proof_1$. The value 2 is obtained in the big jug. For measuring three liters with $\exists x\, J(x, 3) \vee \exists y\, J(3, y)$, the $Proof_3$ has 7 steps. The value 3 is obtained in the small jug.

$Proof_3$	#answer(Action)	#answer(State)
1	"init state"	$J(0, 0)$
2	"fill the small jug"	$J(j_1, 0)$
3	"empty the small jug into the big jug"	$J(0, 4 + 0)$
4	"fill the small jug"	$J(j_1, 4)$
5	"empty the small jug into the big jug"	$J(0, 4 + 4)$
6	"fill the small jug"	$J(j_1, 8)$
7	"small into big, until full"	$J(4 - (j_2 - 8), j_2)$

$\exists x\, J(x, 5) \vee \exists y\, J(5, y)$, the $Proof_5$ has only 3 steps. The value 5 is obtained in the large jug.

$Proof_5$	#answer(Action)	#answer(State)
1	"init state"	$J(0, 0)$
2	"fill the big jug"	$J(0, j_2)$
3	"big into small, until full"	$J(j_1, 9 - (j_1 - 0))$

The graphical representation of each proof is obtained with:
```
prover9 -f waterjugs.in | prooftrans xml renumber |
gvizify | dot -Tpdf
```

Solution

For $\exists x \, J(x, 6) \vee \exists y \, J(6, y)$, the $Proof_6$ has 9 steps. The steps in $Proof_6$ represent the first 9 steps in $Proof_2$.

$Proof_6$	#answer(Action)	#answer(State)
1	"init state"	$J(0, 0)$
2	"fill the big jug"	$J(0, j_2)$
3	"big into small, until full"	$J(j_1, 9 - (j_1 - 0))$
4	"empty the small jug"	$J(0, 5)$
5	"big into small, until full"	$J(j_1, 5 - (j_1 - 0))$
6	"empty the small jug"	$J(0, 1)$
7	"empty the big jug into the small jug"	$J(0 + 1, 0)$
8	"fill the big jug"	$J(1, j_2)$
9	"big into small, until full"	$J(j_1, 9 - (j_1 - 1))$

$Proof_7$ continues $Proof_3$ with the following four steps:

$Proof_7$	#answer(Action)	#answer(State)
8	"empty the big jug"	$J(3, 0)$
9	"empty the small jug into the big jug"	$J(0, 3 + 0)$
10	"fill the small jug"	$J(j_1, 3)$
11	"empty the small jug into the big jug"	$J(0, 4 + 3)$

Finally, $Proof_8$ is the same with the first 5 steps in $Proof_3$.
The following relations between proofs are identified:

$$
\begin{aligned}
Proof_1 + [6, 7, 8, 9, 10, 11] &= Proof_2 \\
Proof_8 + [6, 7] &= Proof_3 \\
Proof_3 + [8, 9, 10, 11] &= Proof_7 \\
Proof_6 + [10, 11] &= Proof_2 \\
Proof_5 + [4, 5] &= Proof_1
\end{aligned}
$$

A lattice can be built with the following subsumption relations between proofs:

$$
\begin{aligned}
Proof_8 &\sqsubseteq Proof_3 \\
Proof_5 \sqsubseteq Proof_1 &\sqsubseteq Proof_2 \\
Proof_6 &\sqsubseteq Proof_2
\end{aligned}
$$

The lattice shows that $Proof_5$ will be a good starting point. The proof can be given to Prover9 as a list of hints. These hints are used by Prover9 to search for related proofs. Here, one can prove that a specific value can be measured (e.g. 1), and then add the specific state as an init state. That is, we know already how to prove, for instance $J(5, 1)$. Now the search is more flexible, as the init state contains both $J(0, 0)$, and $J(5, 1)$.

Puzzle 35. Railway routes

The diagram below represents a simplified railway system, and we want to know how many different ways there are of going from A to E, if we never go twice along the same line in any journey. This is a very simple proposition, but practically impossible to solve until you have hit on some method of recording the routes. You see there are many ways of going, from the short route ABDE, taking one of the large loops, up to the long route ABCDBCDBCDE, which takes you over every line on the system and can itself be varied in order in many ways. How many different ways of going are there? (puzzle 424 from Dudeney (2016))

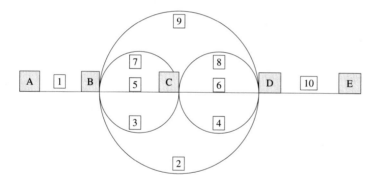

Listing 4.13 Finding all the railway routes between two points

```
1   set(arithmetic).
2   assign(domain_size,11).              % there are 11 edges
3   assign(max_models,−1).
4
5   formulas(demodulators).
6     A = 0.    B = 1.    C = 2.    D = 3.      E = 4.
7     n(10) = E.                         %last node is E
8     f(10) = 10.                        %E is reached only from D (edge 10)
9     g(x)  = f(s(x)).                   % next edge of edge x
10  end_of_list.
11
12  formulas(railway_routes).
13    x < 10 -> s(x) = x + 1.    s(10) = 0.    %succesor
14    x >= 1 -> p(x) = x + −1.   p(0) = 0.    %predecessor
15
16    all x1 all x2 ((f(x1) = f(x2) & f(x1) != 0) -> x1 = x2).
```

```
17
18    all x (n(x)=A | n(x)=B | n(x)=C | n(x)=D | n(x)=E).     %only 5 nodes
19
20    (n(p(x))=A & f(x)=1) -> n(x) = B.                        %A->B (1)
21
22    (n(p(x))=B & (f(x)=5 | f(x)=7 | f(x)=3)) -> n(x)=C.      %B->C (5,7,3)
23
24    (n(p(x))=B & (f(x)=2 | f(x)=9)) -> n(x)=D.     %B->D (2,9)
25
26    (n(p(x))=C & (f(x)=7 | f(x)=5 | f(x)=3)) -> n(x)=B.      %C->B (7,5,3)
27
28    (n(p(x))=C & (f(x)=8 | f(x)=6 | f(x)=4)) -> n(x)=D.      %C->D (8,6,4)
29
30    (n(p(x))=D & (f(x)=9 | f(x)=2)) -> n(x)=B.               %D->B (8,9)
31
32    (n(p(x))=D & (f(x)=8 | f(x)=6 | f(x)=4)) -> n(x)=C.      %D-> C (8,6,4)
33
34    (n(p(x))=D & f(x)=10) -> n(x)=E.                         %D->E (10)
35
36    %zeros of f and n are states not reacheable from any edge
37    %if after using x we get into the 0 edge then current node is A
38    f(x) = 0 -> n(x) = A.
39
40    %zeros of f and n are only in the init state
41    f(x) = 0 -> -(exists y (f(y) != 0 & y < x)).
42
43    %after using edge 1 we cannot use (4,6,8,10)
44    f(x)=1 -> (g(x)!=4 & g(x)!=6 & g(x)!=8 & g(x)!=10).
45
46    %after using edge 2 from D we cannot use (4,6,8,10)
47    (f(x)=2 & n(p(x))=D) -> (g(x)!=10 & g(x)!=6 & g(x)!=8 & g(x)!=4).
48
49    %after using edge 2 from B we cannot use (3,5,7)
50    (f(x)=2 & n(p(x))=B) -> (g(x)!=7 & g(x)!=3 & g(x)!=5).
51
52    %after using edge 1 from B we cannot use (2,9,10)
53    (f(x)=3 & n(p(x))=B) -> (g(x)!=10 & g(x)!=2 & g(x)!=9).
54
55    %after using edge 3 from C we cannot use (4,6,8,10)
56    (f(x)=3 & n(p(x))=C) -> (g(x)!=4 & g(x)!=8 & g(x)!=6 & g(x)!=10).
57
58    %similar as before
59    (f(x)=4 & n(p(x))=D) -> (g(x)!=1  & g(x)!=9 & g(x)!=2  & g(x)!=10).
60    (f(x)=4 & n(p(x))=C) -> (g(x)!=3  & g(x)!=5 & g(x)!=7).
61    (f(x)=5 & n(p(x))=B) -> (g(x)!=9  & g(x)!=2 & g(x)!=1  & g(x)!=10).
62    (f(x)=5 & n(p(x))=C) -> (g(x)!=8  & g(x)!=4 & g(x)!=6  & g(x)!=10).
63    (f(x)=6 & n(p(x))=C) -> (g(x)!=7  & g(x)!=5 & g(x)!=3  & g(x)!=1).
64    (f(x)=6 & n(p(x))=D) -> (g(x)!=9  & g(x)!=2 & g(x)!=10).
65    (f(x)=7 & n(p(x))=B) -> (g(x)!=10 & g(x)!=9 & g(x)!=2  & g(x)!=1).
66    (f(x)=7 & n(p(x))=C) -> (g(x)!=10 & g(x)!=8 & g(x)!=6  & g(x)!=4).
67    (f(x)=8 & n(p(x))=C) -> (g(x)!=1  & g(x)!=7 & g(x)!=5  & g(x)!=3).
68    (f(x)=8 & n(p(x))=D) -> (g(x)!=9  & g(x)!=2 & g(x)!=1  & g(x)!=10).
69    (f(x)=9 & n(p(x))=B) -> (g(x)!=7  & g(x)!=3 & g(x)!=5).
```

```
70    (f(x)=9 & n(p(x))=D)  ->  (g(x)!=8  & g(x)!=4 & g(x)!=10 & g(x)!=6).
71
72    exists y  (y > 0 & y < 9 & (n(y) = 1 & f(y) = 1 &
73              (all  x ((x < y) -> f(x) = 0)))).        % way of 11-y lines
74    end_of_list.
```

Solution

We set the domain size to the number of edges, i.e. 11. Let $n(x) : edge \to node$. That is $n : [0..10] \to \{A, B, C, D, E\}$. Given the edge x, $n(x)$ returns the current node, obtained by using the edge x. For instance, "edge 10 goes only towards node E" is represented with $n(10) = E$ (line 7 in Listing 4.13). "There are only five nodes" is modeled with:

$$\forall x \; n(x) = A \vee n(x) = B \vee n(x) = C \vee n(x) = D \vee n(x) = E \qquad (4.17)$$

Let $A = 0$, $B = 1$, $C = 2$, $D = 3$, $E = 4$. Function $f(x)$ returns the current edge, given the previous edge/step x: $f : [0..10] \to [0..10]$. For instance, $f(x) = 1$ says that the edge 1 has just been used to travel from node x. $f(3) = 2$ means that at step 3 the train uses edge 2 (or after using edge 3 the train uses edge 2).

We define three auxiliary functions: successor $s(x)$, predecessor $p(x)$, and $g(x) = f(s(x))$.

The result is the sum of all ways composed of 2, 3, 4, 5, 6, or 8 steps/edges (we cannot reach E by going through seven edges).

There is an edge y used after edge 1 ($f(y) = 1$) from node A ($n(y) = A$):

$$\exists y \; (y > 0 \wedge y < 9 \wedge n(y) = 1 \wedge f(y) = 1 \wedge (\forall x \; ((x < y) \to f(x) = 0))) \quad (4.18)$$

There are 2,501 ways of going from A to E. Since there is only one route from A to B and also one route from D to E, this value of 2,501 is the same number as going from B to D. Mace4 finds all 2501 solutions, the first one being:

$function(A, \quad [0])$,   function(B, [1]),
$function(C, \quad [2])$,   function(D, [3]),
$function(E, \quad [4])$,   function(c1, [1]),
$function(f(\_), \quad [0,1,2,4,3,5,6,8,7,9,10])$, $function(g(\_), \quad [1,2,4,3,5,6,8,7,9,10,0])$,
$function(p(\_), \quad [0,0,1,2,3,4,5,6,7,8,9])$, $function(s(\_), \quad [1,2,3,4,5,6,7,8,9,10,0])$
$function(n(\_), \quad [0,1,3,2,1,2,3,2,1,3,4])$

The route is stored by the f function. It represents the long route:

$$A \xrightarrow{1} B \xrightarrow{2} D \xrightarrow{4} C \xrightarrow{3} B \xrightarrow{5} C \xrightarrow{6} D \xrightarrow{8} C \xrightarrow{7} B \xrightarrow{9} D \xrightarrow{10} E$$

In the model 2501, $function(f(\_), [0, 0, 0, 0, 0, 0, 0, 0, 1, 9, 10])$ stores the short route:

$$A \xrightarrow{1} B \xrightarrow{9} D \xrightarrow{10} E$$

Puzzle 36. A stolen balsam

Three men robbed a gentleman of a vase containing 24 ounces of balsam. While running away, they met in a forest a glass seller, from whom, in a great hurry, they purchased three vessels. On reaching a place of safety they wished to divide the booty, but they found that their vessels contained 5, 11, and 13 ounces respectively. How could they divide the balsam into equal portions? (puzzle 409 from Dudeney (2016))

Listing 4.14 Finding a proof from the init state $J(24, 0, 0, 0)$ to the goal state $J(8, 8, 8, 0)$

```
1    set(production).
2
3    formulas(usable).
4    all x all y all z all t (J(x,y,z,t) & x+y<=24 --> J(x+y,0,z,t)).%emp(2,1)
5    all x all y all z all t (J(x,y,z,t) & x+z<=24 --> J(x+z,y,0,t)).%emp(3,1)
6    all x all y all z all t (J(x,y,z,t) & x+t<=24 --> J(x+t,y,z,0)).%emp(4,1)
7
8    all x all y all z all t (J(x,y,z,t) & x+y<=11 --> J(0,y+x,z,t)).%emp(1,2)
9    all x all y all z all t (J(x,y,z,t) & y+z<=11 --> J(x,y+z,0,t)).%emp(3,2)
10   all x all y all z all t (J(x,y,z,t) & y+t<=11 --> J(x,y+t,z,0)).%emp(4,2)
11
12   all x all y all z all t (J(x,y,z,t) & z+x<=13 --> J(0,y,z+x,t)).%emp(1,3)
13   all x all y all z all t (J(x,y,z,t) & y+z<=13 --> J(x,0,z+y,t)).%emp(2,3)
14   all x all y all z all t (J(x,y,z,t) & z+t<=13 --> J(x,y,z+t,0)).%emp(4,3)
15
16   all x all y all z all t (J(x,y,z,t) & t+x<=5 --> J(0,y,z,t+x)). %emp(1,4)
17   all x all y all z all t (J(x,y,z,t) & t+y<=5 --> J(x,0,z,t+y)). %emp(2,4)
18   all x all y all z all t (J(x,y,z,t) & z+t<=5 --> J(x,y,0,t+z)). %emp(3,4)
19
20   %from 2 fill 1; from 3 fill 1; from 4 fill 1
21   all x all y all z all t (J(x,y,z,t) & x+y>24 --> J(24,y+ −(24+ −x),z,t)).
22   all x all y all z all t (J(x,y,z,t) & x+z>24 --> J(24,y,z+ −(24+ −x),t)).
23   all x all y all z all t (J(x,y,z,t) & x+t>24 --> J(24,y,z,t+ −(24+ −x))).
24
25   %from 1 fill 2; from 3 fill 2; from 4 fill 2
26   all x all y all z all t (J(x,y,z,t) & y+x>11 --> J(x+ −(11+ −y),11,z,t)).
27   all x all y all z all t (J(x,y,z,t) & y+z>11 --> J(x,11,z+ −(11+ −y),t)).
28   all x all y all z all t (J(x,y,z,t) & y+t>11 --> J(x,11,z,t+ −(11+ −y))).
29
30   %from 1 fill 3; from 2 fill 3; from 4 fill 3
31   all x all y all z all t (J(x,y,z,t) & z+x>13 --> J(x+ −(13+ −z),y,13,t)).
32   all x all y all z all t (J(x,y,z,t) & z+y>13 --> J(x,y+ −(13+ −z),13,t)).
33   all x all y all z all t (J(x,y,z,t) & z+t>13 --> J(x,y,13,t+ −(13+ −z))).
34
35   %from 1 fill 4; from 2 fill 4; from 3 fill 4
36   all x all y all z all t (J(x,y,z,t) & t+x>5 --> J(x+ −(5+ −t),y,z,5)).
37   all x all y all z all t (J(x,y,z,t) & t+y>5 --> J(x,y+ −(5+ −t),z,5)).
38   all x all y all z all t (J(x,y,z,t) & t+z>5 --> J(x,y,z+ −(5+ −t),5)).
39   end_of_list.
```

```
40   formulas(assumptions).
41     J(24,0,0,0).
42   end_of_list.
43
44   formulas(goals).
45     exists x (J(8,8,8,x)).
46   end_of_list.
```

Solution

Let the state $J(x, y, z, t)$ meaning that the first vessel contains $x = 24$ ounces of balsam, the second one can accommodate $y \leq 11$ ounces, the third one $z \leq 13$ and the 4th vessel $t \leq 5$ ounces: $J(24, 11, 13, 5)$. The init state is $J(24, 0, 0, 0)$. Since the last vessel cannot accommodate 8 ounces, there is only one possible goal state to equally split the total amount of 24 ounces:

$$J(8, 8, 8, 0) \tag{4.19}$$

There are two types of actions: (1) emptying a vessel into another one, and (2) filling a vessel.

To exemplify the first category of rules, the rules for emptying the vessels 1 or 2 or 4 in the third one (i.e. 13oz.) are

$$\forall x \, \forall y \, \forall z \, \forall t \; (J(x, y, z, t) \wedge (z + x \leq 13) \rightarrow J(0, y, z + x, t)) \tag{4.20}$$
$$\forall x \, \forall y \, \forall z \, \forall t \; (J(x, y, z, t) \wedge (y + z \leq 13) \rightarrow J(x, 0, z + y, t)) \tag{4.21}$$
$$\forall x \, \forall y \, \forall z \, \forall t \; (J(x, y, z, t) \wedge (z + t \leq 13) \rightarrow J(x, y, z + t, 0)) \tag{4.22}$$

These equations are formalised in lines 12–14 in Listing 4.14.

Similar rules are written for emptying the other vessels (lines 4–18 in Listing 4.14).

To exemplify the second category of rules, the rules to fill the fourth vessel, from 1, 2, and 3 are

$$\forall x \, \forall y \, \forall z \, \forall t \; (J(x, y, z, t) \wedge (t + x > 5) \rightarrow J(x - (5 - t), y, z, 5)) \tag{4.23}$$
$$\forall x \, \forall y \, \forall z \, \forall t \; (J(x, y, z, t) \wedge (t + y > 5) \rightarrow J(x, y - (5 - t), z, 5)) \tag{4.24}$$
$$\forall x \, \forall y \, \forall z \, \forall t \; (J(x, y, z, t) \wedge (t + z > 5) \rightarrow J(x, y, z - (5 - t), 5)) \tag{4.25}$$

Similar rules are written for filling the vessels 1, 2, or 3 (lines 20–33 in Listing 4.14). Prover9 found a five-step solution:

Step	0	1	2	3	4	5
State	$J(24, 0, 0, 0)$	$J(19, 0, 0, 5)$	$J(8, 11, 0, 5)$	$J(8, 11, 5, 0)$	$J(8, 3, 8, 5)$	$J(8, 8, 8, 0)$

The proof is illustrated in Fig. 4.6. With `assign(max_proofs,2)`, one can ask Prover9 to find two solutions.

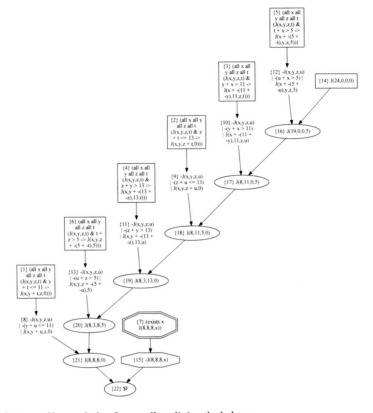

Fig. 4.6 A proof by resolution for equally splitting the balsam

References

Brooke, M. (1963). *150 Puzzles in Crypt-Arithmetic* (Vol. 1039). Dover Publications.

Danesi, M. (2018). *Ahmes' Legacy Puzzles and the Mathematical Mind*. Springer.

Darling, D. (2004). *The Universal Book of Mathematics from Abracadabra to Zeno's Paradoxes*. Wiley.

Dudeney, H. E. (2016). *536 Puzzles and Curious Problems*. Courier Dover Publications.

Gardner, M. (2005). *The Colossal Book of Short Puzzles and Problems*. W.W. Norton.

Gelfond, M., & Przymusinska, H. (1986). Negation as failure: Careful closure procedure. *Artificial Intelligence*, *30*(3), 273–287.

Kordemsky, B. A. (1992). *The Moscow Puzzles: 359 Mathematical Recreations*. Courier Corporation.

Memarsadeghi, N. (2016). NASA computational case study: Golomb rulers and their applications. *Computing in Science & Engineering*, *18*(6), 58–62.

Polash, M. A., Newton, M. H., & Sattar, A. (2017). Constraint-based search for optimal Golomb rulers. *Journal of Heuristics*, *23*(6), 501–532.

Reger, G., Suda, M., & Voronkov, A. (2016). Finding finite models in multi-sorted first-order logic. In *International Conference on Theory and Applications of Satisfiability Testing* (pp. 323–341). Springer.

Chapter 5
Lady and Tigers

Abstract In which we welcome the reader into the world of ladies and tigers created by Smullyan (2009). We formalised here the twelve puzzles appearing in the chapter Ladies or Tigers from Smullyan (2009). The proof displayed by Prover9 provides the reasoning steps that a human agent should follow to find himself the solution. In this way, Prover9 acts as a tool that augments human reasoning capabilities at problem-solving tasks.

$$\exists x \, (interrupt(I, x) \wedge (\forall y \, \neg interrupt(y, I)))$$

Winston S. Churchill

I welcome you here to the world of ladies and tigers built by Smullyan (2009) upon the short story published by F. R. Stockton, The *Lady, or the Tiger?*, 25th century 83, 83–86 (1882). You are a young prince in love with a beautiful princess. Her father, the ruler of the kingdom, asked you to choose between two doors, behind each there is either your beloved one or a killing tiger. There are signs on each door, but you don't know if they are true or false. Your dilemma is: which door to open in order to find the lady, marry her, and get half of the kingdom (as compensation, the wagging tongues would say).

Part of these puzzles has been formalised with Integer Programming by Chlond and Yoase (2003) or with Haskell by van Eijck and Zaytsev (2014).

I formalised here the twelve puzzles appearing in the chapter Ladies or Tigers from Smullyan (2009). Smullyan introduced his puzzles by increasing difficulty, which represents here a pedagogical advantage. By solving increasingly difficult puzzles provides students with motivation to accomplish all twelve levels of logical modelling tasks. Each puzzle is solved both with Mace4 and Prover9. Mace4 is used to find the solution. Prover9 is used to prove the solution found by Mace4. The proof displayed by Prover9 provides the reasoning steps that a human agent should follow to find himself the solution. In this way, Prover9 acts as a tool that augments human reasoning capabilities at problem-solving tasks.

Puzzle 37. Ladies and tigers in jail

Each of the two rooms contains either a lady or a tiger. There could be tigers in both rooms or ladies in both rooms. There is a sign on each room. One of the signs is true and the other is false. Which door to open in order to find the lady, marry her, and get half of the kingdom as compensation? (adapted from Smullyan 2009)

Room₁
In this room there is a lady and
in the other room there is a tiger

Room₂
In one of these rooms there is a lady and
in one of these rooms there is a tiger

Listing 5.1 Finding models when one sign is true and the other is false

```
1   assign(max_models,-1).
2   assign(domain_size,2).
3
4   formulas(day_1).
5     L1 & L2 | L1 & T2 | L2 & T1 | T1 & T2.
6     (L1 -> -T1) & (L2 -> -T2).
7     R1 <-> (L1 & T2).
8     R2 <->  (L1 & T2) | (T1 & L2).
9     (R1 & -R2) | (-R1 & R2).
10  end_of_list.
```

Room₁
In this room there is a lady and
in the other room there is a tiger

Room₂
In one of these rooms there is a lady and
in one of these rooms there is a tiger

Listing 5.2 Formalisation used by Prover9 to prove L_2: lady is in room 2

```
1   formulas(assumptions).
2     L1 & L2 | L1 & T2 | L2 & T1 | T1 & T2.
3     (L1 -> -T1) & (L2 -> -T2).
4     R1 <-> (L1 & T2).
5     R2 <->  (L1 & T2) | (T1 & L2).
6     (R1 & -R2) | (-R1 & R2).
7   end_of_list.
8
9   formulas(goals).
10    L2.
11  end_of_list.
```

Solution

Let the following formalisation: lady is in room 1 (L_1), lady is in room 2 (L_2), tiger is in room 1 (T_1), tiger is in room 2 (T_2), message in room 1 is true (R_1), and message in room 2 is true (R_2). Since all variables are boolean, a domain size of 2 suffices (line 2 in Listing 5.1). With two ladies and two tigers, there are four possible cases:

$$(L_1 \wedge L_2) \vee (L_1 \wedge T_2) \vee (L_2 \wedge T_1) \vee (T_1 \wedge T_2) \qquad (5.1)$$

We know there is a single prisoner in the cell: $(L_1 \to \neg T_1) \wedge (L_2 \to \neg T_2)$. We do not need to state also $T_1 \to \neg L_1$, as this is already inferred by contraposition.
The first message R_1 says there is a lady is in room 1 and a tiger is in room 2:

$$R_1 \leftrightarrow L_1 \wedge T_2 \qquad (5.2)$$

If the second message is true than there is a lady and a tiger in one of the rooms:

$$R_2 \leftrightarrow (L_1 \wedge T_2) \vee (L_2 \wedge T_1) \qquad (5.3)$$

One of the signs is true and the other is false:

$$(R_1 \wedge \neg R_2) \vee (\neg R_1 \wedge R_2) \qquad (5.4)$$

Mace4 outputs a single model, in which the message R_1 is false and R_2 is true:

Model	L_1	L_2	T_1	T_2	R_1	R_2
1	0	1	1	0	0	1

Solution

A proof by resolution for "lady is in room 2" (i.e. L_2) is obtained with:
```
prover9 -f day1p.in | prooftrans xml renumber | gvizify |
dot -Tpdf
```
Here the file *day*1.*p* appears in Listing 5.2. In Fig. 5.1, the theorem to prove appears in the hexagonal node with double edges ({5}). Its negated form appears in the hexagonal node with one edge ({13}). The input sentences appear in rectangular nodes: {1}, {2}, {3}, {4}. In the first step, these input sentences are translated to their clausal form, for instance: {6}, {7}, {8}, {9}, {10}, {11}, {12}. The reasoning steps during resolution are depicted with oval nodes. The theorem is proved by finding a contradiction (node {22}).
For the given puzzle, the prover starts by negating the goal, that is clause {13} $\neg L_2$. From {13} and {6} the system deduces that there is a tiger in room 2 (i.e. clause {15} T_2). With a tiger in room 2 the system infers {16}: either R_2 is true or there is no lady in room 1. Based on clause {18}, the message on door 2 should be true (i.e. {19} R_2). By considering clause {12}, the message on the first door is false (i.e. {21} $\neg R_1$). Since the message on door 2 is true from {19}, the prover uses clause {14} to deduce that a lady is in room 1 (i.e. {20} L_1). By applying hyper-resolution on {17}, {20} and {21}, a contradiction is identified (signalled with $F in clause {22}). Therefore, the initial assumption $\neg L_2$ was wrong. That is, the opposite is considered true: lady is in room 2.

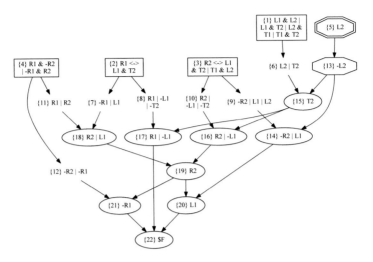

Fig. 5.1 In the first day lady is in room 2: L_2

Puzzle 38. The trials of the second day

Each of the two rooms contains either a lady or a tiger. There could be tigers in both rooms or ladies in both rooms. There is a sign on each room. The signs are either both true or both false. Which door to open in order to find the lady, marry her, and get half of the kingdom as compensation? (Smullyan 2009)

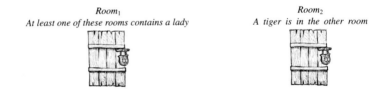

Room₁
At least one of these rooms contains a lady

Room₂
A tiger is in the other room

Listing 5.3 Both messages are either true or false

```
1    assign(max_models,−1).
2    assign(domain_size,2).
3
4    formulas(assumptions).
5      L1 & L2 | L1 & T2 | L2 & T1 | T1 & T2.
6      (L1 −> −T1) & (L2 −> −T2).
7      R1 <−> (L1 | L2).
8      R2 <−> T1.
9
10     (R2 & R1) | (−R2 & −R1).
11   end_of_list.
```

Solution

There are four situations: $(L_1 \wedge L_2) \vee (L_1 \wedge T_2) \vee (L_2 \wedge T_1) \vee (T_1 \wedge T_2)$. We know there is a single prisoner in the cell: $(L_1 \rightarrow \neg T_1) \wedge (L_2 \rightarrow \neg T_2)$.

The first message says there is a lady in room 1 or room 2:

$$R_1 \leftrightarrow L_1 \vee L_2 \tag{5.5}$$

If R_1 is false then its content is also false. Hence, we used the equivalence between the message and its content. Similarly, if the second message is true then there is a tiger in the first room: $R_2 \leftrightarrow T_1$.

Differently now, we learn that the signs are either both true or both false.

$$(R_1 \wedge R_2) \vee (\neg R_1 \wedge \neg R_2) \tag{5.6}$$

(Other modelling option would be $R_1 \leftrightarrow R_2$).

Mace4 computes a single model, in which both messages R_1 and R_2 are true:

Model	L_1	L_2	T_1	T_2	R_1	R_2
1	0	1	1	0	1	1

Room₁
At least one of these rooms contains a lady

Room₂
A tiger is in the other room

Solution

The human agent can follow the reasoning step by step by asking Prover9 for a demonstration (see Fig. 5.2). When adding, for instance, the goal L_2 to Listing 5.3, the reasoning steps during resolution are: The negated goal ($\neg L_2$ in clause {15}) is combined with the clausal form of implication {3} to infer based on the deduced clause {19} that L_1 is true ({20}). The clause {8} inferred from the input clause {2} is then used to support clause {22}, i.e. there is no tiger in room 1. By using hyper-resolution among three clauses {11}, {21} and {22}, a contradiction is found: $R_2 \rightarrow T_1$, R_2, and $\neg T_1$ cannot be simultaneously true. It means that the initial assumption $\neg L_2$ was wrong. Similar proofs can be obtained by adding theorems like T_1, $L_2 \wedge T_1$, or $R_1 \wedge R_2$.

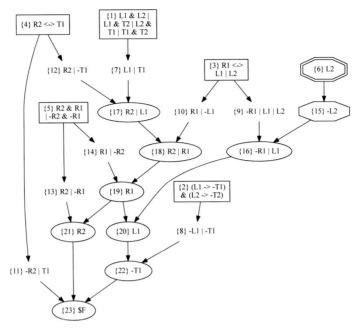

Fig. 5.2 In the second day lady stays in room 2: L_2

Puzzle 39. Ladies only

Each of the two rooms contains either a lady or a tiger. There could be tigers in both rooms or ladies in both rooms. There is a sign on each room. The signs are either both true of both false. Which door to open in order to find the lady, marry her, and get half of the kingdom as compensation? (Smullyan 2009)

$Room_1$
Either a tiger is in this room
or a lady is in the other room

$Room_2$
A lady is in the other room

Listing 5.4 There can be tigers or ladies in both rooms

```
1   assign(max_models, −1).
2   assign(domain_size, 2).
3
4   formulas(assumptions).
5       L1 & L2 | L1 & T2 | L2 & T1 | T1 & T2.
```

6	(L1 -> –T1) & (L2 -> –T2).
7	
8	R1 <-> (T1 | L2).
9	R2 <-> L1.
10	
11	(R2 & R1)	(–R2 & –R1).
12	end_of_list.

Solution

There are four situations: $(L_1 \land L_2) \lor (L_1 \land T_2) \lor (L_2 \land T_1) \lor (T_1 \land T_2)$. There is also a single prisoner in the cell: $(L_1 \rightarrow \neg T_1) \land (L_2 \rightarrow \neg T_2)$.

The first message R_1 says either a tiger is in this room or a lady is in the other room: $R_1 \leftrightarrow T_1 \lor L_2$. The second message R_2 says there is a lady in the other room: $R_2 \leftrightarrow L_1$. The signs are either both true of both false: $(R_1 \land R_2) \lor (\neg R_1 \land \neg R_2)$.

Mace4 finds a single model in which both messages are true:

Model	L_1	L_2	T_1	T_2	R_1	R_2
1	1	1	0	0	1	1

Room₁
Both rooms contain ladies

Room₂
Both rooms contain ladies

Solution

The human agent can follow the reasoning step by step by asking Prover9 for a demonstration (see Fig. 5.3). When adding, for instance, the goal L_1 to Listing 5.4, the reasoning steps during resolution are: The negated goal ($\neg L_1$ in clause {10}) is combined with the clausal form of implication {3} to infer that the second message is false (clause {11} $\neg R_2$). The same negated goal $\neg L_1$ appears positive in clause {6}, allowing resolution to deduce that the tiger is in room 1 (clause {12} $\neg T_1$). Since R_2 is false and given clause {9}, the system deduces that R_1 is also false. By using hyper-resolution among three clauses {12}, {13} and {7}, a contradiction is found: $\neg R_1 \rightarrow \neg T_1$, $\neg R_1$, and T_1 cannot be simultaneously true. It means that the initial assumption $\neg L_1$ was wrong. Similar proofs can be obtained by adding theorems like L_2 or $L_1 \land L_2$.

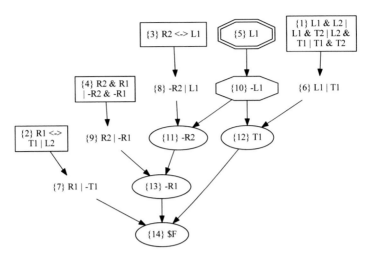

Fig. 5.3 In the third day lady stays in room 1: L_1

Room₁
Both rooms contain ladies

Room₂
Both rooms contain ladies

Listing 5.5 Fourth day: ladies are honest, but tigers are liars

```
1   assign(max_models,−1).
2   assign(domain_size,2).
3
4   formulas(assumptions).
```

```
 5 │  L1 & L2 | L1 & T2 | L2 & T1 | T1 & T2.
 6 │  (L1 -> -T1) & (L2 -> -T2).
 7 │  (R2 & R1) | (-R2 & -R1).
 8 │
 9 │  L1 -> R1.
10 │  T1 -> -R1.
11 │  L2 -> -R2.
12 │  T2 ->  R2.
13 │
14 │  R1 <-> (L1 & L2).
15 │  R2 <-> (L1 & L2).
16 │ end_of_list.
```

Room₁
Both rooms contain ladies

Room₂
Both rooms contain ladies

Solution

There are four situations: $(L_1 \wedge L_2) \vee (L_1 \wedge T_2) \vee (L_2 \wedge T_1) \vee (T_1 \wedge T_2)$. Similarly, there is a single prisoner in the cell: $(L_1 \to \neg T_1) \wedge (L_2 \to \neg T_2)$. We learned already that the signs are either both true of both false: $(R_1 \wedge R_2) \vee (\neg R_1 \wedge \neg R_2)$. There are four cases:

1. If a lady is in room 1, then the sign on the door is true: $L_1 \to R_1$.
2. If a tiger is in room 1, then the sign on the door is false: $T_1 \to \neg R_1$.
3. If a lady is in room 2, then the sign on the door is false: $L_2 \to \neg R_1$.
4. If a tiger is in room 2, then the sign on the door is true: $T_2 \to R_1$.

Both messages say that both rooms contain ladies:

$$(R_1 \leftrightarrow L_1 \wedge L_2) \wedge (R_2 \leftrightarrow L_1 \wedge L_2) \tag{5.7}$$

Mace4 finds a single model in which both messages R_1 and R_2 are false:

Model	L_1	L_2	T_1	T_2	R_1	R_2
1	0	1	1	0	0	0

A proof by resolution for "lady is in room 2" (i.e. L_2) is depicted in the illustration (Fig. 5.4).

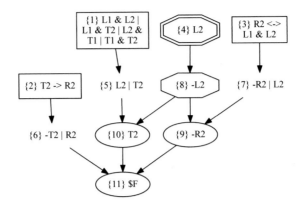

Fig. 5.4 In the fourth day lady stays in room 2: L_2

Puzzle 41. Fifth day

Each of the two rooms contains either a lady or a tiger. There could be tigers in both
rooms or ladies in both rooms. There is a sign on each room. If a lady is in room 1,
then the sign on the door is true. If a tiger is in room 1, then the sign on the door is
false. For room 2, the situation is opposite. If a lady is in room 2, then the sign on the
door is false. If a tiger is in room 2, then the sign on the door is true. The signs are
either both true of both false. Which door to open in order to find the lady, marry her,
and get half of the kingdom as compensation? (Smullyan 2009)

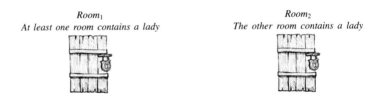

Listing 5.6 Fifth day: the signs are either both true of both false

```
1   assign(max_models,  −1).
2   assign(domain_size,  2).
3
4   formulas(assumptions).
5     L1 & L2  |  L1 & T2  |  L2 & T1  |  T1 & T2.
6
7     (L1 -> –T1) & (L2 -> –T2).
```

```
8
9       (R2 & R1) | (–R2 & –R1).
10
11      L1 –> R1.
12      T1 –> –R1.
13      L2 –> –R2.
14      T2 –> R2.
15
16      R1 <–> (L1 | L2).
17      R2 <–> L1.
18      end_of_list.
```

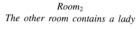

Room₁
At least one room contains a lady

Room₂
The other room contains a lady

Solution

With two rooms available, there are four possibilities: $(L_1 \land L_2) \lor (L_1 \land T_2) \lor (L_2 \land T_1) \lor (T_1 \land T_2)$. Each cell contains a single prisoner: $(L_1 \to \neg T_1) \land (L_2 \to \neg T_2)$. We know that signs are either both true of both false: $(R_1 \land R_2) \lor (\neg R_1 \land \neg R_2)$. The four constraints are formalised as follows:

1. If a lady is in room 1, then the sign on the door is true: $L_1 \to R_1$.
2. If a tiger is in room 1, then the sign on the door is false: $T_1 \to \neg R_1$.
3. If a lady is in room 2, then the sign on the door is false: $L_2 \to \neg R_1$.
4. If a tiger is in room 2, then the sign on the door is true: $T_2 \to R_1$.

The first message says that at least one room contains a lady: $R_1 \leftrightarrow L_1 \lor L_2$.
The second messages say that lady is in the other room: $R_2 \leftrightarrow L_1$.
Mace4 finds a single model in which both messages R_1 and R_2 are true.

Model	L_1	L_2	T_1	T_2	R_1	R_2
1	1	0	0	1	1	1

The human agent can learn a solution by following the proof in Fig. 5.5. Here, the theorem given to Prover9 was "lady is in room 1" (i.e. L_1).

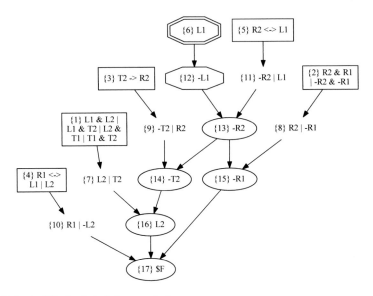

Fig. 5.5 In the fifth day lady is in room 1: L_1

Puzzle 42. Sixth day

Each of the two rooms contains either a lady or a tiger. There could be tigers in both rooms or ladies in both rooms. There is a sign on each room. If a lady is in room 1, then the sign on the door is true. If a tiger is in room 1, then the sign on the door is false. For room 2, the situation is opposite. If a lady is in room 2, then the sign on the door is false. If a tiger is in room 2, then the sign on the door is true. The signs are either both true of both false. Which door to open in order to find the lady, marry her, and get half of the kingdom as compensation? (Smullyan 2009)

Room₁
It makes no difference which room you pick

Room₂
The other room contains a lady

Listing 5.7 Lady is honest in one room but a liar in the other room

```
1   assign(max_models,-1).
2   assign(domain_size,2).
3
4   formulas(assumptions).
5     L1 & L2 | L1 & T2 | L2 & T1 | T1 & T2.
6     (L1 -> -T1) & (L2 -> -T2).
7     (R2 & R1) | (-R2 & -R1).
8
9     L1 -> R1.
10    T1 -> -R1.
11    L2 -> -R2.
12    T2 -> R2.
13
14    R1 <-> (L1 & L2) | (T1 & T2).
15    R2 <-> L1.
16  end_of_list.
```

Room₁
It makes no difference which room you pick

Room₂
The other room contains a lady

Solution

There are four situations to fill two rooms: $(L_1 \wedge L_2) \vee (L_1 \wedge T_2) \vee (L_2 \wedge T_1) \vee (T_1 \wedge T_2)$. Of course, there is a single prisoner in the cell: $(L_1 \rightarrow \neg T_1) \wedge (L_2 \rightarrow \neg T_2)$. The signs are either both true of both false: $(R_1 \wedge R_2) \vee (\neg R_1 \wedge \neg R_2)$. The four constraints are:

1. If a lady is in room 1, then the sign on the door is true: $L_1 \rightarrow R_1$.
2. If a tiger is in room 1, then the sign on the door is false: $T_1 \rightarrow \neg R_1$.
3. If a lady is in room 2, then the sign on the door is false: $L_2 \rightarrow \neg R_1$.
4. If a tiger is in room 2, then the sign on the door is true: $T_2 \rightarrow R_1$.

The first message says that there is no difference which room you pick:

$$R_1 \leftrightarrow (L_1 \wedge L_2) \vee (T_1 \wedge T_2) \tag{5.8}$$

The second messages say that the lady is in the other room: $R_2 \leftrightarrow L_1$. Mace4 finds a single model in which both messages are false.

Model	L_1	L_2	T_1	T_2	R_1	R_2
1	0	1	1	0	0	0

The human agent can learn how to solve the puzzle by following the proof computed by Prover9. Here the proof is for the theorem "lady is in room 2" L_2 (Fig. 5.6).

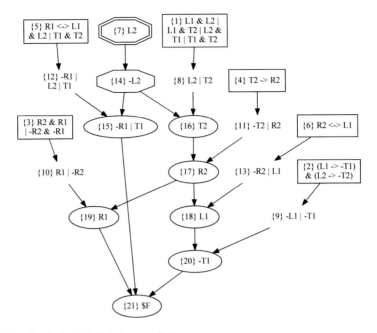

Fig. 5.6 In the sixth day lady is in room 2: L_2

Each of the two rooms contains either a lady or a tiger. There could be tigers in both
rooms or ladies in both rooms. There is a sign on each room. If a lady is in room 1,
then the sign on the door is true. If a tiger is in room 1, then the sign on the door is
false. For room 2, the situation is opposite. If a lady is in room 2, then the sign on the
door is false. If a tiger is in room 2, then the sign on the door is true. The signs are
either both true of both false. Which door to open in order to find the lady, marry her,
and get half of the kingdom as compensation? (Smullyan 2009)

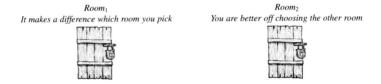

Listing 5.8 Seventh day: it makes no difference which room you pick

```
1   assign(max_models, -1).
2   assign(domain_size, 2).
3
4   formulas(assumptions).
5     L1 & L2 | L1 & T2 | L2 & T1 | T1 & T2.
6
7     (L1 -> -T1) & (L2 -> -T2).
8
9     (R2 & R1) | (-R2 & -R1).
10
11    L1 -> R1.
12    T1 -> -R1.
13    L2 -> -R2.
14    T2 -> R2.
15
16    R1 <-> (L1 & T2) | (T1 & L2).
17    R2 <-> L1 & T2.
18  end_of_list.
```

Room₁
It makes a difference which room you pick

Room₂
You are better of choosing the other room

<u>Solution</u>

There are four situations: $(L_1 \wedge L_2) \vee (L_1 \wedge T_2) \vee (L_2 \wedge T_1) \vee (T_1 \wedge T_2)$. There is a single prisoner in the cell: $(L_1 \rightarrow \neg T_1) \wedge (L_2 \rightarrow \neg T_2)$. The signs are either both true of both false: $(R_1 \wedge R_2) \vee (\neg R_1 \wedge \neg R_2)$.

1. If a lady is in room 1, then the sign on the door is true: $L_1 \rightarrow R_1$.
2. If a tiger is in room 1, then the sign on the door is false: $T_1 \rightarrow \neg R_1$.
3. If a lady is in room 2, then the sign on the door is false: $L_2 \rightarrow \neg R_1$.
4. If a tiger is in room 2, then the sign on the door is true: $T_2 \rightarrow R_1$.

The first message says that there is indeed a difference between which room you pick:

$$R_1 \leftrightarrow (L_1 \wedge T_2) \vee (T_1 \wedge L_2) \tag{5.9}$$

The second message warns that it is better to choose the other room: $R_2 \leftrightarrow L_1 \wedge T_2$. Mace4 finds a single model in which both messages are true.

Models	L_1	L_2	T_1	T_2	R_1	R_2
1	1	0	0	1	1	1

A proof by resolution for "lady is in room 1" (i.e. L_1) is depicted in Fig. 5.7.

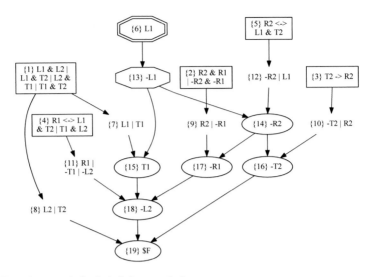

Fig. 5.7 In the seventh day lady is in room 1: L_1

Puzzle 44. Eighth day

Each of the two rooms contains either a lady or a tiger. There could be tigers in both rooms or ladies in both rooms. There is a sign on each room. If a lady is in room 1, then the sign on the door is true. If a tiger is in room 1, then the sign on the door is false. For room 2, the situation is opposite. If a lady is in room 2, then the sign on the door is false. If a tiger is in room 2, then the sign on the door is true. Now, the problem is that signs are not already put on the doors. That is, the two messages can appear either on the first door or on the second one. The signs are either both true of both false. Which door to open in order to find the lady, marry her, and get half of the kingdom as compensation? (Smullyan 2009)

Room$_1$ or Room$_2$?
This room contains a tiger

Room$_1$ or Room$_2$?
Both rooms contain tigers

Listing 5.9 Eighth day: this room contains a tiger

```
1   assign(max_models, −1).
2   assign(domain_size, 2).
3
4   formulas(assumptions).
5     L1 & L2 | L1 & T2 | L2 & T1 | T1 & T2.
6
7     (L1 −> −T1) & (L2 −> −T2).
8
9     (R2 & R1) | (−R2 & −R1).
10
11    L1 −>  R1.
12    T1 −> −R1.
13    L2 −> −R2.
14    T2 −>  R2.
15
16    (R1 <−> T1) | (R2 <−> T2).
17    (R2 <−> T1 & T2) | (R1 <−> T1 & T2).
18  end_of_list.
```

Room₁
Both rooms contain tigers

Room₂
This room contains a tiger

Solution

There are four situations: $(L_1 \wedge L_2) \vee (L_1 \wedge T_2) \vee (L_2 \wedge T_1) \vee (T_1 \wedge T_2)$. Of course, there is a single prisoner in the cell: $(L_1 \rightarrow \neg T_1) \wedge (L_2 \rightarrow \neg T_2)$. The signs are either both true of both false: $(R_1 \wedge R_2) \vee (\neg R_1 \wedge \neg R_2)$.

1. If a lady is in room 1, then the sign on the door is true: $L_1 \rightarrow R_1$.
2. If a tiger is in room 1, then the sign on the door is false: $T_1 \rightarrow \neg R_1$.
3. If a lady is in room 2, then the sign on the door is false: $L_2 \rightarrow \neg R_1$.
4. If a tiger is in room 2, then the sign on the door is true: $T_2 \rightarrow R_1$.

The first message says that there is a tiger in the room. There are two situations: the message can be on the first door or on the second one: $R_1 \leftrightarrow T_1 \vee R_2 \leftrightarrow T_2$. There are also two situations for the second message: $(R_1 \leftrightarrow T_1 \wedge T_2) \vee (R_2 \leftrightarrow T_1 \wedge T_2)$. Mace4 computes a single model, in which both messages should be false:

Model	L_1	L_2	T_1	T_2	R_1	R_2
1	0	1	1	0	0	0

A proof by resolution for "lady is in room 2" (i.e. L_2) is depicted in Fig. 5.8.

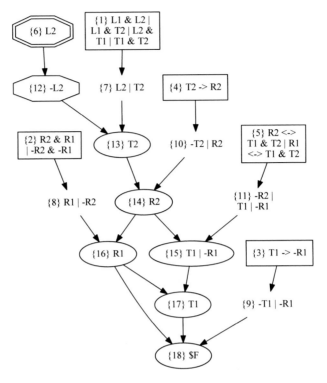

Fig. 5.8 In the eighth day lady is in room 2: L_2

Puzzle 45. Ninth day: three rooms

One room contains a lady and the other two contain tigers. At most one of the three signs is true. Which door to open in order to find the lady, marry her, and get half of the kingdom as compensation? (Smullyan 2009)

$Room_1$	$Room_2$	$Room_3$
A tiger is in this room	A lady is in this room	A tiger is in room 2

```
1   assign(max_models, −1).
2   assign(domain_size, 2).
3
4   formulas(assumptions).
5     L1 | L2 | L3.
6     (L1 -> –L2 & –L3) & (L2 -> –L1 & –L3) & (L3 -> –L1 & –L2).
7     (T1 & T2) | (T1 & T3) | (T2 & T3).
8
9     (L1 -> –T1) & (L2 -> –T2) & (L3 -> –T3).
10
11    (R1 & –R2 & –R3) | ( R2 & –R1 & –R3) |
12    (R3 & –R1 & –R2) | (–R1 & –R2 & –R3).
13
14    R1 <-> T1.              R2 <-> L2.              R3 <-> T2.
15  end_of_list.
```

Room₁
A tiger is in this room

Room₂
A lady is in this room

Room₃
A tiger is in room 2

Solution

One room contains a lady and the other two rooms contain tigers. We learn from this that there is exactly one lady and two tigers. "Exactly one lady" is formalised with two sentences:

1. There is at least one lady: $L_1 \vee L_2 \vee L_3$.
2. Lady is unique:
   $(L_1 \to (\neg L_2 \wedge \neg L_3)) \wedge (L_2 \to (\neg L_1 \wedge \neg L_3)) \wedge (L_3 \to (\neg L_1 \wedge \neg L_2))$.

Two tigers can be arranged in three rooms in three ways: $(T_1 \wedge T_2) \vee (T_1 \wedge T_3) \vee (T_2 \wedge T_3)$.

Each cell accommodates a single guest: $(L_1 \to \neg T_1) \wedge (L_2 \to \neg T_2) \wedge (L_3 \to \neg T_3)$. We do not need to state also the opposite implications (e.g. $T_1 \to \neg L_1$), since the system learns them by contraposition. Next we formalise that at most one of the messages is true:

$$(R_1 \wedge \neg R_2 \wedge \neg R_3) \vee (\neg R_1 \wedge R_2 \wedge \neg R_3) \vee (\neg R_1 \wedge \neg R_2 \wedge R_3) \vee (\neg R_1 \wedge \neg R_2 \wedge \neg R_3) \tag{5.10}$$

The three messages are formalised as follows: $(R_1 \leftrightarrow T_1) \wedge (R_2 \leftrightarrow L_2) \wedge (R_3 \leftrightarrow T_2)$. Mace4 computes a single model in which only the third message is true. Hence, there is a tiger in room 2 and 3, while the lady is in the first room.

Model	L_1	L_2	L_3	T_1	T_2	T_3	R_1	R_2	R_3
1	1	0	0	0	1	1	0	0	1

A proof by resolution for "lady is in room 1" (i.e. L_1) is depicted in Fig. 5.9.

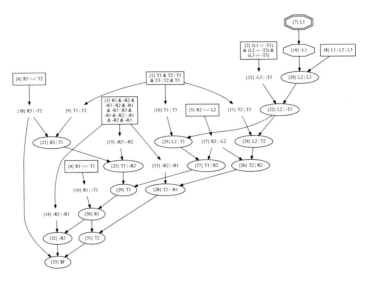

Fig. 5.9   In the ninth day lady stays in room 1: L_1

Puzzle 46. A lady and two tigers

One room contains a lady and the other two contain tigers. The sign on the door containing the lady is true and at least one of the other two signs is false. Which door to open in order to find the lady, marry her, and get half of the kingdom as compensation? (Smullyan 2009)

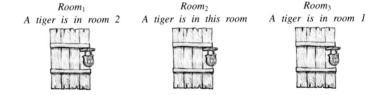

$Room_1$	$Room_2$	$Room_3$
A tiger is in room 2	*A tiger is in this room*	*A tiger is in room 1*

Listing 5.10   A lady and two tigers

```
1    assign(max_models,-1).
2    assign(domain_size,2).
3
4    formulas(assumptions).
5      L1 | L2 | L3.
6      (L1 -> -L2 & -L3) & (L2 -> -L1 & -L3) & (L3 -> -L1 & -L2).
7      (T1 & T2) | (T1 & T3) | (T2 & T3).
8      (L1 -> -T1) & (L2 -> -T2) & (L3 -> -T3).
9
10     (L1 -> R1) & (L2 -> R2) & (L3 -> R3).
11     (L1 -> -R2 | -R3) & (L2 -> -R1 | -R3) & (L3 -> -R1 | -R2).
12
13     (R1 <-> T2) & (R2 <-> T2) & (R3 <-> T1).
14   end_of_list.
```

Solution

One room contains a lady and the other two contain tigers. We learn from this that there is exactly one lady and two tigers. "Exactly one lady" is formalised with two sentences:

1. There is at least one lady: $L_1 \vee L_2 \vee L_3$.
2. Lady is unique:
$(L_1 \rightarrow (\neg L_2 \wedge \neg L_3)) \wedge (L_2 \rightarrow (\neg L_1 \wedge \neg L_3)) \wedge (L_3 \rightarrow (\neg L_1 \wedge \neg L_2))$.

$Room_1$
A tiger is in room 2

$Room_2$
A tiger is in this room

$Room_3$
A tiger is in room 1

Solution

Two tigers can be arranged in three rooms in three ways: $(T_1 \wedge T_2) \vee (T_1 \wedge T_3) \vee (T_2 \wedge T_3)$.

Each cell accommodates a single guest: $(L_1 \rightarrow \neg T_1) \wedge (L_2 \rightarrow \neg T_2) \wedge (L_3 \rightarrow \neg T_3)$.

Next, we formalise that the sign on the door containing the lady is true:

$$(L_1 \rightarrow R_1) \wedge (L_2 \rightarrow R_2) \wedge (L_3 \rightarrow R_3) \tag{5.11}$$

We know also that at least one of the other two signs is false:

$$(L_1 \rightarrow \neg R_2 \vee \neg R_3) \wedge (L_2 \rightarrow \neg R_1 \vee \neg R_3) \wedge (L_3 \rightarrow \neg R_1 \vee \neg R_2) \tag{5.12}$$

The three messages are formalised with: $(R_1 \leftrightarrow T_2) \wedge (R_2 \leftrightarrow T_2) \wedge (R_3 \leftrightarrow T_1)$.

Mace4 computes a single model with lady in room 1 and tigers in rooms 2 and 3. The message R_1 in room 1 is true, and also one sign from the other rooms (i.e. R_2).

Model	L_1	L_2	L_3	T_1	T_2	T_3	R_1	R_2	R_3
1	1	0	0	0	1	1	1	1	0

A human agent can learn how to solve the puzzle by following the reasoning steps in Fig. 5.10.

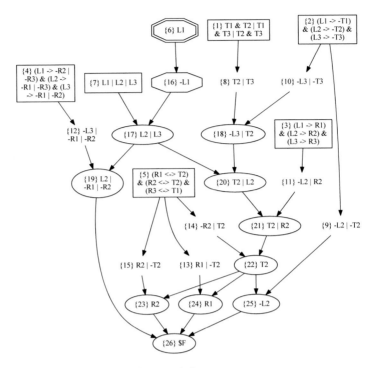

Fig. 5.10 In the tenth day lady stays in room 1: L_1

Puzzle 47. A lady, a tiger, and an empty room

One room contains a lady, another a tiger, and the third room is empty. The sign of the door containing a lady is true. The sign of the door containing a tiger is false. The sign of the empty room can be either true or false. Which door to open in order to find the lady, marry her, and get half of the kingdom as compensation? (Smullyan 2009)

Room₁
Room 3 is empty

Room₂
The tiger is in room 1

Room₃
This room is empty

Listing 5.11 A lady, a tiger, and an empty room

```
1   assign(max_models,−1).
2   assign(domain_size,2).
3
4   formulas(assumptions).
5     (L1 & T2 & E3) | (L1 & E2 & T3) | (T1 & L2 & E3) |
6     (T1 & E2 & L3) | (E1 & L2 & T3) | (E1 & T2 & L3) .
7
8     (L1 −> −T1 & −E1) & (L2 −> −T2 & −E2) & (L3 −> −T3 & −E3).
9     (T1 −> −L1 & −E1) & (T2 −> −L2 & −E2) & (T3 −> −L3 & −E3).
10
11    (L1 −> R1) & (L2 −> R2) & (L3 −> R3).
12    (T1 −> −R1) & (T2 −> −R2) & (T3 −> −R3).
13
14    R1 <−> E3.      R2 <−> T1.      R3 <−> E3.
15  end_of_list.
```

Solution

There are six situations with one lady, one tiger, and one empty room:

$$(L_1 \wedge T_2 \wedge E_3) \vee (L_1 \wedge E_2 \wedge T_3) \vee (T_1 \wedge L_2 \wedge E_3)$$
$$\vee (T_1 \wedge E_2 \wedge L_3) \vee (E_1 \wedge L_2 \wedge T_3) \vee (E_1 \wedge T_2 \wedge L_3)$$

We use two statements to formalise that there is only one prisoner in each room:

$$(L_1 \to \neg T_1 \wedge \neg E_1) \wedge (L_2 \to \neg T_2 \wedge \neg E_2) \wedge (L_3 \to \neg T_3 \wedge \neg E_3) \qquad (5.13)$$
$$(T_1 \to \neg L_1 \wedge \neg E_1) \wedge (T_2 \to \neg L_2 \wedge \neg E_2) \wedge (T_3 \to \neg L_3 \wedge \neg E_3) \qquad (5.14)$$

Based on these two equations, the system deduces that an empty room does not contain a tiger or a lady, for instance, $E_1 \to \neg L_1 \wedge \neg T_1$. We know that the sign on the door containing a lady is true: $(L_1 \to R_1) \wedge (L_2 \to R_2) \wedge (L_3 \to R_3)$. The sign on the door containing a tiger is false: $(T_1 \to \neg R_1) \wedge (T_2 \to \neg R_2) \wedge (T_3 \to \neg R_3)$. The sign on the empty room can be either true or false. This is a tautology.

	Room₂	*Room₃*
Room₁	*The tiger is in room 1*	*This room is empty*
Room 3 is empty		

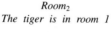

Solution

The three messages are formalised with: $(R_1 \leftrightarrow E_3) \wedge (R_2 \leftrightarrow T_1) \wedge R_3 \leftrightarrow E_3$.
Mace4 computes a single model, with lady in room 1, tiger in room 2 and room 3 is
empty. The messages are indeed true in room 1 (R_1) and false in room 2 ($\neg R_2$).

Model	L_1	L_2	L_3	T_1	T_2	T_3	E_1	E_2	E_3	R_1	R_2	R_3
1	1	0	0	0	1	0	0	0	1	1	0	1

A proof by resolution for "lady is room 1" (i.e. L_1) is depicted in Fig. 5.11.

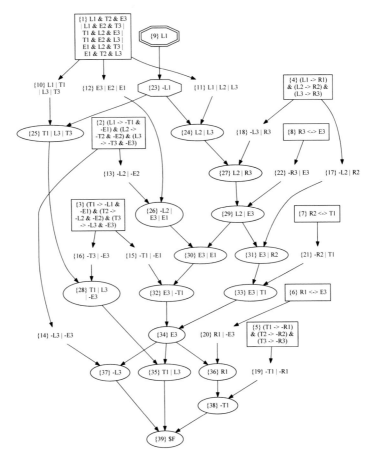

Fig. 5.11   In the eleventh day lady stays in room 1: L_1

There are nine rooms. Only one room contains the lady. Each of the other eight contains a tiger or they are empty. The sign on the door containing a lady is true. The signs on the doors containing tigers are false. The signs on the empty rooms can be either true or false.

You studied the situation for a long while: "The problem is unsolvable!" "I know" laughed the king. "Now, at least give me a decent clue: is Room 8 empty or not?". The king was decent enough to tell whether Room 8 was empty or not, and you are then able to deduce where the lady is. Which door to open in order to find the lady, marry her, and get half of the kingdom as compensation? (Smullyan 2009)

Room₁
Lady is an odd numbered room

Room₂
This room is empty

Room₃
Either sign 5 is right or sign 7 is wrong

Room₄
Sign 1 is wrong

Room₅
Either sign 2 or sign 4 is right

Room₆
Sign 3 is wrong

Room₇
Lady is not in room 1

Room₈
This room contains a tiger and room 9 is empty

Room₉
This room contains a tiger and sign 6 is wrong

Listing 5.12 A logical labyrinth in propositional logic

```
1    assign(max_models,−1).
2    assign(domain_size, 2).
3
4    formulas(assumptions).
5     L1 | L2 | L3 |L4 | L5 | L6 |L7 |L8 |L9.   %at least one lady
6     %T1 | T2 | T3 |T4 | T5 | T6 |T7 |T8 |T9.  %at least one tiger
7     %E1 | E2 | E3 |E4 | E5 | E6 |E7 |E8 |E9.  %at least one empty
8
9    %at most one lady
10    L1 −> −(L2 | L3 | L4 | L5 | L6 | L7 | L8 | L9).
11    L2 −> −(L1 | L3 | L4 | L5 | L6 | L7 | L8 | L9).
12    L3 −> −(L2 | L1 | L4 | L5 | L6 | L7 | L8 | L9).
13    L4 −> −(L2 | L3 | L1 | L5 | L6 | L7 | L8 | L9).
14    L5 −> −(L2 | L3 | L4 | L1 | L6 | L7 | L8 | L9).
15    L6 −> −(L2 | L3 | L4 | L5 | L1 | L7 | L8 | L9).
16    L7 −> −(L2 | L3 | L4 | L5 | L6 | L1 | L8 | L9).
17    L8 −> −(L2 | L3 | L4 | L5 | L6 | L7 | L1 | L9).
18    L9 −> −(L2 | L3 | L4 | L5 | L6 | L7 | L8 | L1).
19
20    %lady is true
21    (L1 −> R1) & (L2 −> R2) & (L3 −> R3) & (L4 −> R4) & (L5 −> R5).
22    (L6 −> R6) & (L7 −> R7) & (L8 −> R8) & (L9 −> R9).
23
24    %tiger is false
25    (T1 −> −R1) & (T2 −> −R2) & (T3 −> −R3) & (T4 −> −R4) & (T5 −> −R5).
26    (T6 −> −R6) & (T7 −> −R7) & (T8 −> −R8) & (T9 −> −R9).
27
28    %lady, tiger or empty
29    (L1 | T1 | E1) & (L2 | T2 | E2) & (L3 | T3 | E3) & (L4 | T4 | E4) &
30    (L5 | T5 | E5) & (L6 | T6 | E6) & (L7 | T7 | E7) & (L8 | T8 | E8) &
31    (L9 | T9 | E9).
32
33    %single occupant
34    (L1 −> −T1 & −E1) & (L2 −> −T2 & −E2) & (L3 −> −T3 & −E3).
35    (L4 −> −T4 & −E4) & (L5 −> −T5 & −E5) & (L6 −> −T6 & −E6).
36    (L7 −> −T7 & −E7) & (L8 −> −T8 & −E8) & (L9 −> −T9 & −E9).
37
38    (T1 −> −L1 & −E1) & (T2 −> −L2 & −E2) & (T3 −> −L3 & −E3).
39    (T4 −> −L4 & −E4) & (T5 −> −L5 & −E5) & (T6 −> −L6 & −E6).
40    (T7 −> −L7 & −E7) & (T8 −> −L8 & −E8) & (T9 −> −L9 & −E9).
41
42    (E1 −> −L1 & −T1) & (E2 −> −L2 & −T2) & (E3 −> −L3 & −T3).
43    (E4 −> −L4 & −T4) & (E5 −> −L5 & −T5) & (E6 −> −L6 & −T6).
44    (E7 −> −L7 & −T7) & (E8 −> −L8 & −T8) & (E9 −> −L9 & −T9).
45
46    R1 <−> L1 | L3 | L5 | L7 | L9.
47    R2 <−> E2.
48    R3 <−> (R5 | −R7).
49    R4 <−> −R1.
50    R5 <−> (R2 | R4).
51    R6 <−> −R3.
52    R7 <−> −L1.
53    R8 <−> T8 & E9.
54    R9 <−> T9 & −R6.
55
56    −E8. %extra clue − first case; L7 is true in all 8 models
57    %E8. %extra clue − second case
58    %L9. %L8. %L7. %L6. %L5. %L4. %L3. %L2. %L1.
59    end_of_list.
```

> **Solution**
>
> In line with the previous puzzles, we keep the formalisation in propositional logic. There are two alternative clues: room 8 is empty or not.
>
> For the first case, assume that the king says that room 8 is not empty $\neg E_8$ (step 0). It implies that the lady or the tiger is in room 8 ($L_8 \vee T_8$). If L_8 then the sign in room 8 is true. It follows that T_8 is also true, which means both the lady and the tiger are in the same room, which is impossible. It means that L_8 is not true and therefore the cell is occupied by a tiger (step 1). T_8 implies that its message R_8 should be false. $T_8 \wedge E_9$ is false iff room 9 is not empty $\neg E_9$ (step 2).
>
> $\neg E_9$ implies lady or tiger in room 9 ($L_9 \vee T_9$). If L_9 then the sign on her room is true (R_9). R_9 says T_9, which cannot be as the cell already contains the lady. With no lady in room 9, we learn that T_9 (step 3). With tiger in room 9, it follows that sign R_9 is false. That is sign R_6 is right. Therefore R_3 is false. It implies that $\neg R_5 \wedge R_7$ (step 4).
>
> From R_7 we learn that the lady is not in room 1. From $\neg R_5$ we know that $\neg R_2 \wedge \neg R_4$ (step 5). R_4 is false means that sign 1 is true. We learn now that lady is in 3, 5, 7, or 9. R_2 is false means room 2 is not empty, hence it contains a tiger (step 6).

Step	What we know	Messages
0	$\neg E_8$	
1	$\neg E_8, \neg L_8, T_8$	$\neg R_8$
2	$\neg E_8, \neg L_8, T_8 \ \neg E_9$	
3	$\neg E_8, \neg L_8, T_8 \ \neg E_9, \neg L_9, T_9$	$\neg R_8, \neg R_9$
4	$\neg E_8, \neg L_8, T_8 \ \neg E_9, \neg L_9, T_9$	$\neg R_8, \neg R_9, R_6, \neg R_3, \neg R_5, R_7$
5	$\neg E_8, \neg L_8, T_8 \ \neg E_9, \neg L_9, T_9, \neg L_1$	$\neg R_8, \neg R_9, R_6, \neg R_3, \neg R_5, R_7, \neg R_4, \neg R_2$
6	$\neg E_8, \neg L_8, T_8 \ \neg E_9, \neg L_9, T_9, \neg L_1, T_2$	$\neg R_8, \neg R_9, R_6, \neg R_3, \neg R_5, R_7, \neg R_4, \neg R_2, R_1$

At this moment, we know all signs if they are true or not:

R_1	R_2	R_3	R_4	R_5	R_6	R_7	R_8	R_9
1	0	0	0	0	1	1	0	0

Since only the signs R_1, R_6, and R_7 are true, princess can be only in one of these cells. We have learned in step 5 that $\neg L_1$. Since R_1 says the princess stays in odd rooms the only solution is L_7.

Based on the formalisation in Listing 5.12, Mace4 computes 8 models. In all these models only L_7 is true. Thereby, we can ask Prover9 to prove that lady is in room 7 in all models. Recall that this was only for the first case in which that the king said that room 8 is not empty (line 56).

For the second case (i.e. the king said that room 8 is empty), we investigate what would happen when the extra clue is E_8. Mace4 outputs 60 models:

Lady	L_1	L_2	L_3	L_4	L_5	L_6	L_7	L_8	L_9
Models (60)	24	0	4	8	4	0	20	0	0

These results were obtained by replacing $\neg E_8$ with E_8 in Listing 5.12 (line 57). Here we need to run Mace4 nine times, for each lady L_i, i=[1...9] (line 58). Since the lady can be in room 1 (24 models), room 3 (4 models), room 4 (8 models), room 5 (4 models), or room 7 (20 models), the agent cannot find which room to open.

We learn from this that what the king actually said is that the room 8 is not empty. For this case, Mace4 found eight models with L_7 true in all of them. Hence, Prover9 can prove that L_7 is true.

References

Chlond, M. J., & Toase, C. M. (2003). IP modeling and the logical puzzles of Raymond Smullyan. *INFORMS Transactions on Education, 3*(3), 1–12.

Smullyan, R. M. (2009). *The lady or the tiger?: and other logic puzzles*. Courier Corporation.

van Eijck, J., & Zaytsev, V. (2014). A course in Haskell-based software testing. Technical report, University of Amsterdam.

Chapter 6
Einstein Puzzles

Abstract In which, we provide formalisations of the various popular Einstein or zebra puzzles collected from Dudeney Dudeney (2016), Clessa Clessa (1996), Kordemsky Kordemsky (1992), or from the websites Math is fun and Brainzilla. The models are formalised in equational FOL and are given to Mace4. With each clue modelled in Mace4, the interested reader can see how the number of interpretation model decreases. From the modelling perspective, the task is to squeeze all the given clues until a single model remains. The twelve zebra-like puzzles modelled here equip students with more examples and modelling techniques behind the classical Einstein's riddle.

$$\exists x\, \exists y\, ((x \neq y \wedge (\forall z\, wayToLive(z) \leftrightarrow (z = x \vee z = y))) \wedge (wayToLive(x) \rightarrow$$
$$(\neg\exists u\, miracle(u))) \wedge (wayToLive(y) \rightarrow (\forall u\, miracle(u))))$$

Albert Einstein

The famous riddle attributed to Albert Einstein, the root of the so-called *zebra puzzles*, is considered as a textbook example Szeredi (2003) and it has been given some consideration in literature as a combinatorial optimisation problem Yeomans (2003). The Einstein's riddle asks you to identify five men of different nationalities and with different jobs, pets, and favourite drinks, and who live in consecutive houses on a street. To check your solution you have to answer the question: Who owns the zebra? With five nationalities, there are five possible nationality assignments to the first house. For each of these, there are four remaining nationalities that can be assigned to the second house (i.e. 4×3 combinations). For each of these, there are three nationalities left for the third house, two nationalities for the fourth house, and one for the fifth house. The total number of nationality assignments to houses is therefore $5 \times 4 \times 3 \times 2 \times 1 = 5!$. Assume that we pick one of the 5! nationality assignments. For this selection, we can choose 5! assignments for the colour of the houses. Following the same line of reasoning, there are $(5!)^5$ possible assignments. That is almost 25 billion assignments.

A plethora of approaches have been proposed for modelling or finding solution to zebra puzzles. For instance Yeoman has used the spreadsheet model of Microsoft Excel and its Solver Yeomans (2003), Bright et al. has used SAT solvers Bright et al. (2019), or logic programming Gregor et al. (2015).

The puzzles proposed in this chapter follow the zebra-like pattern and were collected from Dudeney Dudeney (2016), Clessa Clessa (1996), Kordemsky Kordemsky (1992), or from the websites Math is fun[1] and Brainzilla[2]. The models are formalised in equational FOL and are given to Mace4. With each clue modelled in Mace4, the interested reader can see how the number of interpretation model decreases. From the modelling perspective, the task is to squeeze all the given clues until a single model remains. Having more than one model signals that something hasn't been modelled from the given clues. In this case, Mace4 acts as a debugging tool, signalling that the problem hasn't been solved completely.

The 12 zebra-like puzzles modelled here equip students with more examples and modelling techniques behind the classical Einstein's riddle.

Puzzle 49. The magisterial bench

A bench of magistrates consists of two Englishmen, two Scotsmen, two Welshmen, one Frenchman, one Italian, one Spaniard, and one American. The Englishmen will not sit beside one another, the Scotsmen will not sit beside on another, and the Welshmen also object to sitting together. In how many different ways may the ten men sit in a straight line so that no two men of the same nationality shall ever be next to one another? (puzzle 447 from Dudeney (2016))

E_1 E_2 S_1 S_2 W_1 W_2 F I S A

Listing 6.1 The magisterial bench

```
1   set(arithmetic).
2   assign(domain_size,11).
3   assign(max_models,-1).
4
5   list(distinct).
6     [0,E1,E2,S1,S2,W1,W2,I,S,A,F].
7   end_of_list.
8
9   formulas(bench).
10    right_neighbor(x,y) <-> x+1 = y.
11    neighbors(x,y) <-> right_neighbor(x,y) | right_neighbor(y,x).
12
13    %-neighbors(E1,E2).    -neighbors(S1,S2).    -neighbors(W1,W2).
14    %neighbors(E1,E2) & neighbors(S1,S2) & neighbors(W1,W2).  %40,320 models
15    neighbours(E1,E2) & neighbors(S1,S2).
16  end_of_list.
```

---

[1] www.mathisfun.com.

[2] www.brainzilla.com.

> **Solution**

Let two Englishmen E_1 and E_2, two Scotsmen S_1 and S_2, two Welshmen W_1 and W_2 and the remaining persons F, I, S, A. These persons are distinct are different of 0 (line 6 in Listing 6.1). With ten people on the bench, we use a domain size of 11. A human agent can reason as follows. Without any constraints, ten persons can sit in $10! = 3,628,800$ ways. From this number, we subtract the cases that do not satisfy the puzzle constraints:

First, the arrangements in which the Englishmen stay next to each other, the Scotsmen stay next to each other, and also the Welshmen (i.e. $(E_i E_j)(S_i S_j), (W_i W_j), i \neq j$) are $7! \times 2^3$ Indeed, for these three constraints:

$$neighbours(E_1, E_2) \wedge neighbours(S_1, S_2) \wedge neighbours(W_1, W_2) \tag{6.1}$$

If line 13 in Listing 6.1 is used instead of line 12, Mace4 correctly identifies $7! \times 2^3 = 40,320$ models.

Second, the arrangements $(E_i E_j)(S_i S_j)$ (where the W_1 and W_2 are not bracketed but free) occur in $8! \times 2^2$ cases. Indeed, for the two constraints:

$$neighbours(E_1, E_2) \wedge neighbours(S_1, S_2) \tag{6.2}$$

Mace4 correctly identifies $8! \times 2^2 = 161,280$ models (line 15). The same number of models is identified for the two Scotsmen S_1 and S_2 unbracketed $(E_i E_j)(W_i W_j)$, or when the two Englishmen E_i and E_2 unbracketed $(S_i S_j), (W_i W_j)$. Hence, the cases in which two men from the same nationality are unbracked are $3 \times 161,280 = 724,560$. Third, we may have $(E_i E_j)S_i S_j W_i W_j F I S A$, where both Scotsmen and both Welshmen are unbracketed. This gives $9! \times 2$ cases. Similarly, for only S_1 and S_2 bracketed or W_1 and W_2 bracketed.

When adding these three cases together, we get 1,733,760 models:

$$40,320 + 3 \times 161,280 + 3 \times 724,560 = 1.733.760 \tag{6.3}$$

which deducted from the number first given (10!=3,628,800 ways) leaves us with 1,895,040 solutions.

The software agent (i.e. Mace4) also finds 1,895,040 solutions based on the constraint:

$$\neg neighbours(E_1, E_2) \wedge \neg neighbours(S_1, S_2) \wedge \neg neighbours(W_1, W_2) \tag{6.4}$$

Some models are

Model	E_1	E_2	S_1	S_2	W_1	W_2	A	F	I	S
1	1	3	2	4	5	7	6	8	9	10
2	1	3	2	4	5	7	6	8	10	9
3	3	1	5	3	4	8	6	7	9	10
4	1	3	2	4	8	5	6	7	9	10
5	1	3	5	2	4	8	6	7	10	9
6	10	8	9	7	6	4	5	3	2	1

We can see here a way to use Mace4 to validate the human reasoning. Mace4 can validate the number of models estimated by the human agent in each case.

Puzzle 50. Ships

There are five ships in a port:

1. The Greek ship leaves at six and carries coffee.

2. The Ship in the middle has a black exterior.

3. The English ship leaves at nine.

4. The French ship with blue exterior is to the left of a ship that carries coffee.

5. To the right of the ship carrying cocoa is a ship going to Marseille.

6. The Brazilian ship is heading for Manila.

7. Next to the ship carrying rice is a ship with a green exterior.

8. A ship going to Genoa leaves at five.

9. The Spanish ship leaves at 7 and is to the right of the ship going to Marseille.

10. The ship with a red exterior goes to Hamburg.

11. Next to the ship leaving at seven is a ship with a white exterior.

12. The ship on the border carries corn.

13. The ship with a black exterior leaves at eight.

14. The ship carrying corn is anchored next to the ship carrying rice.

15. The ship to Hamburg leaves at six.

Which ship goes to Port Said? Which ship carries tea? (taken from Math is fun—www.mathisfun.com)

Listing 6.2 Five ships in a port

```
1    set ( arithmetic ).
2    assign ( domain_size ,5).
3    assign ( max_models , −1).
4
5    list ( distinct ).
6        [ Greek , English , French , Brazilian , Spanish ].
7        [ Black , Blue , Green , Red , White ].
8        [ Nine , Five , Seven , Eight , Six ].
9        [ Hamburg , Genoa , Marseille , Manila , Port_Said ].
10       [ Tea , Coffee , Cocoa , Rice , Corn ].
11   end_of_list .
12
13   formulas ( utils ).
14       right_neighbor (x,y) <−> x + 1 = y.
15       left_neighbor (x,y)  <−> x = y + 1.
16       neighbor (x,y) <−> right_neighbor (x,y) | left_neighbor (x,y).
17       middle (x) <−> x = 2.
18       border (x) <−> x = 4 | x = 0.
19   end_of_list .
20
21   formulas ( ships ).
22       Greek = Six & Greek = Coffee                              #label ( Clue1 ).
23       middle ( Black )                                          #label ( Clue2 ).
24       English = Nine                                            #label ( Clue3 ).
25       French = Blue & left_neighbor ( Coffee , French )         #label ( Clue4 ).
26       right_neighbor ( Cocoa , Marseille )                      #label ( Clue5 ).
27       Brazilian = Manila                                       #label ( Clue6 ).
28       neighbor ( Rice , Green )                                 #label ( Clue7 ).
29       Genoa = Five                                             #label ( Clue8 ).
30       Spanish = Seven & right_neighbor ( Marseille , Spanish )  #label ( Clue9 ).
```

```
32        Hamburg = Red                                        #label(Clue10).
33        neighbor(Seven , White)                              #label(Clue11).
34        border(Corn)                                         #label(Clue12).
35        Black = Eight                                        #label(Clue13).
36        neighbor(Corn , Rice)                                #label(Clue14).
37        Hamburg = Six                                        #label(Clue15).
38   end_of_list.
```

Solution

There are five distinct nationalities, colours, hours, cities, or drinks (lines 5–11 in Listing 6.2). We need to define auxiliary predicates: $neighbour(x, y)$, x has neigbour y; $middle(x)$, x is in the middle; or $border(x)$, x is on the border. Given that the domain size is 5, their corresponding definitions appear in lines 13–20. The 15 clues are formalised in lines 22–38. We rely here on functions instead of predicates. Mace4 finds a single model:

Domain value	Country	Time	Carry	Colour	Destination
0	French	5:00	tea	blue	Genoa
1	Greek	6:00	coffee	red	Hamburg
2	Brazilian	8:00	cocoa	black	Manila
3	English	9:00	rice	white	Marseille
4	Spanish	7:00	corn	green	Port Said

Puzzle 51. The ladies of the committee

Six ladies are eligible for the offices of Captain, Vice-captain, and Treasurer (in descending order of seniority), in the local ladies' golf club (puzzle 33 in Clessa (1996)).

1. Audrey won't serve if Elaine is Captain, or if Freda is Treasurer.

2. Betty won't be Treasurer if Cynthia is one of the officials.

3. Audrey won't serve with both Betty and Elaine.

4. Freda won't serve if Elaine is also an official.

5. Betty refuses to be Vice-captain.

6. Freda won't serve if she outranks Audrey.

7. Cynthia won't serve with Audrey or Betty unless she is Captain.

8. Doris won't serve unless Betty is Captain.

9. Betty won't serve with Doris unless Elaine is also an official.

10. Elaine won't serve unless she or Audrey is Captain.

How can the three offices be filled?

Solution

With six ladies, we need a domain size of 6. Our assumptions are: First, the six ladies
are distinct (line 5 in Listing 6.3). Second, there are only three positions a person can
serve. Hence, we used the closed predicate:

$$\forall x \ (serve(x) \leftrightarrow captain(x) \lor vice(x) \lor treasurer(x)) \tag{6.5}$$

Third, a person can have only one job:

$$\forall x \ \forall y \ (captain(x) \land captain(y) \rightarrow x = y). \tag{6.6}$$

$$\forall x \ \forall y \ (vice(x) \land vice(y) \rightarrow x = y). \tag{6.7}$$

$$\forall x \ \forall y \ (treasurer(x) \land treasurer(y) \rightarrow x = y). \tag{6.8}$$

Listing 6.3 Ladies in the committee

```
1    assign(domain_size ,6).
2    assign(max_models , -1).
3
4    list(distinct).
5      [Audrey , Elaine , Betty , Freda , Cynthia , Doris ].
6    end_of_list.
7
8    formulas(assumptions).
9      all  x  (serve(x) <-> captain(x) | vice(x) | treasurer(x)).
10
11     all  x  (captain(x)    -> -vice(x)    & -treasurer(x)).
12     all  x  (vice(x)       -> -captain(x) & -treasurer(x)).
13     all  x  (treasurer(x) -> -captain(x) & -vice(x)).
14
15     all  x  all  y  (captain(x)    & captain(y)    -> x=y).
16     all  x  all  y  (vice(x)       & vice(y)       -> x=y).
17     all  x  all  y  (treasurer(x) & treasurer(y) -> x=y).
18
19     all  x  all  y  (outranks(x,y) <-> captain(x) | (vice(x) & treasurer(y))).
20
21     exists  x  (captain(x)) & exists  x  (vice(x)) & exists  x  (treasurer(x)).
22   end_of_list.
23
24   formulas(puzzle_clues).
25     -serve(Audrey) <- (captain(Elaine) | treasurer(Freda))    #label(Clue1 ).
26     -treasurer(Betty) <- serve(Cynthia)                        #label(Clue2 ).
27     -serve(Audrey) <- serve(Betty) & serve(Elaine)            #label(Clue3 ).
28     -serve(Freda) <- serve(Elaine)                             #label(Clue4 ).
29     -vice(Betty)                                               #label(Clue5 ).
30     outranks(Freda ,Audrey) -> -serve(Freda)                  #label(Clue6 ).
31     (serve(Audrey) | serve(Betty)) & -captain(Cynthia)
32            -> -serve(Cynthia)                                  #label(Clue7 ).
33     -captain(Betty) -> -serve(Doris)                          #label(Clue8 ).
34     -serve(Elaine) -> -(serve(Betty) & serve(Doris))          #label(Clue9 ).
35     -(captain(Elaine) | captain(Audrey)) -> -serve(Elaine)    #label(Clue10 ).
36   end_of_list.
```

Solution

There is an order relation between these jobs. Captain outranks vice and vice outranks treasurer: $\forall x\ \forall y\ (outranks(x, y) \leftrightarrow captain(x) \lor (vice(x) \land treasurer(y)))$. Each job should be fulfilled by a lady: $\exists x\ captain(x) \land \exists x\ vice(x) \land \exists x\ treasurer(x)$. Note there are three distinct variables (with the same name x). Given these assumptions, the clues are formalised as follows:

1. Audrey won't serve if Elaine is Captain, or if Freda is Treasurer:

$$captain(Elaine) \lor treasurer(Freda) \rightarrow \neg serve(Audrey) \qquad (6.9)$$

2. Betty won't be Treasurer if Cynthia is one of the officials:

$$serve(Cynthia) \rightarrow \neg treasurer(Betty) \qquad (6.10)$$

   Note that based on contraposition, the clue can be written as
   $treasurer(Betty) \rightarrow \neg serve(Cynthia)$
3. Audrey won't serve with both Betty and Elaine:

$$serve(Betty) \land serve(Elaine) \rightarrow \neg serve(Audrey) \qquad (6.11)$$

4. Freda won't serve if Elaine is also an official:
   $serve(Elaine) \rightarrow \neg serve(Freda)$.
5. Betty refuses to be Vice-captain: $\neg vice(Betty)$.
6. Freda won't serve if she outranks Audrey (we need the predicate $outranks$):

$$outranks(Freda, Audrey) \rightarrow \neg serve(Freda) \qquad (6.12)$$

7. Cynthia won't serve with Audrey or Betty unless she is Captain:

$$(serve(Audrey) \lor serve(Betty)) \land \neg captain(Cynthia) \rightarrow \neg serve(Cynthia)$$
$$(6.13)$$

8. Doris won't serve unless Betty is Captain: $\neg captain(Betty) \rightarrow \neg serve(Doris)$.
   Based on contraposition, this is the same as: $serve(Doris) \rightarrow captain(Betty)$).
9. Betty won't serve with Doris unless Elaine is also an official:

$$\neg serve(Elaine) \rightarrow \neg(serve(Betty) \land serve(Doris)) \qquad (6.14)$$

10. Elaine won't serve unless she or Audrey is Captain:

$$\neg(captain(Elaine) \lor captain(Audrey)) \rightarrow \neg serve(Elaine) \qquad (6.15)$$

The implementation in Listing 6.3 outputs a single model in which $captain(Audrey)$, $vice(Freda)$, and $treasurer(Betty)$.

Audrey: 0   Betty: 1   Cynthia: 2   Doris: 3   Elaine: 4   Freda: 5

c1: 0   c2: 5   c3: 1

captain:

0	1	2	3	4	5
1	0	0	0	0	0

serve:

0	1	2	3	4	5
1	1	0	0	0	1

treasurer:

0	1	2	3	4	5
0	1	0	0	0	0

outranks:

	0	1	2	3	4	5
0	1	1	1	1	1	1
1	0	0	0	0	0	0
2	0	0	0	0	0	0
3	0	0	0	0	0	0
4	0	0	0	0	0	0
5	0	1	0	0	0	0

vice:

0	1	2	3	4	5
0	0	0	0	0	1

Puzzle 52. Cocktail party

Six people meet at a cocktail party. Their names are Annie, Brian, Celia, Don, Erica, and Frank, and their professions, though not necessarily, respectively, are Teacher, Engineer, Programmer, Doctor, Accountant, and Solicitor.

1. Frank and the Teacher both vote Tory.
2. Don and the Engineer both vote Liberal.
3. Annie and the Programmer both vote Labour.
4. Celia and Erica are both Scots. The programmer is Welsh.
5. The Accountant is older than Frank.
6. The Solicitor is older than Annie.
7. Celia and the Teacher both adore classical music.
8. Annie and the Accountant both hate the classics but love jazz.

What is each person's respective profession? (puzzle 65 from Clessa (1996))

Listing 6.4 Modelling the cocktail party in propositional logic

```
1    assign(domain_size ,6).
2    assign(max_models , -1).
3
4    list(distinct).
5      [Annie , Brian , Celia , Don , Erica , Frank ].
6    end_of_list.
7
8    formulas(assumptions).
9      Teacher = 0.    Engineer    = 1.    Programmer  = 2. %remove isomorphisms
10     Doctor  = 3.    Accountant  = 4.    Solicitator = 5.
11
12     Frank != Teacher.        %Clue 1
13     Don   != Engineer.       %Clue 2
14     Frank != Engineer.       %Clue 1 and 2
15     Don   != Teacher.        %Clue 1 and 2
16     Annie != Programmer.     %Clue 3
17     Frank != Programmer.     %Clue 3 and 1
18     Don   != Programmer.     %Clue 3 and 2
19     Annie != Teacher.        %Clue 3 and 1
```

```
20    Annie != Engineer.          %Clue  3  and  2
21    Erica != Programmer.        %Clue  4
22    Celia != Programmer.        %Clue  4
23    Frank != Accountant.        %Clue  5
24    Annie != Solicitator.       %Clue  6
25    Celia != Teacher.           %Clue  7
26    Annie != Accountant.        %Clue  8
27    Celia != Accountant.        %Clue  8  and  7
28    Annie != Teacher.           %Clue  8  and  7
29  end_of_list.
```

Solution

We use here functions, instead of predicates. Since, there are six persons and six professions, we need a domain size of 6. To avoid isomorphic models, we fix each profession to a specific value in the domain. For instance, the teacher is set to 0 (see line 9 in Listing 6.4).

The task is to assign names to each of these fixed professions. Of course, each name represents a distinct value from the domain (line 5). That is, each name has a different profession. Hence, we do not have to state explicitly this constraint. Also, we do not need to state that each name has at least one job. Since, the names will be assigned to distinct values, all the six values from the domain will be assigned.

What we need to model are only the clues (see lines 12–28). For instance, from the first clue, we learn that Franks is distinct of the teacher (line 12). Since Frank vote Tory (clue 1) and Engineer vote Liberal (clue 2), we learn that Frank is not the engineer. Given the constraints in lines 12–28, Mace4 finds a single model: Annie was the Doctor, Brian the Programmer, Celia the Engineer, Don the Accountant, Erica the Teacher, Frank the Solicitor.

Puzzle 53. Borrowed books

Four children, borrowed a specific book from the library. Each book has a unique subject and number of pages. Find out in which month each child got the book and their number of pages. You know that:

1. If Janet got a History book, then the Fantasy book has 350 pages.

2. The Thriller is 50 pages larger than the book borrowed in May, but 50 pages shorter than Anthony's choice.

3. A girl borrowed a book in May, the other girl, Meredith, got a History book.

4. Janet didn't get the shortest book of 250 pages.

5. Exactly one of the customers shares the initial in the name and in the month he/she got a book.

6. The fantasy is 50 pages larger than Mark's book.

7. In June a Fantasy book wasn't borrowed, nor a Romance one.

8. The largest book and the Romance one have been borrowed in consecutive months.

9. If Mark borrowed a book in the first month (i.e. April), then the Fantasy book is the largest one with 400 pages long.

(taken from Brainzilla—www.brainzilla.com)

┌─ **Solution** ───┐

By reading the clues, we learn that the children names are: Antony (clue 2), Meredith (clue 3), Janet (clue 4), Mark (clue 6). The subjects are history (clue 1), fantasy
(clue 1), romance (clue 7), thriller (clue 2). The number of pages are: 250 (clue 4),
350 (clue 1), 400 (clue 9), and 300 (inferred by us). There are three mentioned months:
April (clue 9), May (clue 3), June (clue 7). As they are consecutive, the fourth one can
be either March or July. March is not the case since clue 8 mentions April as the first
month of the story.

Since there are four children, books, types, pages and months, we set the domain size
to 4. We state also in lines 6–9 that the individuals are distinct.

└──┘

┌─ **Solution** ───┐

Then, we can fix one of the variables (e.g. number of pages) in order to remove some
isomorphic models: $p250 = 0 \land p300 = 1 \land p350 = 2 \land p400 = 3$.

Next, we define two predicates that appear in the clues: $50plus(x, y)$ for "book x has
50 pages more than book y", and $next(x, y)$ for "month x is consecutive to month y".
For the predicate $50plus(x, y)$, it is enough to specify that two books do not differ by
50 pages if their difference is not one:

$$\neg 50plus(x, y) \leftrightarrow x \neq y + 1 \tag{6.16}$$

└──┘

Listing 6.5 Borrowed books

```
1    set ( arithmetic ).
2    assign ( domain_size ,4 ).
3    assign ( max_models , −1 ).
4
5    list ( distinct ).
6       [ Janet , Anthony , Mark , Meredith ].
7       [ Fantasy , Romance , Thriller , History ].
8       [ April , May , June , July ].
9       [ p250 , p300 , p350 , p400 ].
10   end_of_list .
11
12   formulas ( utils ).
13      p250 = 0 & p300 = 1 & p350 = 2 & p400 = 3.              %avoid  isomorphisms
14      −50plus ( x , y ) <−> x != y + 1.
15      next ( x , y ) <−> next ( y , x ).
16      −next ( x , x ).
17      next ( April , May )   &   next ( May , June )   &   next ( June , July ).
18      −next ( April , June ) & −next ( April , July ) & −next ( May , July ).
19   end_of_list .
20
21   formulas ( clues ).
22      ( Janet = History )  −>  Fantasy = p350.                        %Clue 1
23      50plus ( Thriller , May ) & 50plus ( Anthony , Thriller ).      %Clue 2
24      Janet = May & Meredith = History.                              %Clue 3
25      Janet != p250.                                                 %Clue 4
26      Janet = June                                                   %Clue 5
27        <−> ( Janet != July & Meredith != May & Mark != May & Anthony != April ).
28      Janet = July
29        <−> ( Janet != June & Meredith != May & Mark != May & Anthony != April ).
30      Meredith = May
31        <−> ( Janet != June & Janet != July & Mark != May & Anthony != April ).
32      Mark = May  <−> ( Janet != June &
33             Janet != July & Meredith != May & Anthony != April ).
34      Anthony = April <−> ( Janet != June &
35             Janet != July & Meredith != May & Mark != May ).
36      50plus ( Fantasy , Mark ).                                     %Clue 6
37      June != Fantasy & June != Romance.                             %Clue 7
38      next ( p400 , Romance ).                                       %Clue 8
39      ( Mark = April )  −>  Fantasy = p400.                          %Clue 9
40   end_of_list .
```

Solution

The consecutive months predicate $next(x, y)$ is symmetric and irreflexive:

$$next(x, y) \leftrightarrow next(y, x) \wedge \neg next(x, x). \tag{6.17}$$

Then we specify which months are consecutive or not:

$$next(April, May) \wedge next(May, June) \wedge next(June, July) \tag{6.18}$$
$$\neg next(April, June) \wedge \neg next(April, July) \wedge \neg next(May, July) \tag{6.19}$$

The information above together with properties in equation 6.17 are enough to completely determine the predicate $next(x, y)$. Given these assumptions, we can start formalising the clues.

1. If Janet got a History book, then the Fantasy book has 350 pages:

$$Janet = History \rightarrow Fantasy = p350 \tag{6.20}$$

2. The Thriller is 50 pages larger than the book borrowed in May, but 50 pages shorter than Anthony's choice:

$$50plus(Thriller, May) \wedge 50plus(Anthony, Thriller) \tag{6.21}$$

3. A girl borrowed a book in May, the other girl, Meredith, got a History book:

$$Janet = May \wedge Meredith = History. \tag{6.22}$$

4. Janet didn't get the shortest book of 250 pages: $Janet \neq p250$.
5. Exactly one of the customers shares the initial in the name and in the month he/she got a book.

   $$Janet = June \vee Janet = July \vee Meredith = May \vee Mark = May \vee Anthony = April.$$

   $$Janet = June \rightarrow (Janet \neq July \wedge Meredith \neq May \wedge Mark \neq May \wedge Anthony \neq April)$$

   First, we specify that at least one child shares the initials. Then we have to specify that the other children do not share this. Equation 6.23 states this at most constraints for Janes only.
6. The fantasy is 50 pages larger than Mark's book: $50plus(Fantasy, Mark)$
7. In June a Fantasy book wasn't borrowed, nor a Romance one.

$$June \neq Fantasy \wedge June \neq Romance \tag{6.23}$$

8. The largest book and the Romance one have been borrowed in consecutive months: $next(p400, Romance)$.
9. If Mark borrowed a book in the first month (i.e. April), then the Fantasy book is the largest one with 400 pages long: $Mark = April \rightarrow Fantasy = p400$.

Solution

Given the formalisation in Listing 6.5, Mace4 finds 2 models:

Domain value	Pages	Name	Book	Month		next:	0	1	2	3
0	250	Meredith	History	June		0	0	1	1	0
1	300	Janet	Romance	May		1	1	0	0	1
2	350	Mark	Thriller	July		2	1	0	0	0
3	400	Anthony	Fantasy	April		3	0	1	0	0

Domain value	Pages	Name	Book	Month		next:	0	1	2	3
0	250	Meredith	History	July		0	0	0	1	0
1	300	Janet	Romance	May		1	0	0	1	1
2	350	Mark	Thriller	June		2	1	1	0	0
3	400	Anthony	Fantasy	April		3	0	1	0	0

There are only two differences between these two models. In the first model, $June = 0$ and $July = 2$, while in the second model, the other way around.

Puzzle 54. Baseball coach dilemma

A baseball coach recalls that:

1. Charles played 5 more games than the player who wore number 28.
2. Jorge wore number 3.
3. Martin or who played shortstop, one wore number 21 and the other played 11 games.
4. Armando either wore number 32 or number 21.
5. Jorge either played 13 games or played first base.
6. The player who played 13 games didn't wear number 29.
7. Armando didn't play shortstop.
8. The right field player played 1 more game than the center field player.
9. Pedro played somewhat fewer games than Martin.
10. Neither Russell nor the player who played 13 games was number 21.
11. Benny didn't play second base.
12. Russell was either the boy who played right field or played 12 games.
13. Number 35 played somewhat fewer games than number 28.
14. Neither number 3 nor the player who played 13 games was Armando.
15. Number 29 was either the player who played 8 games or the person who played third base.
16. Armando didn't play 10 games.

Could you help the coach to figure out the number of games played by each player, their numbers and positions? (taken from Math is fun—www.mathisfun.com)

Solution

Let the seven players denoted by their initials. The seven positions are: central field
(p_{cf}), right field (p_{rf}), left field (p_{lf}), first base (p_{fb}), second base (p_{sb}), shortstop
(p_{ss}). Each player played between 8 and 14 games (g_8–g_{14}), while numbers on the t-
shirt are denoted with n (e.g. n_{32}).

Listing 6.6 Baseball coach dilemma

```
1    set(arithmetic).
2    assign(domain_size ,7).
3    assign(max_models , -1).
4
5    list(distinct).
6      [A,B,C,J,M,P,R].              [p_cf , p_rf , p_fb , p_tb , p_sb , p_lf , p_ss ].
7      [g8,g9,g10,g11,g12,g13,g14].  [n35 , n28 , n3 , n29 , n32 , n7 , n21 ].
8    end_of_list.
9
10   formulas(clues).
11     A = 0 & B = 1 & C = 2 & J = 3 & M = 4 & P = 5 & R = 6. %-isomorphisms
12
13     C != n28 & ((C = g13 & n28 = g8) | (C = g14 & n28 = g9)).     %Clue 1
14     J = n3.                                                        %Clue 2
15     M != p_ss & ((M=g11 & p_ss = n21) | (M = n21 & p_ss = g11)).   %Clue 3
16     (A = n32) | (A = n21).                                         %Clue 4
17     (J = g13) | (J = p_fb).                                        %Clue 5
18     g13 != n29.                                                    %Clue 6
19     A != p_ss.                                                     %Clue 7
20
21     (p_cf = g8  & p_rf = g9)  | (p_cf = g9  & p_rf = g10) |         %Clue 8
22     (p_cf = g10 & p_rf = g11) | (p_cf = g11 & p_rf = g12) |
23     (p_cf = g12 & p_rf = g13) | (p_cf = g13 & p_rf = g14).
24
25     M != g8 & P != g14.                                            %Clue 9
26     (M = g14 & (P = g13 | P = g12 | P = g11 | P = g10 | P = g9 | P = g8))|
27     (M = g13 & (P = g12 | P = g11 | P = g10 | P = g9 | P = g8))|
28     (M = g12 & (P = g11 | P = g10 | P = g9 | P = g8))|
29     (M = g11 & (P = g10 | P = g9 | P = g8))|
30     (M = g10 & (P = g9 | P = g8))|
31     (M = g9  & P = g8).
32
33     (g13 != n21) & (R != n21) & (R != g13).                        %Clue 10
34     B != p_sb.                                                     %Clue 11
35     (R = p_rf) | (R = g12).                                        %Clue 12
36
37     (n28=g14 & (n35=g13 | n35=g12 | n35=g11 | n35=g10 | n35=g9 | n35=g8 ))|
38     (n28=g13 & (n35=g12 | n35=g11 | n35=g10 | n35=g9 | n35=g8 ))|
39     (n28=g12 & (n35=g11 | n35=g10 | n35=g9 | n35=g8 ))|
40     (n28=g11 & (n35=g10 | n35=g9 | n35=g8 ))|
41     (n28=g10 & (n35=g9 | n35=g8 ))|
42     (n28=g9  & (n35=g9 | n35=g8 )).                                %Clue 13
43
44     (A != n3) & (A != g13) & n3 != g13.                            %Clue 14
45     g8 != p_tb.                                                    %Clue 15
46     (n29 = g8 & n29 != p_tb) | (n29 = p_tb & (n29 != g8 )).
47     A != g10.                                                      %Clue 16
48   end_of_list.
```

Solution

1. Charles played 5 more games than number 28 implies that Charles is not the number 28 (i.e. $C \neq n_{28}$) and he played either g_{13} or g_{14} games. There are two cases. First, if he played g_{13} games, than number 28 played g_8 games. Second, if he played g_{14} games, than number 28 played g_9 games: $(C = g_{13} \wedge n_{28} = g_8) \vee (C = g_{14} \wedge n_{28} = g_9)$.

2. Jorge wore number 3: $J = n_3$.

3. Of the boy who played p_{ss} and Martin, one wore number 21 and the other played 11 games: implies that Martin did not play as shortstop ($M \neq p_{ss}$) and $(M = g_{11} \wedge p_{ss} = n_{21}) \vee (M = n_{21} \wedge p_{ss} = g_{11})$.

4. Armando wore either number 32 or 21: $A = n_{32} \vee A = n_{21}$.

5. Jorge played 13 games or played first base: $J = g_{13} \vee J = p_{fb}$.

6. The player who played 13 games didn't wear number 29: $g13 \neq n29$.

7. Armando didn't play shortstop: $A \neq p_{ss}$.

8. The right field player played 1 more game than the center field player—gives six possibilities:
$(p_{cf} = g_8 \wedge p_{rf} = g_9) \vee (p_{cf} = g_9 \wedge p_{rf} = g_{10}) \vee (p_{cf} = g_{10} \wedge p_{rf} = g_{11}) \vee$
$(p_{cf} = g_{11} \wedge p_{rf} = g_{12}) \vee (p_{cf} = g_{12} \wedge p_{rf} = g_{13}) \vee (p_{cf} = g_{13} \wedge p_{rf} = g_{14})$.

9. Pedro played somewhat fewer games than Martin—we infer that $M \neq g_8$ and $P \neq g_{14}$ and we need to enumerate all possibilities (lines 25–31).

10. Neither R nor the player with 13 games was wore number 21: we infer that $R \neq g_{13}$) and also $g_{13} \neq n_{21} \wedge R \neq n_{21}$).

11. B didn't play second base: $B \neq p_{sb}$.

12. R played right field or played 12 games: $R = p_{rf} \vee R = g_{12}$.

13. Number 35 played fewer games than number 28: requires the enumeration of all possibilities (lines 37–42).

14. Neither the player who wore number 3 nor the player who played 13 games was A: it indirectly means that $n_3 \neq g_{13}$ and also $A \neq n_3 \wedge A \neq g_{13}$.

15. The player who wore number 29 was either the player who played 8 games or the person who played third base: first we learn tat $g_8 \neq p_{tb}$, and second $(n_{29} = g_8 \wedge n_{29} \neq p_{tb}) \vee (n_{29} = p_{tb} \wedge (n_{29} \neq g_8))$.

16. A didn't play 10 games: $g_{10} \neq A$.

Mace4 finds a single model:

Domain value	0	1	2	3	4	5	6
Player	A	B	C	J	M	P	R
Games	g_{12}	g_{13}	g_{14}	g_{10}	g_{11}	g_8	g_9
Number	n_{32}	n_7	n_{21}	n_3	n_{29}	n_{35}	n_{28}
Position	p_{sb}	p_{lf}	p_{ss}	p_{fb}	p_{tb}	p_{cf}	p_{rf}

Susan's perfect man has black hair, brown eyes, and is tall and slim. Susan knows four men: Arthur, Bill, Charles, and Dave. Only one of them has all the characteristics that Susan requires.

1. Arthur and Bill have the same eye colour.

2. Only one of the men has both black hair and brown eyes.

3. Bill and Charles have the same hair colour.

4. Only two of the men are both tall and slim.

5. Charles and Dave are of differing build.

6. Only two of the men are both tall and dark-haired.

7. Dave and Arthur are the same height.

8. Only three of the men are both slim and brown-eyed.

Who is Susan's perfect man? (taken from Clessa (1996))

Solution

Since there are four candidates, we set the domain size to 4 and we avoid some of the isomorphisms by fixing $arthur = 0$, $bill = 1$, $charles = 2$ and $dave = 3$. We are going to define the perfect man in first-order logic, by using four predicates:

$$\forall x \ (perfectMan(x) \leftrightarrow (tall(x) \wedge dark(x) \wedge slim(x) \wedge brownEyed(x))) \quad (6.24)$$

Obviously, there is a single perfect man:

$$\exists x \ perfectMan(x) \quad (6.25)$$
$$perfectMan(x) \wedge perfectMan(y) \rightarrow x = y \quad (6.26)$$

Solution

Then, it follows the clues. Clue$_1$ says that Arthur and Bill have the same eye colour: $brownEyed(arthur) \leftrightarrow brownEyed(bill)$. Clue$_2$ needs two constraints to formalise that exactly one man has both black hair and brown eyes:

$$\exists x \ (brownEyed(x) \wedge dark(x)) \quad (6.27)$$
$$(brownEyed(x) \wedge dark(x) \wedge brownEyed(y) \wedge dark(y)) \rightarrow x = y \quad (6.28)$$

Listing 6.7 Finding the perfect man

```
1   assign(domain_size , 4).
2   assign(max_models , -1).
3
4   formulas(perfect_man).
5     arthur = 0 & bill = 1 & charles = 2 & dave = 3.      %avoid isomorphisms
6
7     all x (perfectMan(x) <-> (tall(x) & dark(x) & slim(x) & brownEyed(x))).
8
9     exists x perfectMan(x).                              %at least one perfect man
10    perfectMan(x) & perfectMan(y) -> x = y.              %at most one perfect man
11
12    brownEyed(arthur) <-> brownEyed(bill).                            %Clue1
13    exists x (brownEyed(x) & dark(x)).                               %Clue2
14    (brownEyed(x) & dark(x) & brownEyed(y) & dark(y)) -> x = y.
15    dark(bill) <-> dark(charles).                                    %Clue3
16
17    exists x exists y (tall(x) & slim(x) & tall(y) & slim(y) & x != y).
18    (tall(x) & slim(x) & tall(y) & slim(y) & tall(z) & slim(z) &
19                    x != y) -> (z = x | z = y).                      %Clue4
20
21    slim(charles) <-> -slim(dave).                                   %Clue5
22
23    exists x exists y (tall(x) & dark(x) & tall(y) & dark(y) & x != y).
24    (tall(x) & dark(x) & tall(y) & dark(y) & tall(z) & dark(z) &
25                    x != y) -> (z = x | z = y).                      %Clue6
26
27    tall(dave) <-> tall(arthur).                                     %Clue7
28
29    exists x -(brownEyed(x) & slim(x)).                              %Clue8
30    (-(brownEyed(x) & slim(x)) & -(brownEyed(y) & slim(y))) -> x = y.
31  end_of_list.
32
33  formulas(test).
34    %perfectMan(1) | perfectMan(2) | perfectMan(3).
35    perfectMan(0).
36  end_of_list.
```

Solution

We learn from the Clue₃ that Bill and Charles have the same hair colour:
$dark(bill) \leftrightarrow dark(charles)$. Clue₄ also requires two constraints: "at least two" and "at most two":

$$\exists x\, \exists y\, (tall(x) \wedge slim(x) \wedge tall(y) \wedge slim(y) \wedge x \neq y) \qquad (6.29)$$

$$(tall(x) \wedge slim(x) \wedge tall(y) \wedge slim(y) \wedge tall(z) \wedge slim(z) \wedge x \neq y) \to (z = x \vee z = y) \qquad (6.30)$$

Here, if two distinct men x and y are both slim and tall, than the third one z should be one of them. Next, we learn from the Clue₅ that Charles and Dave are of different build. For instance, if one is slim the other is not: $slim(charles) \leftrightarrow \neg slim(dave)$. Clue₆ needs two constraints:

$$\exists x\, \exists y\, (tall(x) \wedge dark(x) \wedge tall(y) \wedge dark(y) \wedge x \neq y) \qquad (6.31)$$

$$(tall(x) \wedge dark(x) \wedge tall(y) \wedge dark(y) \wedge tall(z) \wedge dark(z) \wedge x \neq y) \to (z = x \vee z = y) \qquad (6.32)$$

Clue$_7$ states that Dave and Arthur are either both tall or not: $tall(dave) \leftrightarrow tall(arthur)$. The last Clue$_8$ says there are exactly 3 men both slim and brown-eyed. Since there are 4 men, there is exactly one man that is not both slim and brown-eyed:

$$\exists x \; \neg(brownEyed(x) \wedge slim(x)) \qquad (6.33)$$

$$(\neg(brownEyed(x) \wedge slim(x)) \wedge \neg(brownEyed(y) \wedge slim(y))) \rightarrow x = y \qquad (6.34)$$

Mace4 found 8 models. In all of these models, only $perfectMan(0)$ (that is Arthur) is true. We can test this by adding the disjunction:

$$perfectMan(1) \vee perfectMan(2) \vee perfectMan(3) \qquad (6.35)$$

With this additional formula, Mace4 fails to find a model. That is no one from 1, 2, or 3 is the perfect man. By adding $perfectMan(0)$ instead of Eq. (6.35), Mace4 still outputs eight models. Hence, all these eight possible worlds for Susan life, only Arthur has all the characteristics that Susan requires.

Puzzle 56. Four bikers

There are four bikers that are riding their own bikes. Each bike has a brand and each motorcyclist is wearing a helmet.

1. Exactly one person is riding a bike by a brand with the same initial of his/her name. That's not the boy with the yellow helmet.

2. A girl is riding a Groovers.

3. Herold is on the Yamada or has a XL helmet.

4. Yoshi's helmet is one size larger than Anne's. None of them has the white one.

5. Greta's helmet is one size larger than a boy's one, but one size smaller than the other boy's one.

6. The blue helmet is Small or Large. However, it doesn't belong to the person on Aprily.

7. Herold's helmet is XL or white.

8. The person on the Honshu has Medium or Large helmet.

9. If the green Helmet is Large, then the white helmet is on a Grooves. However, the green helmet is not on the Aprily.

Who is riding the Aprily bike? (adapted from Brainzilla—www.brainzilla.com)

Listing 6.8 Four bikers

```
1    assign(domain_size ,4).
2    assign(max_models , -1).
3    set(arithmetic).
4
5    list(distinct).
```

```
 6       [Aprily , Grooves , Honshu , Yamada ].
 7       [Anne , Greta , Harold , Yoshi ].
 8       [Blue , Green , White , Yellow ].
 9    end_of_list .
10
11    formulas ( bikers ).
12      S = 0 & M = 1 & L = 2 & XL = 3.                              %avoid  isomorphisms
13
14      Anne = Aprily  |  Greta = Grooves  |  Harold = Honshu  |  Yoshi = Yamada.
15
16      Anne = Aprily  -> ( Greta != Grooves & Harold != Honshu & Yoshi != Yamada ).
17      Greta = Grooves -> ( Anne != Aprily & Harold != Honshu & Yoshi != Yamada ).
18      Harold = Honshu -> ( Anne != Aprily & Greta != Grooves & Yoshi != Yamada ).
19      Yoshi = Yamada  -> ( Anne != Aprily & Greta != Grooves & Harold != Honshu ).
20
21      Yellow = Harold  -> Harold != Honshu .
22      Yellow = Yoshi   -> Yoshi != Yamada .
23      Harold = Yellow  |  Yoshi = Yellow .
24
25      Grooves = Anne   |  Grooves = Greta .                         %Clue2
26      Harold = Yamada  |  Harold = XL.                              %Clue3
27      ( Anne != White & Yoshi != White ) & ( Yoshi + -1 = Anne ).   %Clue4
28      Greta + -1 = Harold  |  Greta + -1 = Yoshi .                  %Clue5
29      Greta +  1 = Harold  |  Greta +  1 = Yoshi .
30      ( Blue = S  |  Blue = L ) & ( Blue != Aprily ).               %Clue6
31      Harold = XL  |  Harold = White .                             %Clue7
32      Honshu = M  |  Honshu = L .                                  %Clue8
33      ( Green = L  -> White = Grooves ) & ( Green != Aprily ).     %Clue9
34    end_of_list .
```

Solution

With four bikers we need a domain size of 4. Their names, helmets colours or sizes, and bikes producers are distinct. Let the size of the helmets $S = 0$, $M = 1$, $L = 2$, and $XL = 3$. This notation avoids some isomorphic models but also is useful when formalising Clue$_4$ and Clue$_5$.

Clue$_1$ provides several pieces of information. First, we know that at least one biker drives a bike with the same initial:

$$Anne = Aprily \lor Greta = Grooves \lor Harold = Honshu \lor Yoshi = Yamada \tag{6.36}$$

Second, we know that at most one biker drives a bike with the same initial:

$$Anne = Aprily \rightarrow (Greta \neq Grooves \land Harold \neq Honshu \land Yoshi \neq Yamada) \tag{6.37}$$

$$Greta = Grooves \rightarrow (Anne \neq Aprily \land Harold \neq Honshu \land Yoshi \neq Yamada) \tag{6.38}$$

$$Harold = Honshu \rightarrow (Anne \neq Aprily \land Greta \neq Grooves \land Yoshi \neq Yamada) \tag{6.39}$$

$$Yoshi = Yamada \rightarrow (Anne \neq Aprily \land Greta \neq Grooves \land Harold \neq Honshu) \tag{6.40}$$

Third, that's not the boy with the yellow helmet. We deduce that a boy has the yellow helmet $Harold = Yellow \lor Yoshi = Yellow$ but also that

$$(yellow = Harold \rightarrow Harold \neq Honshu) \land (yellow = Yoshi \rightarrow Yoshi \neq Yamada) \tag{6.41}$$

The rest of clues are formalised in lines 25–33. Given the formalisation in Listing 6.8, Mace4 computes three models. In all of them, Harold drives an Aprily.

Model 1	0	1	2	3
Helmet size	S	M	L	XL
Biker	Anne	Yoshi	Greta	Harold
Bike	Yamada	Honshu	Grooves	Aprily
Helmet colour	Green	Yellow	Blue	White

Model 2	0	1	2	3
Helmet size	S	M	L	XL
Biker	Anne	Yoshi	Greta	Harold
Bike	Yamada	Honshu	Grooves	Aprily
Helmet colour	Blue	Green	White	Yellow

Model 3	0	1	2	3
Helmet size	S	M	L	XL
Biker	Anne	Yoshi	Greta	Harold
Bike	Grooves	Yamada	Honshu	Aprily
Helmet colour	Blue	Green	White	Yellow

Puzzle 57. Movies night

Four boys are at home to watch some movies. Figure out which is the favourite kind of movie of each one.

1. Joshua is in one of the ends.
2. The boy wearing the Black shirt is somewhere to the left of the youngest boy.
3. Joshua likes Horror movies.
4. The 14-year-old boy is in the third position.
5. The boy wearing the Red shirt is somewhere between the 13-year-old boy and the one who likes Action movies, in that order.
6. Daniel likes Thriller movies.
7. The boy who is going to eat Cookies is in one of the ends.
8. The boy wearing the Black shirt is exactly to the left of the one who likes Thriller movies.
9. The boy who is going to eat Crackers is exactly to the right of the boy who likes Comedy movies.
10. The boy wearing the Red shirt is somewhere between the boy who is going to eat Popcorn and Nicholas, in that order.
11. In one of the ends is the boy who likes Thriller movies.
12. Nicholas is somewhere between Joshua and Daniel, in that order.
13. At first position is the boy wearing the Green shirt.

(taken from Brainzilla—www.brainzilla.com)

Listing 6.9 Equations for the movies night

```
1    set ( arithmetic ).
2    assign ( domain_size , 4 ).
3    assign ( max_models , −1 ).
4
5    list ( distinct ).
6      [ Green , Black , Red , Blue ].
7      [ Joshua , Ryan , Nicholas , Daniel ].
8      [ Thriller , Horror , Comedy , Action ].
9      [ Popcorn , Chips , Crackers , Cookies ].
10     [ 14 years , 13 years , 12 years , 11 years ].
11   end_of_list .
12
13   formulas ( assumptions ).
14     Joshua = 0  I  Joshua = 3 .              %Clue  1
15     Black < 11 years .                       %Clue  2
16     Joshua = Horror .                        %Clue  3
17     14 years = 2 .                           %Clue  4
18     Red > 13 years & Red < Action .          %Clue  5
19     Daniel = Thriller .                      %Clue  6
20     Cookies = 0  I  Cookies = 3 .            %Clue  7
21     Black+1 = Thriller .                     %Clue  8
22     Crackers =  Comedy + 1 .                 %Clue  9
23     Popcorn < Red & Red < Nicholas .         %Clue  10
24     Thriller = 0  I  Thriller = 3 .          %Clue  11
25     Joshua < Nicholas & Nicholas < Daniel .  %Clue  12
26     Green = 0 .                              %Clue  13
27   end_of_list .
```

Solution

With four boys we need a domain size of 4. Their names, shirts, movie preferences, ages, or beverages are distinct (lines 6–10 in Listing 6.9. We consider the position most to the left has value 0, and the right most position has value 4. Hence, the predicate left corresponds to the "<" operator. The clues are straightforward formalised in lines 14–26.

Mace4 computes a single model:

Domain Value	0	1	2	3
Name	*Joshua*	*Ryan*	*Nicholas*	*Daniel*
Shirt	*Green*	*Red*	*Black*	*Blue*
Movie	*Horror*	*Comedy*	*Action*	*Thriller*
Snack	*Popcorn*	*Chips*	*Crackers*	*Cookies*
Age	13	12	14	11

Listing 6.10 Auxiliary predicates and clues formalisation for the secret Santa

```
1    set(arithmetic).
2    assign(domain_size,5).
3    assign(max_models,-1).
4
5    list(distinct).
6      [Black, Blue, Green, Red, White].
7      [Cody, Jason, Riley, Steven, Tyler].
8      [Book, Chocolate, Mug, Notepad, Tie].
9      [HR, IT, Marketing, R_and_D, Sales].
10     [Age_23, Age_28, Age_35, Age_41, Age_50].
11     [Coffee, Juice, Soft_drink, Tea, Water].
12   end_of_list.
13
14   formulas(utils).
15     right_neighbor(x,y) <-> x+1 = y.
16     left_neighbor(x,y) <-> y+1 = x.
17     neighbors(x,y) <-> right_neighbor(x,y) | left_neighbor(x,y).
18     between(x,y,z) <-> (x+1=y | x+2=y | x+3=y) & ( y+1=z | y+2=z | y+3=z).
19     somewhereLeft(x,y) <-> x+1=y | x+2=y | x+3=y | x+4=y.
20   end_of_list.
21
22   formulas(santa_clues).
23     Cody = Age_23.                     %Clue  1
24     Book = Riley.                      %Clue  2
25     Juice = 4.                         %Clue  3
26     neighbors(Riley,Age_41).           %Clue  4
27     Age_35 = 0 | Age_35 = 4.           %Clue  5
28     between(Mug,Red,Soft_drink).       %Clue  6
29     left_neighbor(Notepad,Coffee).     %Clue  7
30     right_neighbor(Blue,Tea).          %Clue  8
31     neighbors(Green,Age_28).           %Clue  9
32     right_neighbor(Cody,Steven).       %Clue  10
33     Water = 1.                         %Clue  11
34     R_and_D = 2.                       %Clue  12
35     Tyler = Mug.
36     Age_50 = 4.                        %Clue  14
37     Soft_drink = 2.                    %Clue  15
38     neighbors(Riley,Tie).              %Clue  16
39     between(Water,Age_23,Age_50).      %Clue  17
40     right_neighbor(Black,Jason).       %Clue  18
41     neighbors(Cody,Soft_drink).        %Clue  19
42     somewhereLeft(Blue,Sales).         %Clue  20
43     IT = Notepad.                      %Clue  21
44     Tea = 3.                           %Clue  22
45     Chocolate = HR.                    %Clue  23
46     Tie = Green.                       %Clue  24
47   end_of_list.
```

Puzzle 58. Secret Santa

Five employees are side by side at their company secret santa. Find out what each one is drinking, which department they work and what the gift they got was.

1. Cody is the youngest employee.

2. The person gifted with a Book is Riley.

3. In the fifth position is the person drinking Juice.

4. Riley is next to the 41-year-old employee.

5. The 35-year-old employee is at one of the ends.

6. The man wearing the Red shirt is somewhere between the one who received a Mug and the one drinking Soft drink, in that order.

7. The one drinking Coffee is exactly to the left of who got a Notepad as a gift.

8. The man drinking Tea is exactly to the right of the man wearing the Blue shirt.

9. The employee wearing the Green shirt is next to the 28-year-old.

10. Steven is exactly to the right of Cody.

11. In the second position is the person drinking Water.

12. The employee that works at the R&D department is at the third position.

13. Tyler's gift was a Mug.

14. The oldest employee is at the fifth position.

15. The person drinking Soft drink is at the third position.

16. Riley is next to the person who got a Tie as a gift.

17. The youngest employee is somewhere between the person drinking Water and the oldest person, in that order.

18. Jason is exactly to the right of the man wearing the Black shirt.

19. Cody is next to the one drinking Soft drink.

20. The man wearing Blue is somewhere to the left of who works at the Sales department.

21. The employee that works at the IT department was gifted a Notepad.

22. At the fourth position is the one drinking Tea.

23. The person gifted with chocolate is working at the HR department.

24. The man wearing the Green shirt is the person gifted with a tie.

(taken from Brainzilla - www.brainzilla.com)

Solution

By analysing the clues, we learn that there are five positions, employees (Cody, Jason, Riley, Steven, Tyler), shirts (black, blue, green, red, white), gifts (book, chocolate, mug, notepad, tie), departments (HR, IT, marketing, R&D, sales), ages (23, 28, 35, 41, 50), and drinks (coffee, juice, soft drink, tea, water). Hence, the domain size is set to 5. We start by defining some auxiliary predicates. First, we need predicates signalling that two persons are neighbours to the right or left:

$$right\_neighbor(x, y) \leftrightarrow x + 1 = y \quad (6.42)$$

$$left\_neighbor(x, y) \leftrightarrow y + 1 = x \quad (6.43)$$

$$neighbors(x, y) \leftrightarrow right\_neighbor(x, y) \vee left\_neighbor(x, y) \quad (6.44)$$

Second, the predicate $between(x, y, z)$ returns true when y is somewhere between x and z:

$$between(x, y, z) \leftrightarrow (x + 1 = y \vee x + 2 = y \vee x + 3 = y) \wedge$$

$$(y + 1 = z \vee y + 2 = z \vee y + 3 = z) \quad (6.45)$$

Note that y is not necessarily neighbour of x or z. Third, the predicate $somewhereLeft(x, y)$ returns true one x is to the left of y, but not necessarily neighbour of y:

$$somewhereLeft(x, y) \leftrightarrow (x + 1 = y) \vee (x + 2 = y) \vee (x + 3 = y) \vee (x + 4 = y) \quad (6.46)$$

The formalisation of each clue is straightforward (see Listing 6.10). Mace4 computes the following unique model:

Position	0	1	2	3	4
Employee	Tyler	Jason	Cody	Riley	Steven
Shirt	Black	Red	Blue	Green	White
Gift	Mug	Notepad	Book	Tie	Chocolate
Department	Marketing	IT	R&D	Sales	HR
Drink	Coffee	Water	Soft drink	Tea	Juice
Age	35	41	28	23	50

Puzzle 59. Seating the party

As the Crackham family were taking their seats on starting out on their tour, Dora asked in how many different ways they could all be seated, as there were six of them and six seats-one beside the driver, two with their backs to the driver, and two behind, facing the driver-if no two of the same sex are ever to sit side by side? As the Colonel, Uncle Jabez, and George were the only ones who could drive, it required just a little thinking out. Perhaps the reader will like to work out the answer concerning on which they all agreed at the end of the day. (puzzle 465 from Dudeney (2016))

---

s_1	s_2	s_3
s_4	s_5	s_6

Listing 6.11 Six seats in a car

```
1   assign(domain_size ,6).
2   assign(max_models ,-1).
3
4   formulas(assumptions).
5     neq(a,b).  neq(a,c).  neq(a,d).  neq(a,e).  neq(a,f).
6     neq(b,c).  neq(b,d).  neq(b,e).  neq(b,f).
7     neq(c,d).  neq(c,e).  neq(c,f).
8     neq(d,e).  neq(d,f).
9     neq(e,f).
10    neq(x, y) -> neq(y, x).
11
12    %each person stays somewhere
13    s1(a) | s2(a) | s3(a) | s4(a) | s5(a) | s6(a).
14    s1(b) | s2(b) | s3(b) | s4(b) | s5(b) | s6(b).
15    s1(c) | s2(c) | s3(c) | s4(c) | s5(c) | s6(c).
16    s1(d) | s2(d) | s3(d) | s4(d) | s5(d) | s6(d).
17    s1(e) | s2(e) | s3(e) | s4(e) | s5(e) | s6(e).
18    s1(f) | s2(f) | s3(f) | s4(f) | s5(f) | s6(f).
19
20    %two persons can not share the same seat
21    s1(x) & s1(y) -> -neq(x,y).     s2(x) & s2(y) -> -neq(x,y).
22    s3(x) & s3(y) -> -neq(x,y).     s4(x) & s4(y) -> -neq(x,y).
23    s5(x) & s5(y) -> -neq(x,y).     s6(x) & s6(y) -> -neq(x,y).
24
25    man(a) & man(b) & man(c) & woman(d) & woman(e) & woman(f).
26    man(x) -> -woman(x).
27
28    driver(a) & driver(b) & driver(c) &-driver(d) &-driver(e) &-driver(f).
29    s1(x) -> driver(x).
30    s1(x) & man(x) & s2(y) -> woman(y).
31    s3(x) & man(x) & s4(y) -> woman(y).
32    s5(x) & man(x) & s6(y) -> woman(y).
33  end_of_list.
```

Solution

With six persons and six seats, we set the domain size to 6. Let the predicate $s_i(x)$ returning true when the person x stays on the sit s_i, $i \in [1..6]$. Let the six persons denoted by a, b, c, d, e, f. First, we specify that the persons are distinct. For this we introduce in lines 5–10 the predicate not equal $neq(x, y)$. Second, we state that each person stays somewhere in the car (lines 13–18 in Listing 6.11). For instance, person a can stay on the places s_i:

$$s_1(a) \lor s_2(a) \lor s_3(a) \lor s_4(a) \lor s_5(a) \lor s_6(a) \tag{6.47}$$

Third, two persons cannot share the same seat (lines 21–23). For instance, for the first seat, if two individuals x and y stay there, they are the same person:

$$s_1(x) \land s_1(y) \rightarrow \neg neq(x, y) \tag{6.48}$$

Note that based on the facts that: 1) a person stays somewhere and 2) no two persons stay on the same seat, one can deduce that a person cannot stay in two seats. Fourth, we assume three men and three women (lines 25–26).

Next, we formalise the clues: i) only men can drive (line 28); ii) the first seat is the driver's seat (line 29); iii) no two of the same sex are ever to sit side by side (lines 30–32).

Mace4 finds 144 models. One such model is

	a	b	c	d	e	f
s_1	1	0	0	0	0	0
s_2	0	0	0	1	0	0
s_3	0	1	0	0	0	0
s_4	0	0	0	0	1	0
s_5	0	0	1	0	0	0
s_6	0	0	0	0	0	1

This model is depicted in Figure 6.1, where the women occupy the pink seats and the blue seat represents the driver.

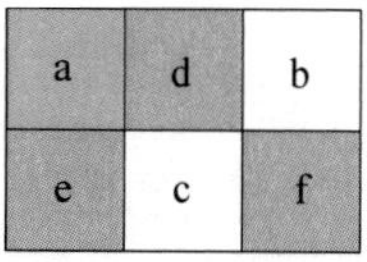

Fig. 6.1 One model among 144 solutions found by Mace4

Puzzle 60. Passengers in a railroad compartment

Six passengers sharing a compartment are from Moscow, Leningrad, Tula, Kiev, Kharkov, Odessa.

1. A and the man from Moscow are physicians.

2. E and the Leningrader are teachers.

3. The man from Tula and C are engineers.

4. B and F are WWII veterans, but the man from Tula has never served in the army.

5. The man from Kharkov is older than A.

6. The man from Odessa is older than C.

7. At Kiev, B and the man from Moscow got off.

8. At Vinnitsa, C and the man from Kharkov got off.

Match initials, professions, and cities. Also, are these facts both sufficient and necessary to solve the problem? (puzzle 256 from Kordemsky (1992))

Listing 6.12 Passengers in a railroad compartment

```
1   assign(max_models,-1).
2   assign(domain_size,6).
3
4   formulas(assumptions).
5     A=0 & B=1 & C=2 & D=3 & E=4 & F=5.                    %remove isomorphic
6
7   all x ((  Mosc(x) & -Leni(x) & -Tula(x) & -Kiev(x) & -Khar(x) & -Odes(x))
8          | (-Mosc(x) &  Leni(x) & -Tula(x) & -Kiev(x) & -Khar(x) & -Odes(x))
9          | (-Mosc(x) & -Leni(x) &  Tula(x) & -Kiev(x) & -Khar(x) & -Odes(x))
10         | (-Mosc(x) & -Leni(x) & -Tula(x) &  Kiev(x) & -Khar(x) & -Odes(x))
11         | (-Mosc(x) & -Leni(x) & -Tula(x) & -Kiev(x) &  Khar(x) & -Odes(x))
12         | (-Mosc(x) & -Leni(x) & -Tula(x) & -Kiev(x) & -Khar(x) &  Odes(x))).
13
14  all x ((  physician(x) & -teacher(x) & -engineer(x) & -armForce(x))
15         | (-physician(x) &  teacher(x) & -engineer(x) & -armForce(x))
16         | (-physician(x) & -teacher(x) &  engineer(x) & -armForce(x))
17         | (-physician(x) & -teacher(x) & -engineer(x) &  armForce(x))).
18
19    Mosc(x) & Mosc(y) -> x=y.          Leni(x) & Leni(y) -> x=y.
20    Tula(x) & Tula(y) -> x=y.          Kiev(x) & Kiev(y) -> x=y.
21    Khar(x) & Khar(y) -> x=y.          Odes(x) & Odes(y) -> x=y.
22
23    physician(A) & -Mosc(A) & (Mosc(x) -> physician(x)).       %Clue 1
24    teacher(E)   & -Leni(E) & (Leni(x) -> teacher(x)).         %Clue 2
25    engineer(C)  & -Tula(C) & (Tula(x) -> engineer(x)).        %Clue 3
26    -Tula(B)     & -Tula(F) & (Tula(x) -> -armForce(x)).       %Clue 4
27    -Khar(A).                                                  %Clue 5
28    -Odes(C).                                                  %Clue 6
29    -Mosc(B).                                                  %Clue 7
30    -Khar(C) & -Khar(A).                                       %Clue 8
31  end_of_list.
```

Solution

With six passengers we need a domain of six elements. Since we do not care whether a person is associated with one value or the other, we can fix the values for each person (line 5 in Listing 6.12). We have three assumptions. First, a person cannot be from two distinct cities (lines 7–12). For instance, if someone is from Moscow, that person cannot be from the other cities:

$$\forall x \ (Mosc(x) \land \neg Leni(x) \land \neg Tula(x) \land \neg Kiev(x) \land \neg Khar(x) \land \neg Odes(x)) \tag{6.49}$$

Second, a person cannot have two professions (lines 14–17). For instance, if someone is a physician, that person cannot have a different profession:

$$\forall x \ (physician(x) \land \neg teacher(x) \land \neg engineer(x) \land \neg armForce(x)) \tag{6.50}$$

Third, we assume there are no two persons from the same city (lines 19–21). For instance, if x and y are from Moscow, then x and y represent the same person:

$$Mosc(x) \land Mosc(y) \rightarrow x = y \tag{6.51}$$

1. A and the man from Moscow are physicians:

$$physician(A) \land \neg Mosc(A) \land \forall x \ (Mosc(x) \rightarrow physician(x)) \tag{6.52}$$

2. E and the Lenier are teachers: $teacher(E) \land \forall x \ (Leni(x) \rightarrow teacher(x))$.
3. The man from Tula and C are engineers:
   $engineer(C) \land \forall x \ (Tula(x) \rightarrow engineer(x))$.
4. B and F are World War II veterans, but the man from Tula has never served in the army forces: $\neg Tula(B) \land \neg Tula(F) \land \forall x \ (Tula(x) \rightarrow \neg armForce(x))$. The assumption here is that WWII does not necessarily implies $armForce$.
5. The man from Kharkov is older than A: $\neg Kharkov(A)$.
6. The man from Odessa is older than C: $\neg Odessa(C)$.
7. At Kiev, B and the man from Moscow got off: $\neg Moscow(B) \land Kiev(B)$.
8. At Vinnitsa, C and the man from Khar got off: $\neg Kharkov(C) \land \neg Kharkov(A)$.

Mace4 finds the following model:

	A	B	C	D	E	F
Moscow	0	0	0	0	0	1
Leningrad	0	1	0	0	0	0
Kiev	0	0	1	0	0	0
Tula	0	0	0	1	0	0
Odessa	1	0	0	0	0	0
Kharkov	0	0	0	0	1	0

	A	B	C	D	E	F
ArmForce:	0	0	0	0	0	0
Engineer	0	0	1	1	0	0
Physician	1	0	0	0	0	1
Teacher	0	1	0	0	1	0

There is more than one passenger for some jobs, because there are four jobs for six people. Note that not all clues are necessary. For instance, both Clue 5 and Clue 8 state that A is not from Kharkov. One can test this redundancy simply by commenting Clue 5 (line 27 in Listing 6.12)

References

Bright, C., Gerhard, J., Kotsireas, I., and Ganesh, V. (2019). Effective problem solving using SAT solvers. In *Maple Conference*, pages 205–219. Springer.

Clessa, J. (1996). *Math and Logic Puzzles for PC Enthusiasts*. Courier Corporation.

Dudeney, H. E. (2016). 536 Puzzles and curious problems. Courier Dover Publications.

Gregor, M., Zábovská, K., and Smataník, V. (2015). The zebra puzzle and getting to know your tools. In *2015 IEEE 19th International Conference on Intelligent Engineering Systems (INES)*, pages 159–164. IEEE.

Kordemsky, B. A. (1992). *The Moscow puzzles: 359 mathematical recreations*. Courier Corporation.

Szeredi, P. (2003). Teaching constraints through logic puzzles. In *International Workshop on Constraint Solving and Constraint Logic Programming*, pages 196–222. Springer.

Yeomans, J. S. (2003). Solving Einstein's Riddle using spreadsheet optimization. *INFORMS Transactions on Education, 3*(2), 55–63.

Chapter 7
Island of Truth

Abstract In which we visit the island of truth, an island also discovered by Smullyan (2011). The island is populated by knights, knaves, or spies. The knights are telling the truth, while the knaves are telling lies. Occasionally, you find other inhabitants whom you don't know if they are telling the truth or not. But you do not have to worry since you have been mastering for some chapters a very powerful weapon: reasoning in FOL. And this reasoning weapon will help you distinguish between truth and lie on this island. You are fully equipped with two technologies. In one hand you have Mace4 that finds all the interpretation models of the sentences conveyed by the inhabitans. In the other hand you have Prover9 that explains to you all the reasoning steps as to why a statement is true or false.

$$(\forall x\; learns(x) \wedge \neg thinks(x) \rightarrow lost(x)) \wedge (\forall x\; thinks(x) \wedge \neg learns(x) \rightarrow great\, Danger(x))$$

Confucius

We visit here the island of truth, an island also discovered by Smullyan (2011). The island is populated by knights, knaves, or spies. The knights are telling the truth, while the knaves are telling lies. Occasionally, you find other inhabitants whom you don't know if they are telling the truth or not. But you do not have to worry since you have been mastering for some chapters a very powerful weapon: reasoning in FOL. And this reasoning weapon will help you distinguish between truth and lie on this island. You are fully equipped with two technologies. In one hand you have Mace4 that finds all the interpretation models of the sentences conveyed by the inhabitans. In the other hand you have Prover9 that explains to you all the reasoning steps as to why a statement is true or false.

This strange island has been visited by explorers using various languages to communicate with the inhabitants: For instance, in 1995, Baumgartner et al. explored the island using model elimination and logic programming (Baumgartner et al. 1995). In 2001, Aszalos explored the shore of the island (that is, only one puzzle) using propositional logic (Aszalós 2001). In 2013, Lie and Wang built a new language on

© The Author(s), under exclusive license to Springer Nature Switzerland AG 2021 131
A. Groza, *Modelling Puzzles in First Order Logic*,
https://doi.org/10.1007/978-3-030-62547-4_7

top of *public announcement logic*, dedicated to explore islands where agents made public announcements that can be true or not (Ciraulo and Maschio 2020). Recently, in 2020, Ciraulo and Maschio have changed the exploration strategy: instead of logic reasoning they use algebra-based computations. They did this by translating the language of the inhabitants into *commutative algebra* (i.e. Boolean rings) equations. Since each logic operand is translated into a polynomial form, the puzzles are modelled with equations only.

Now it is your turn! Dare to visit the island and meet its first inhabitants.

Puzzle 61. We are both knaves

On the island of knights and knaves, knights always tell the truth, while knaves always lie. You are approached by two people. The first one says: "We are both knaves". What are they actually? (Smullyan 2011)

Listing 7.1 Island of truth: finding models based on the single message "We are both knaves"

```
1   assign(domain_size,2).
2   assign(max_models,-1).
3
4   formulas(island_of_truth).
5      all x (inhabitant(x) -> knight(x) | knave(x)).
6      all x ((knight(x) -> -knave(x)) & (knave(x) -> -knight(x))).
7      knight(x) -> m(x).
8      knave(x)  -> -m(x).
9   end_of_list.
10
11  formulas(puzzle).
12     inhabitant(a) & inhabitant(b).
13     m(a) <-> knave(a) & knave(b).
14  end_of_list.
```

Listing 7.2 Proving *knave*(*a*) ∧ *knight*(*b*) based on "We are both knaves"

```
1   formulas(assumptions).
2      all x (inhabitant(x) -> knight(x) | knave(x)).
3      all x ((knight(x) -> -knave(x)) & (knave(x) -> -knight(x))).
4      knight(x) -> m(x).            knave(x)  -> -m(x).
5
6      inhabitant(a) & inhabitant(b).
7      m(a) <-> knave(a) & knave(b).
8   end_of_list.
9
10  formulas(goals).
11     knave(a) & knight(b).
12  end_of_list.
```

Solution

Let start by formalising the assumptions regarding the island of truth (lines 5–8 in Listing 7.1). First, the inhabitants are either knights or knaves:

$$\forall x \ (inhabitant(x) \rightarrow knight(x) \lor knave(x)) \qquad (7.1)$$

Second, one cannot be a knight and a knave in the same time:

$$\forall x \ ((knight(x) \rightarrow \neg knave(x)) \land (knave(x) \rightarrow \neg knight(x))) \qquad (7.2)$$

Third, a message $m(x)$ said by a knight x is always true: $knight(x) \rightarrow m(x)$. Fourth, a message $m(x)$ said by a knave x is always false: $knave(x) \rightarrow \neg m(x)$. This model will be reused in the following puzzles from the same island of truth.

Next, we formalise the first puzzle on the island. We learn that there are two inhabitants, let's say a and b: $inhabitant(a) \land inhabitant(b)$ (line 12). We need a domain size of two individuals (line 1). The message of inhabitant a is: $m(a) \leftrightarrow knave(a) \land knave(b)$.

Mace4 finds a single model:

$$a = [0] \qquad\qquad b = [1]$$
$$inhabitant(\_) = [1, 1] \quad m(\_) = [0, 1]$$
$$knight(\_) = [0, 1] \qquad knave(\_) = [1, 0]$$

Here, the inhabitant a gets value 0 from the domain, while the inhabitant b gets value 1. The predicate *inhabitant* is true for both a and b, as it is explicitly stated in line 12. Mace4 deduces that the message conveyed by the agent with value 0 (i.e. inhabitant a) is false. The predicate *knave* has value 1 for the first inhabitant ($a = 0$), and value 0 for the second one ($b = 1$). Quite the opposite, the predicate *knight* has value 0 for the first inhabitant ($a = 0$), and value 1 for the second one ($b = 1$).

The above model is consistent with the one found by the following line of reasoning a human agent might have: If the message is true, then it had to be conveyed by a knight, i.e. *knight*(*a*). But the message says *knave*(*a*). Hence the message must be false. That means that it was conveyed by a knave, i.e. *knave*(*a*). Since the message is false, it also means that $\neg knave(a) \lor \neg knave(b)$. In other words, we have *knight*(*a*) ∨ *knight*(*b*). Hence b is a knight.

We can ask the software agent for its own line of reasoning. Given the code in Listing 7.2, Prover9 finds the proof in Fig. 7.1 for the goal $knave(a) \land knight(b)$. Here,
the prover finds a contradiction based on three pieces of knowledge. First, the negated
conclusion $\{12\}$ is equivalent to $\{18\}$: $knave(b) \lor \neg knave(a)$. Second, the agent
deduces $\{17\}$: a is a knave or the message is true. Further, based on $\{15\}$, the system deduces that a is a knave (clause $\{22\}$). Third, the prover infers $\{20\}$ based on
the input clause $\{5\}$ (knaves are liars) and the given message $\{3\}$. Using resolution
between $\{20\}$ and $\{22\}$, b is not a knave. By combining these three pieces of knowledge ($\{18\}, \{22\}, \{23\}$) through hyper-resolution, the software agent signals a contradiction. Hence, theorem $\{6\}$ is proved.

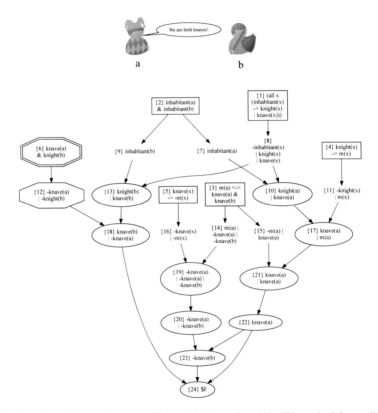

Fig. 7.1 Proof for $\{6\}$: a is a knave and b is a knight based on $\{3\}$: "We are both knaves"

Puzzle 62. At least one of us is a knave

On the island of knights and knaves, knights always tell the truth, while knaves always lie. You are approached by two people. The first one says: "At least one of us is a knave". What are they actually? (puzzle 28 from Smullyan (2011))

a   b

Listing 7.3 Finding models based on the single message "At least one of us is a knave"

```
1   assign(domain_size ,2).
2   assign(max_models , -1).
3
4   formulas(island_of_truth).
5     all x (inhabitant(x) -> knight(x) | knave(x)).
6     all x ((knight(x) -> -knave(x)) & (knave(x) -> -knight(x))).
7     knight(x) -> m(x).
8     knave(x)  -> -m(x).
9   end_of_list.
10
11  formulas(puzzle).
12    inhabitant(a) & inhabitant(b).
13    m(a) <-> knave(a) | knave(b).
14  end_of_list.
```

Listing 7.4 Proving $knight(a) \wedge knave(b)$ based on "At least one of us is a knave"

```
1   formulas(assumptions).
2     all x (inhabitant(x) -> knight(x) | knave(x)).
3     all x ((knight(x) -> -knave(x)) & (knave(x) -> -knight(x))).
4     knight(x) -> m(x).            knave(x)  -> -m(x).
5
6     inhabitant(a) & inhabitant(b).
7     m(a) <-> knave(a) | knave(b).
8   end_of_list.
9
10  formulas(goals).
11    knight(a) & knave(b).
12  end_of_list.
```

Solution

The puzzle is similar with puzzle 61, but the message is different. The current message
is: $m(a) \leftrightarrow knave(a) \vee knave(b)$ (line 13 in Listing 7.3). Mace4 finds a single model:

$$a = [0] \qquad\qquad b = [1]$$
$$inhabitant(\_) = [1, 1] \quad m(\_) = [0, 1]$$
$$knight(\_) = [0, 1] \qquad knave(\_) = [1, 0]$$

Here the inhabitant a gets the value 0 from the domain, while the inhabitant b gets the
value 1. The function $inhabitant$ is true for both a and b, as it is explicitly stated in
line 12. Mace4 deduces that the message conveyed by the element with value 0 (i.e.
inhabitant a) is true. The function $knave$ has value 0 for the first inhabitant ($a = 0$),
and value 1 for the second one ($b = 1$). Quite the opposite, the function $knight$ has
value 1 for the first inhabitant ($a = 0$), and value 0 for the second one ($b = 1$).
The above model is consistent with the one found by the following line of reasoning:
Suppose the message is false. Hence, the following holds: $\neg knave(a) \wedge \neg knave(b)$.
This is equivalent to $knight(a) \wedge knight(b)$. But if the message is false, it was conveyed by a knave, i.e. $knave(a)$, which is a contradiction. Now, suppose the message is
true. Then it was conveyed by a knight, i.e. $knight(a)$. The message is true only when
b is a knave.
Based on Listing 7.4, Prover9 finds the proof for $knight(a) \wedge knave(b)$ as illustrated
in Fig. 7.2. First, based on the negated goal $\{11\}$ and the rules on the island of truth,
the software agent deduces $\{16\}$: a is knave or b is not knave. Second, based on the
current message $\{5\}$ and the information that knaves are liars ($\{3\}$), the agent infers
$\{18\}$: a is not a knave. Third, based on the current message $\{5\}$ and the deduced clause
$\{15\}$ (either a is a knave or the message is true), the clause $\{19\}$ is computed. By resolution between $\{19\}$ and $\{18\}$, the agent learns that b is a knave. By combining these
three pieces of knowledge ($\{16\}, \{18\}, \{20\}$) through hyper-resolution, the software
agent signals the contradiction $\{21\}$. Hence, theorem $\{6\}$ is proved.

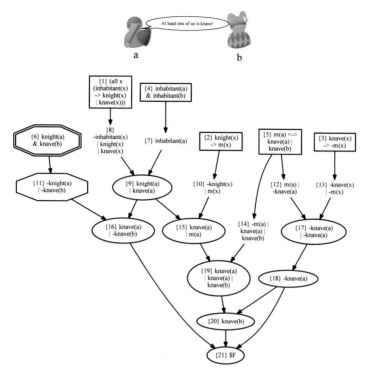

Fig. 7.2 Proof for {6}: *a* is a knight and *b* is a knave based on {5}: "At least one of us is knave"

Puzzle 63. Either I am a knave or *b* is a knight

On the island of knights and knaves, knights always tell the truth, while knaves always lie. You are approached by two people. The first one says: "Either I am a knave or B is a knight.". What are they actually? (puzzle 29 from Smullyan (2011))

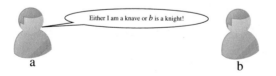

Listing 7.5 Finding models based on "Either I am a knave or *b* is a knight"

```
1   assign(domain_size ,2).
2   assign(max_models , -1).
3
4   formulas(island_of_truth).
5     all x (inhabitant(x) -> knight(x) | knave(x)).
6     all x ((knight(x) -> -knave(x)) & (knave(x) -> -knight(x))).
7     knight(x) -> m(x).
8     knave(x)  -> -m(x).
9   end_of_list.
10
11  formulas(puzzle).
12    inhabitant(a) & inhabitant(b).
13    m(a) <-> (knave(a) | knight(b)).
14    a=0 & b=1.  %remove isomorphisms
15  end_of_list.
```

Listing 7.6 Proving $knight(a) \land knave(b)$ based on "Either I am a knave or *b* is a knight"

```
1   formulas(assumptions).
2     all x (inhabitant(x) -> knight(x) | knave(x)).
3     all x ((knight(x) -> -knave(x)) & (knave(x) -> -knight(x))).
4     knight(x) -> m(x).              knave(x)  -> -m(x).
5
6     inhabitant(a) & inhabitant(b).
7     m(a) <-> knave(a) | knight(b).
8   end_of_list.
9
10  formulas(goals).
11    knight(a) & knight(b).
12  end_of_list.
```

Solution

The current message is: $m(a) \leftrightarrow knave(a) \lor knight(b)$. Note that the message is formalised as an "inclusive or", and not as an exclusive one. Note also in line 14 in Listing 7.5 that the values of *a* and *b* are fixed in order to avoid isomorphic models. Mace4 computes a single model:

$$a = [0] \qquad\qquad b = [1]$$
$$inhabitant(\_) = [1, 1] \quad m(\_) = [1, 1]$$
$$knight(\_) = [1, 1] \qquad knave(\_) = [0, 0]$$

Here the message is true and both inhabitants are knights.

To follow the reasoning steps of the software agent we can ask for a proof for this statement: $knight(a) \land knight(b)$.

Given the code in Listing 7.6, Prover9 finds the proof in Fig. 7.3 for the goal $knight(a) \land knight(b)$. Here, the software agents find a contradiction based on three pieces of knowledge. First, by classifying the input sentence {2}, the system knows that the message is true or x is not a knight. Second, based on the current message {5}, the agent learns clause {18}: a is a knight or the message is true. Further, based on {17} the system infers through resolution that a is a knight ({19}). Third, based on this new knowledge {19} and the negated goal {14}, b is not a knight ({20}). Therefore the message conveyed by a is false ({21}). By combining these three pieces of knowledge ({2}, {19}, {21}) through hyper-resolution, the prover signals the contradiction {22}. Hence, theorem {6} is proved.

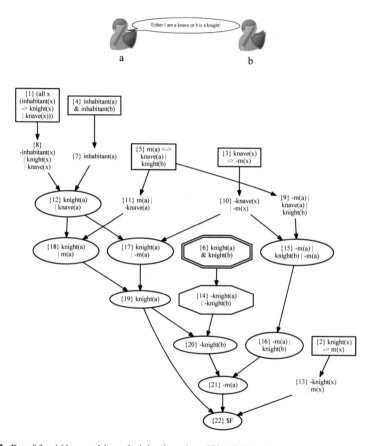

Fig. 7.3 Proof for {6}: a and b are knights based on {5}: "Either I am a knave or b is a knight"

Puzzle 64. We are both the same

On the island of knights and knaves, knights always tell the truth, while knaves always
lie. You are approached by two people. The first one says: "We are the same". Is the
other person a knight or a knave? (adapted from Smullyan (2011))

Listing 7.7 Finding models based on the single message "We are both the same"

```
1    assign(domain_size ,2).
2    assign(max_models ,-1).
3
4    formulas(island_of_truth).
5      all  x (inhabitant(x) -> knight(x) | knave(x)).
6      all  x ((knight(x) -> -knave(x)) & (knave(x) -> -knight(x))).
7      knight(x)  ->  m(x).
8      knave(x)   ->  -m(x).
9    end_of_list.
10
11   formulas(puzzle).
12     inhabitant(a) & inhabitant(b).
13     m(a) <-> ((knave(a) & knave(b)) | (knight(a) & knight(b))).
14     a=0 & b=1. %remove isomorphism
15   end_of_list.
```

Listing 7.8 Proving *knight*(*b*) based on "We are both the same"

```
1    formulas(assumptions).
2      all  x (inhabitant(x)  -> knight(x) | knave(x)).
3      all  x ((knight(x)  -> -knave(x)) & (knave(x) -> -knight(x))).
4      knight(x)  ->  m(x).
5      knave(x)   -> -m(x).
6      inhabitant(a) & inhabitant(b).
7      m(a) <->  ((knave(a) & knave(b)) | (knight(a) & knight(b))).
8    end_of_list.
9
10   formulas(goals).
11     knight(b).
12   end_of_list.
```

Solution

The message of the first inhabitant (i.e. a) states that a and b are both knaves or both knights:

$$m(a) \leftrightarrow (knave(a) \wedge knave(b)) \vee (knight(a) \wedge knight(b)) \qquad (7.3)$$

For $a = 0$ and $b = 1$, Mace computes two models:

Model 1	Model 2
relation(knave(_), [1, 0])	relation(knave(_), [0, 0])
relation(knight(_), [0, 1])	relation(knight(_), [1, 1])
relation(m(_), [0, 1])	relation(m(_), [1, 1])

Note that in both models b is a knight. This can be demonstrated with Prover9 (see Fig. 7.4).

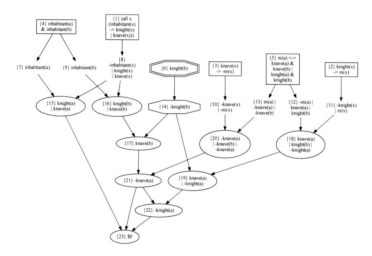

Fig. 7.4 Proof for $\{7\}$: b is a knight based on the message: "We are both the same"

Puzzle 65. Three inhabitants and two messages

Again we have three people, A, B, C, each of them is either a knight or a knave. A and B make the following statements: A: "All of us are knaves". B: Exactly one of us is a knight. What are A, B, C? (puzzle 31 from Smullyan (2011))

Listing 7.9 Finding models for three inhabitants with two messages

```
1   assign(domain_size ,3).
2   assign(max_models , -1).
3
4   formulas(island_of_truth).
5     all  x  (inhabitant(x)  ->  knight(x)  |  knave(x)).
6     all  x  ((knight(x)  ->  -knave(x))  &  (knave(x)  ->  -knight(x))).
7     knight(x)  ->   m(x).
8     knave(x)   ->  -m(x).
9   end_of_list.
10
11  formulas(puzzle).
12    inhabitant(a) & inhabitant(b) & inhabitant(c).
13    a=0 & b=1 & c=2.                          %remove  isomorphisms
14    m(a) <-> (knave(a) & knave(b) & knave(c)).
15    m(b) <-> (knight(a) & knave(b) & knave(c)) |
16             (knave(a) & knight(b) & knave(c)) |
17             (knave(a) & knave(b) & knight(c)).
18  end_of_list.
```

Listing 7.10 Proving *knave*(*a*), *knight*(*b*) and *knave*(*c*) based on two messages *m*(*a*) and *m*(*b*)

```
1   formulas(assumptions).
2     all  x  (inhabitant(x)  ->  knight(x)  |  knave(x)).
3     all  x  ((knight(x)  ->  -knave(x))  &  (knave(x)  ->  -knight(x))).
4     knight(x)  ->   m(x).        knave(x)   ->  -m(x).
5
6     inhabitant(a) & inhabitant(b) & inhabitant(c).
7     a != b  & b !=c   & c !=a.
8     m(a) <-> (knave(a) & knave(b) & knave(c)).
9     m(b) <-> (knight(a) & knave(b) & knave(c)) |
10             (knave(a) & knight(b) & knave(c)) |
11             (knave(a) & knave(b) & knight(c)).
12  end_of_list.
13
14  formulas(goals).
15   % knave(a) & knight(b) & knave(c).
16   % knave(a).
17   % knight(b).
18    knave(c).
19  end_of_list.
```

> **Solution**
>
> With three inhabitants we need a domain size of 3. To avoid isomorphisms we fix
> $a = 0, b = 1, c = 2$ (line 13 in Listing 7.9). The first message is that all inhabitants are
> knaves:
>
> $$m(a) \leftrightarrow (knave(a) \wedge knave(b) \wedge knave(c)) \tag{7.4}$$
>
> The second message says two points: there is knight, and that knight is unique, that can
> be modelled with $m(b) \leftrightarrow ((\exists x\ knight(x)) \wedge (\forall x\ \forall y\ (knight(x) \wedge knight(y) \rightarrow x =$
> $y)))$ The problem with this formalisation is that it introduces three Skolem constants
> that generate isomorphic models. Instead, we use an explicit modelling of the three
> possible cases:
>
> $$m(b) \leftrightarrow (knight(a) \wedge knave(b) \wedge knave(c) \vee$$
> $$(knave(a) \wedge knight(b) \wedge knave(c) \vee$$
> $$(knave(a) \wedge knave(b) \wedge knight(c) \vee \tag{7.5}$$
>
> Mace4 finds a single model which includes the following relevant relations:
>
> $$relation(knave(\_), [\ 1, 0, 1\ ])$$
> $$relation(knight(\_), [\ 0, 1, 0\ ])$$
> $$relation(m(\_), [\ 0, 1, 0\ ])$$
>
> Here, the first message $m(a)$ is false and $m(b)$ is true. Also b is a knight, while a and
> c are knaves. One can follow the Prover9's line of reasoning to prove this theorem
> $knave(a) \wedge knight(b) \wedge knave(c)$ (line 15 in Listing 7.10). Since the proof is large,
> we can prove separately each statement (lines 14–16).

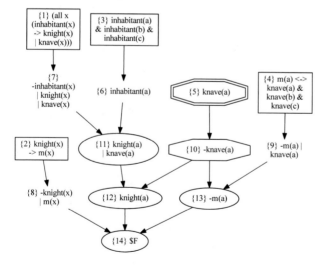

Fig. 7.5 Proof for {5}: a is a knave based on message {4}: "All of us are knaves"

Solution

For instance, we can start to prove that a is a knave (see Fig. 7.5). First, based on the negated conclusion {10} and the fact the $m(a)$ is false or a is a knave (i.e. clause {9}), Prover9 infers that the message $m(a)$ is false. Second, following the clause {10}, a must be a knight (i.e. clause {12}). Third, if a is a knight its message $m(a)$ is true (i.e. clause {8}). Applying hyper-resolution among {13}, {12} and {8} leads to a contradiction. Thus, theorem {5} is proved. Note that the message $m(b)$ was not needed within the proof. Next we can prove that c is knave (line 18 in Listing 7.10). The proof appears in Fig. 7.6. Here, the prover finds a different reasoning chain to prove that a is a knave (clause {25}. Based on message $m(b)$ and assuming that c is a knight, the prover deduces {26}: $m(b)$ is true or b is not a knave (clause {26}. Hence, the message $m(b)$ is true. This is a contradiction, since we know that knaves do not convey true messages.

Similarly, one can ask for the proof for the last term $knight(b)$ (line 17 in Listing 7.10).

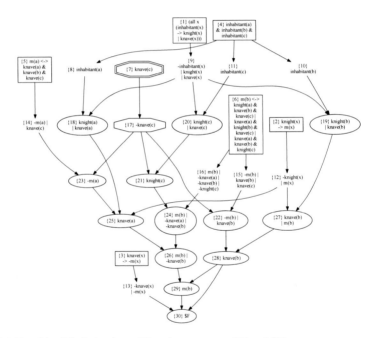

Fig. 7.6 Proof for {7}: "c is a knave" based on messages {5} and {6}

Puzzle 66. Three inhabitants and not enough information

Suppose instead, *A* and *B* say the following: *A*: "All of us are knaves". *B*: "Exactly one of us is a knave". Can it be determined what *B* is? Can it be determined what *C* is? (puzzle 32 from Smullyan (2011))

Listing 7.11 Finding models for three inhabitants and not enough information

```
1   assign(domain_size,3).
2   assign(max_models,-1).
3
4   formulas(island_of_truth).
5     all x (inhabitant(x) -> knight(x) | knave(x)).
6     all x (knight(x) -> -knave(x)).
7     all x (knight(x) -> m(x)) & (knave(x) -> -m(x)).
8   end_of_list.
9
10  formulas(puzzle).
11    inhabitant(a) & inhabitant(b) & inhabitant(c).
12    a=0 & b=1 & c=2.                %remove isomorphisms
13    m(a) <-> (knave(a) & knave(b) & knave(c)).
14    m(b) <-> (knave(a) & knight(b) & knight(c)) |
15            (knight(a) & knave(b) & knight(c)) |
16            (knight(a) & knight(b) & knave(c)).
17  end_of_list.
```

Listing 7.12 Proving *knight*(*c*) based on two messages *m*(*a*) and *m*(*b*)

```
1   formulas(assumptions).
2     all x (inhabitant(x) -> knight(x) | knave(x)).
3     all x ((knight(x) -> -knave(x)) & (knave(x) -> -knight(x))).
4     knight(x) -> m(x).      knave(x)  -> -m(x).
5
6     inhabitant(a) & inhabitant(b) & inhabitant(c).
7     a != b & b !=c & c !=a.
8     m(a) <-> (knave(a) & knave(b) & knave(c)).
9     m(b) <-> (knave(a) & knight(b) & knight(c)) |
10            (knight(a) & knave(b) & knight(c)) |
11            (knight(a) & knight(b) & knave(c)).
12  end_of_list.
13
14  formulas(goals).
15    knight(c).
16  end_of_list.
```

Solution

With three inhabitants we need a domain size of 3. To avoid isomorphisms, we fix $a = 0$, $b = 1$, $c = 2$ (line 13 in Listing 7.11). The first message is that all inhabitants are knaves:

$$m(a) \leftrightarrow (knave(a) \wedge knave(b) \wedge knave(c)) \tag{7.6}$$

The second message "there is exactly one knave" can be modelled by enumerating the three possible cases:

$$
\begin{aligned}
m(b) \leftrightarrow (knave(a) \wedge knight(b) \wedge knight(c) \vee \\
(knight(a) \wedge knave(b) \wedge knight(c) \vee \\
(knight(a) \wedge knight(b) \wedge knave(c) \vee
\end{aligned}
\tag{7.7}
$$

Mace4 finds two models:

	Model 1	Model 2
relation(knave(_), [1, 0, 0])		relation(knave(_), [1, 1, 0])
relation(knight(_), [0, 1, 1])		relation(knight(_), [0, 0, 1])

Observe that in both models, a is knave (i.e. first position in relation knave is 1 in both models). Similarly, in both models c is a knight (i.e. third position in relation knight is 1 in both models). Hence, one can prove $knave(a) \wedge knight(c)$, since this sentence is true in all models. Differently, b (position 2 in each relation) is a knight in the first model, and a knave in the second model. Hence, there is not enough information to prove what b is (Fig. 7.7).

Prover9 can be used to explain the reasoning for the theorem $knave(a) \wedge knight(c)$. Recall from Fig. 7.5 that $knave(a)$ can be proved only based on the first message. Since the first message is the same here, the proof is identical. The proof for $knight(c)$ is illustrated in Fig. 7.8.

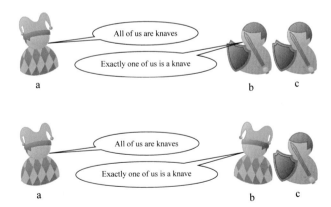

All of us are knaves

Exactly one of us is a knave

a   b   c

All of us are knaves

Exactly one of us is a knave

a   b   c

Fig. 7.7 Two models found: b can be either a knave or a knight

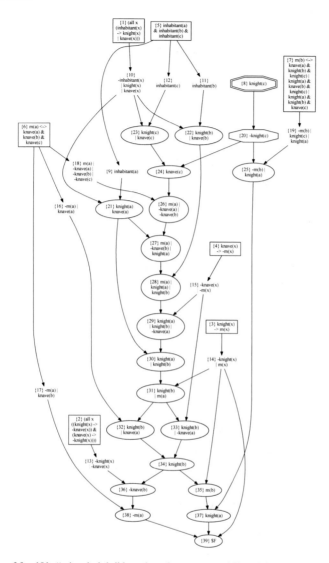

Fig. 7.8 Proof for {8}: "*c* is a knight" based on the messages {6} and {7}

Puzzle 67. Three inhabitants and two of the same type

Suppose instead, *A* and *B* say the following: *A*: "*B* is a knave". *B*: "*A* and *C* are of the same type". What is *C*? (puzzle 34 from Smullyan (2011))

Listing 7.13 Finding models for three inhabitants and two of the same type

```
1   assign(domain_size ,3).
2   assign(max_models , -1).
3
4   formulas(island_of_truth).
5     all x (inhabitant(x) -> knight(x) | knave(x)).
6     all x ((knight(x) -> -knave(x)) & (knave(x) -> -knight(x))).
7     knight(x) -> m(x).    knave(x)  -> -m(x).
8   end_of_list.
9
10  formulas(puzzle).
11    inhabitant(a) & inhabitant(b) & inhabitant(c).
12    a=0 & b=1 & c=2.                    %remove isomorphisms
13    m(a) <-> knave(b).
14    m(b) <-> ((knave(a)  & knave(c)) | (knight(a) & knight(c))).
15  end_of_list.
```

Solution

With three inhabitants we need a domain size of 3. To avoid isomorphisms we fix $a = 0, b = 1, c = 2$ (line 13 in Listing 7.13). The first message is $m(a) \leftrightarrow knave(b)$. The second message implies two cases: $m(b) \leftrightarrow (((knave(a) \wedge knave(c)) \vee (knight(a) \wedge knight(c)))$. Mace4 finds two models:

	Model 1	Model 2
	relation(knave(_), [1, 0, 1])	relation(knave(_), [0, 1, 1])
	relation(knight(_), [0, 1, 0])	relation(knight(_), [1, 0, 0])

Observe that in both models (see Fig. 7.9), c is knave (i.e. the third position corresponding to $c = 2$ in relation knave is 1 in both models). For the theorem $knave(c)$, Prover9 finds the demonstration in Fig. 7.10.

Fig. 7.9 Two models found: in both of them c is a knave

Listing 7.14 Proving *knave*(*c*) based on two messages *m*(*a*) and *m*(*b*): *A* and *C* are of the same type

```
1    formulas(assumptions).
2      all x (inhabitant(x) -> knight(x) | knave(x)).
3      all x ((knight(x) -> -knave(x)) & (knave(x) -> -knight(x))).
4      knight(x) -> m(x).       knave(x) -> -m(x).
5
6      inhabitant(a) & inhabitant(b) & inhabitant(c).
7      a != b  & b !=c  & c !=a.
8      m(a) <-> knave(b) .
9      m(b) <-> ((knave(a)  & knave(c)) | (knight(a) & knight(c))).
10   end_of_list.
11
12   formulas(goals).
13     knave(c).
14   end_of_list.
```

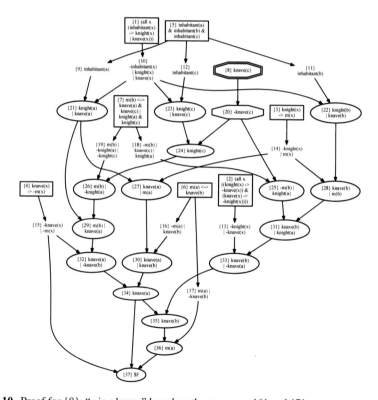

Fig. 7.10 Proof for {8}: "*c* is a knave" based on the messages {6} and {7}

Puzzle 68. Jim, Jon, and Joe

On the island of knights and knaves, knights always tell the truth, while knaves always lie. You are approached by three people: Jim, Jon, and Joe. Jim says: *Joe is a knave or I am a knight*. Jon says, *"Jim could claim that I am a knave"*. Joe says, *"Neither Jim nor Jon are knights"*. What are they actually? (taken from Popular mechanics - www. popularmechanics.com/science/math)

Listing 7.15 Jim, Jon, and Joe

```
1   formulas(assumptions).
2     all x (inhabitant(x) -> knight(x) | knave(x)).
3     all x ((knight(x) -> -knave(x)) & (knave(x) -> -knight(x))).
4     knight(x) -> m(x).
5     knave(x)  -> -m(x).
6
7     inhabitant(jim) & inhabitant(jon) & inhabitant(joe).
8     jim = 0 & joe = 1 & jon = 2.
9     m(jim) <-> (knave(joe) | knight(jim)).
10    m(jon) <-> ((knight(jim) & knave(jon)) | (knave(jim) & knight(jon))).
11    m(joe) <-> (-knight(jim) & -knight(jon)).
12  end_of_list.
```

Solution

Mace4 finds a single model (see Fig. 7.11):

$$relation(knave(\_), [1, 0, 1])$$
$$relation(knight(\_), [0, 1, 0])$$

So $Jim = 0$ (i.e. first position) and $Jon = 2$ (i.e. third position) are knaves, while $Joe = 1$ (i.e second position) is a knight. Since, Prover9 found a large proof for the goal $knave(jim) \wedge knight(joe) \wedge knave(jon)$ (line 14 in Listing 7.16), one can separately prove each subgoal. For instance, the proof for the subgoal $knave(jim)$ is illustrated in Fig. 7.12.

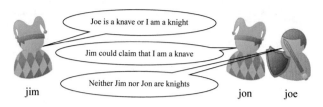

Fig. 7.11 The single model found by Mace4 for Jim, Joe, and Joe

Listing 7.16 Proving $knave(jim) \wedge knight(joe) \wedge knave(jon)$ based on three messages

```
1    formulas ( assumptions ) .
2       all x ( inhabitant(x) -> knight(x) | knave(x)).
3       all x (( knight(x) -> -knave(x)) & (knave(x) -> -knight(x))).
4       knight(x) -> m(x).    knave(x) -> -m(x).
5
6       inhabitant(jim) & inhabitant(joe) & inhabitant(jon).
7       jim != jon.  jon != joe.  jim != joe.
8       m(jim) <-> (knave(joe) | knight(jim)).
9       m(jon) <-> ((knight(jim) & knave(jon)) | (knave(jim) & knight(jon))).
10      m(joe) <-> (-knight(jim) & -knight(jon)).
11   end_of_list.
12
13   formulas ( goals ) .
14      %knave(jon) & knight(joe) & knave(jim).
15      knave(jon).  %knight(joe).  %knave(jim).
16   end_of_list.
```

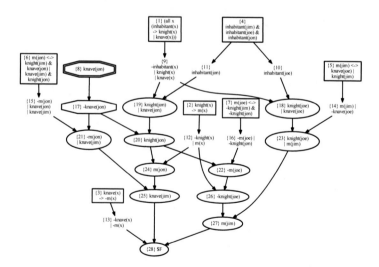

Fig. 7.12 Proof for {8}: "Jon is a knave" based on three messages {5}, {6} and {7}

Puzzle 69. A spy appears

On the island of knights and knaves a new type on inhabitants has settled: spies. Spies
can lie or tell the truth at will. You are approached by three people wearing different
coloured clothes. You know that one is a knight, one is a knave, and one is a spy. They
speak in the following order: The man wearing blue says, "I am a knight". The man
wearing red says, "He speaks the truth". The man wearing green says, "I am a spy".
Who is the knight, who is the knave, and who is the spy? (taken from Popular mechanics - www.popularmechanics.com/science/math)

Listing 7.17 Finding models for knights, knaves, and spies

```
1    assign(domain_size ,3).
2    assign(max_models , -1).
3
4    formulas(island_of_truth_with_spies).
5      all  x  (inhabitant(x) -> knight(x) | knave(x) | spy(x)).
6      all  x  (knight(x) -> -knave(x) & -spy(x)).
7      all  x  (knave(x) -> -knight(x) & -spy(x)).
8      all  x  (spy(x) -> -knight(x) & -knave(x)).
9      knight(x)  ->  m(x).
10     knave(x)   -> -m(x).
11   end_of_list.
12
13   formulas(puzzle).
14     inhabitant(blue) & inhabitant(red) & inhabitant(green).
15     blue = 0 & red = 1 & green = 2.
16     knight(x) & knight(y) -> x=y.
17     knave(x)  & knave(y)  -> x=y.
18     spy(x)    & spy(y)    -> x=y.
19
20     (exists  x  knight(x)) & (exists  x  knave(x)) & (exists  x  spy(x)).
21
22     m(blue)  <-> knight(blue).
23     m(red)   <-> knight(blue).
24     m(green) <-> spy(green).
25   end_of_list.
```

Fig. 7.13 The single model found by Mace4 for Blue, Red, and Green

When a spy appears on the island of truth, our assumptions are:

1. There are 3 types of inhabitants:
   $\forall x \ (inhabitant(x) \rightarrow knight(x) \vee knave(x) \vee spy(x))$

2. The set of knights, knaves and spies are disjoint:

$$\forall x \ (knight(x) \rightarrow \neg knave(x) \wedge \neg spy(x))$$
$$\forall x \ (knave(x) \rightarrow \neg knight(x) \wedge \neg spy(x)) \qquad (7.8)$$
$$\forall x \ (spy(x) \rightarrow \neg knight(x) \wedge \neg knave(x))$$

3. The message $m(x)$ is true if x is a knight and false if x is a knave:

$$(knight(x) \rightarrow (x) \rightarrow m(x)) \wedge (knave(x) \rightarrow \neg m(x)) \qquad (7.9)$$

The puzzle says that:

1. There are three inhabitants: blue, red, and green:

$$inhabitant(blue) \wedge inhabitant(red) \wedge inhabitant(green). \qquad (7.10)$$

   To avoid generation of isomorphic models we fix each individual to a unique value in the domain: $blue = 0, red = 1, green = 2$.

2. There is a single knight, knave or spy:

$$\forall x \ \forall y \ (knight(x) \wedge knight(y) \rightarrow x = y)$$
$$\forall x \ \forall y \ (knave(x) \wedge knave(y) \rightarrow x = y) \qquad (7.11)$$
$$\forall x \ \forall y \ (spy(x) \wedge spy(y) \rightarrow x = y)$$

   Equation 7.11 says there is at most one knight, at most one knave or at most one spy. We need to add the "at least" constraint: $\exists x \ knight(x)$, $\exists x \ knave(x)$, $\exists x \ spy(x)$. Then we formalise the messages:

3. Blue's message is that blue is a knight: $m(blue) \leftrightarrow knight(blue)$.

4. Red's message is that blue is a knight: $m(red) \leftrightarrow knight(blue)$.

5. Green's message is that green is a spy: $m(green) \leftrightarrow spy(green)$.

Given the formalisation in Listing 7.17, Mace4 computes a single model in which $knight(blue)$, $knave(green)$, and $spy(red)$ (see Fig. 7.13). In this model, the spy's message $m(spy)$ is true. The proof for $knight(blue)$ appears in Fig. 7.14. Note that the proof does not use the message of the blue person $m(b)$.

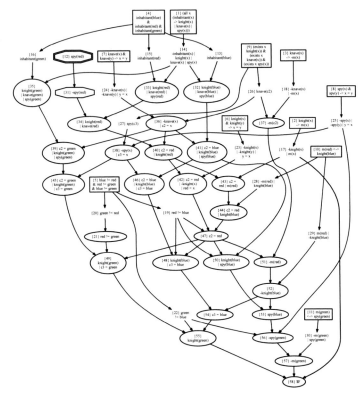

Fig. 7.14 Proof for $spy(red)$ ({12}) based on the messages $m(red)$ ({10}) and $m(green)$ ({11})

Puzzle 70. Who is the spy?

On the island of knights and knaves and spies, you come across three people. One
wears blue, one wears red, and one wears green. You know that one is a knight, one
is a knave, and one is a spy. "Who is the spy?" you ask. The man wearing blue says,
"That man in red is the spy". The man wearing red says, "No, the man in green is the
spy". The man wearing green says, "No, the man in red is in fact the spy". (taken from
Popular mechanics - www.popularmechanics.com/science/math)

Listing 7.18 Finding models with spies

```
1    assign(domain_size ,3).
2    assign(max_models , -1).
3
4    formulas(island_of_truth_with_spies).
5      all x (inhabitant(x) -> knight(x) | knave(x) | spy(x)).
6      all x (knight(x) -> -knave(x) & -spy(x)).
7      all x (knave(x)  -> -knight(x) & -spy(x)).
8      all x (spy(x)    -> -knight(x) & -knave(x)).
9      knight(x)         -> m(x).              knave(x)  -> -m(x).
10   end_of_list.
11
12   formulas(puzzle).
13     inhabitant(blue) & inhabitant(red) & inhabitant(green).
14     blue = 0 & red = 1 & green = 2.
15     (knight(x) & knight(y)) -> x=y.
16     (knave(x)  & knave(y))  -> x=y.
17     (spy(x)    & spy(y))    -> x=y.
18     (exists x knight(x)) & (exists x knave(x)) & (exists x spy(x)).
19
20     m(blue)  <-> spy(red).
21     m(red)   <-> spy(green).
22     m(green) <-> spy(red).
23   end_of_list.
```

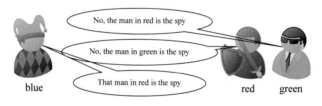

blue   red   green

Fig. 7.15 The single model found by Mace4

Solution

The only difference from the previous puzzle is that the messages differ. We formalise here only the three messages:

1. Blue's message is that blue is a knight: $m(blue) \leftrightarrow spy(red)$.
2. Red's message is that blue is a knight: $m(red) \leftrightarrow spy(green)$.
3. Green's message is that green is a spy: $m(green) \leftrightarrow spy(red)$.

Note that red and green accuse each other of being the spy (Fig. 7.15).
Given the formalisation in Listing 7.17, Mace4 computes a single model in which $knight(red)$, $knave(blue)$, and $spy(green)$ are true. In this model, the spy's message $m(spy)$ is false. For instance, the proof for the subgoal $spy(green)$ appears in Fig. 7.16.

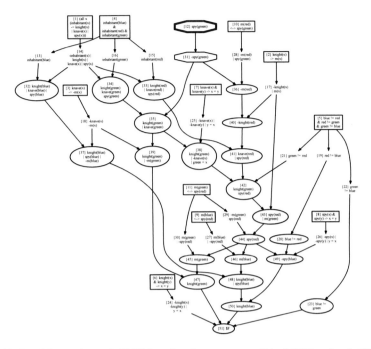

Fig. 7.16 Proof for $spy(green)$ ({12}) based on the messages $m(blue)$ ({9}), $m(red)$ ({10}), and $m(green)$ ({11})

Puzzle 71. The whole truth and nothing but the truth

On a famous island in the Pacific Ocean (whose name I forgot) live three tribes: the *Whites*, who always tell the truth; the *Blacks*, who always tell lies; and the *Greys*, who lie and tell the truth alternatively (although their first answer may be either truth or lie). These natives always gather in groups of three, with one representative of each tribe in the group. I approached such a group of three natives, and had the following conversation with the first native: "Are you the White, the Black, or the Grey?" "I am the Grey". "And what about your friend here?" "He is a Black". "So, your other friend is the White? " "Of course". Was the "other friend" a White, and if not, what was he?

Listing 7.19 White is honest; Black is a liar; Grey alternatively tells truth or lies

```
1    assign(domain_size ,3).
2    assign(max_models ,-1).
3
4    list(distinct).
5      [n1,n2,n3].
6    end_of_list.
7
8    formulas(assumptions).
9      exists x (white(x)) & exists x (black(x)) & exists x (gray(x)).
10     all x (white(x) & white(y) -> x=y). %Each one from a different tribe
11     all x (black(x) & black(y) -> x=y).
12     all x (gray(x)  & gray(y)  -> x=y).
13
14     all x (white(x) -> -black(x) & -gray(x)). %Each one from a single tribe
15     all x (black(x) -> -white(x) & -gray(x)).
16     all x (gray(x)  -> -white(x) & -black(x)).
17   end_of_list.
18
19   formulas(alltruth_puzzle).
20     white(n1) ->    gray(n1) &  black(n2) &  white(n3).   %Case 1: w answers
21     black(n1) ->  -gray(n1) & -black(n2) & -white(n3).   %Case 2: b answers
22     gray(n1)  -> ( gray(n1) & -black(n2) &  white(n3)) | %Case 3: g answers
23                  (-gray(n1) &  black(n2) &  white(n3)).
24   end_of_list.
```

Solution

The functions n_1, n_2, and n_3 represent the three natives. These natives are distinct (line 5 in Listing 7.19). There are two assumptions: (i) Each native belongs to a different tribe (lines 9–12); (ii) A native belongs to a single tribe (lines 14–16). Depending on who provides the answer, there are three cases:

1. white answers → its answer is true:

$$white(n_1) \rightarrow grey(n_1) \wedge black(n_2) \wedge white(n_3) \qquad (7.12)$$

2. black answers → its answer is false:

$$black(n_1) \rightarrow \neg grey(n_1) \wedge \neg black(n_2) \wedge \neg white(n_3) \qquad (7.13)$$

3. grey answers → its answers are true and false alternatively, i.e. there are two subcases: (3.1) the first answer is true, respectively (3.2) the first answer is false:

$$grey(n_1) \rightarrow (grey(n_1) \wedge \neg black(n_2) \wedge white(n_3)) \vee$$
$$(\neg grey(n_1) \wedge black(n_2) \wedge white(n_3)) \qquad (7.14)$$

Mace4 outputs a single model: $n_1=0$ $n_2=1$ $n_3=2$ $c_1=1$ $c_2=0$ $c_3=2$

black:	0 1 2	gray:	0 1 2	white:	0 1 2
	1 0 0		0 0 1		0 1 0

Here, c_1, c_2, and c_3 are the Skolem constants from line 9 in Listing 7.19 (e.g. $\exists x\ white(x)$ for c_1). Hence, n_1 (i.e. the person that answers) is black, n_2 is white, while n_3 is grey (see Fig. 7.17).

Fig. 7.17 The single model found by Mace4 for Black, White, and Grey inhabitants

---

Puzzle 72. Three goddesses

Three goddesses were sitting in an old Indian temple. Their names were Truth (always telling the truth), Lie (always lying), and Wisdom (sometimes lying). A visitor asked the one on the left: "Who is sitting next to you?" "Truth", she answered. Then he asked the one in the middle: "Who are you?" "Wisdom". Last, he asked the one on the right: "Who is your neighbour?" "Lie", she replied. And then it became clear who was who.

Listing 7.20 Finding models of three goddesses in propositional logic

```
1    assign(domain_size,2).
2    assign(max_models,-1).
3
4    formulas(assumptions).
5      (T1 | T2 | T3) & (L1 | L2 | L3) & (W1 | W2 | W3).
6
7      %T, L and W occupies only one position
8      (T1 -> -T2 & -T3) & (T2 -> -T1 & -T3) & (T3 -> -T1 & -T2).
9      (L1 -> -L2 & -L3) & (L2 -> -L1 & -L3) & (L3 -> -L1 & -L2).
10     (W1 -> -W2 & -W3) & (W2 -> -W1 & -W3) & (W3 -> -W1 & -W2).
11
12     %Each position is occupied by one godess.
13     (T1 -> -L1 & -W1) & (T2 -> -L2 & -W2) & (T3 -> -L3 & -W3).
14     (L1 -> -T1 & -W1) & (L2 -> -T2 & -W2) & (L3 -> -T3 & -W3).
15     (W1 -> -L1 & -T1) & (W2 -> -L2 & -T2) & (W3 -> -L3 & -T3).
16
17     (L1 -> -T2) & (T1 -> T2).         % the first  answer
18     (L2 -> -W2) & (T2 -> W2).         % the second answer
19     (T3 -> L2) & (L3 -> -L2).         % the third  answer
20   end_of_list.
```

Solution

We will use propositional logic. Let Truth (T), Wisdom (W), Lie (L) are in left (i.e. 1), middle (i.e. 2) or right (i.e. 3) positions. With three goddesses and three positions, there are nine possible combinations:

$$(T_1 \vee T_2 \vee T_3) \wedge (L_1 \vee L_2 \vee L_3) \wedge (W_1 \vee W_2 \vee W_3) \qquad (7.15)$$

Each goddess occupies only one position. For instance, if Truth is on the first position she cannot be on the other positions:

$$(T_1 \rightarrow \neg T_2 \wedge \neg T_3) \wedge (T_2 \rightarrow \neg T_1 \wedge \neg T_3) \wedge (T_3 \rightarrow \neg T_1 \wedge \neg T_2) \qquad (7.16)$$

Each position is occupied by one goddess. For instance, if there is Truth, there can't be Lie or Wisdom:

$$(T_1 \rightarrow \neg L_1 \wedge \neg W_1) \wedge (T_2 \rightarrow \neg L_2 \wedge \neg W_2) \wedge (T_3 \rightarrow \neg L_3 \wedge \neg W_3) \qquad (7.17)$$

The goddess on the first position says that Truth stays next to her (i.e. position 2, thereby T_2). If she is L then the answer is false ($\neg T_2$), if she is T then the answer is true (T_2), if she is W we cannot infer anything:

$$(L_1 \rightarrow \neg T_2) \wedge (T_1 \rightarrow T_2) \qquad (7.18)$$

The goddess on the second position says she is Wisdom (i.e. W_2). If she is Truth the answer is true, if she is Lie the answer is false:

$$(L_2 \rightarrow \neg W_2) \wedge (T_2 \rightarrow W_2) \qquad (7.19)$$

The goddess on the third position says that Lie is her neighbour (i.e. L_2), which can be true or false:

$$(T_3 \rightarrow L_2) \wedge (L_3 \rightarrow \neg L_2) \qquad (7.20)$$

Mace4 finds a model in which Lie is in the middle, Truth on the right, and Wisdom on the left:

Model	L_1	L_2	L_3	T_1	T_2	T_3	W_1	W_2	W_3
1	0	1	0	0	0	1	1	0	0

This model appears in Fig. 7.18. The proof for the theorem $L_2 \wedge T_3 \wedge W_1$ appears in Fig. 7.19.

Fig. 7.18 The single model found by Mace4 for three goddesses

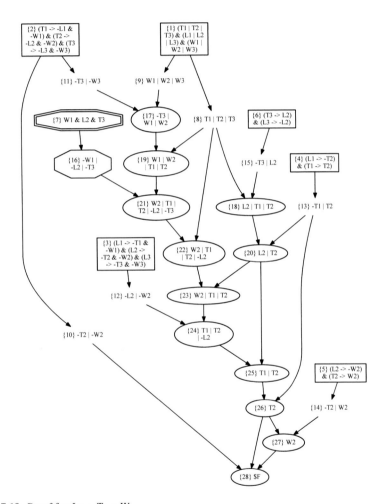

Fig. 7.19 Proof for $L_2 \wedge T_3 \wedge W_1$

References

Aszalós, L. (2001). Automated solution of the riddle of Dracula and other puzzles. In *Proceedings of the International Joint Conference on Automated Reasoning (IJCAR'01)* (pp. 1–10).

Baumgartner, P., Furbach, U., & Stolzenburg, F. (1995). Model elimination, logic programming and computing answers. In *IJCAI* (pp. 335–341).

Ciraulo, F., & Maschio, S. (2020). Solving knights-and-knaves with one equation. *The College Mathematics Journal, 51*(2), 82–89.

Smullyan, R. M. (2011). *What is the name of this book? The riddle of Dracula and other logical puzzles*. Dover Publications.

Chapter 8
Love and Marriage

Abstract In which we enter into the world of interwoven family relationships, love, and marriage. The classical "Looking at unmarried people" puzzle has been adapted here for the television series "Friends". In this puzzle, Monica is looking at Ross, Ross is looking at Rachel. Monica is married; Rachel is not. Is a married person looking at an unmarried person? The answer may seem difficult since we do not know if Ross is married or not. A fan of the "Friends" series would know that even Ross has some difficulties to figure out if he is married or not. You do not have to worry since this is an easy one for Prover9. The chapter continues with puzzles related to lying within mariage, or puzzles asking you to figure out various family relationships.

$$\forall x \ (man(x) \wedge wife(x) = c \rightarrow (\forall z \ \neg attractive(x, z) \leftrightarrow z \neq c))$$

Oscar Wilde

The following puzzles are related to interwoven family relationships, love, and marriage. The classical "Looking at unmarried people" puzzle has been adapted here for the television series "Friends". In this puzzle, Monica is looking at Ross, Ross is looking at Rachel. Monica is married; Rachel is not. Is a married person looking at an unmarried person? The answer may seem difficult since we do not know if Ross is married or not. A fan of the "Friends" series would know that even Ross has some difficulties to figure out if he is married or not. You do not have to worry since this is an easy one for Prover9. The chapter continues with puzzles related to lying within marriage, or puzzles asking you to figure out various family relationships. For instance, the "Family tree" puzzles ask you to figure out a family tree of ten members over three generations, based on some clues. The solution shows how each clue reduces the number of trees (i.e. interpretation models) until a single tree remains.

The chapter can be seen as a review of the modelling techniques that have been introduced until now. It contains truth-telling puzzles in line with Chap. 7, zebra-like puzzles related to marriage in line with Chap. 6, or arithmetic puzzles in line with Chap. 4.

> **Puzzle 73. Looking at unmarried people**
>
> There are three friends staying on the couch in Central Perk: Rachel, Ross, and Monica. Monica is looking at Ross. Ross is looking at Rachel. Monica is married; Rachel is not. Is a married person looking at an unmarried person?

Rachel   Ross   Monica

Listing 8.1 Proving that a married person is looking at an unmarried one

```
1    formulas(assumptions).
2        married(Monica).
3      −married(Rachel).
4        looking(Monica,Ross).
5        looking(Ross,Rachel).
6    end_of_list.
7
8    formulas(goals).
9      exists x exists y (married(x) &−married(y) & looking(x,y)).
10   end_of_list.
```

> **Solution**
>
> A proof by resolution for
>
> $$\exists x \, \exists y \, (married(x) \wedge \neg married(y) \wedge looking(x, y)). \qquad (8.1)$$
>
> is obtained with:
>
> ```
> prover9 -f married.in | prooftrans xml renumber | gvizify
>                       | dot -Tpdf
> ```
>
> The prover starts by negating the goal, that is clause {2} (see Fig. 8.1). On the one hand, resolution introduces clause {4} into {2}, resulting in $\neg married(Ross) \vee married(Rachel)$ (i.e. clause {8}). As we know from {6} that Rachel is not married, the system deduces that Ross is not married (i.e. clause {10}). On the other hand, resolution introduces clause {3} into {2} to infer clause {7}: $\neg married(Monica) \vee married(Ross)$. Knowing that Monica is married (clause {5}), the prover infers that Ross must be married. Obviously, clauses {9} and {10} contradict each other, meaning that the negated goal was a wrong assumption. That is, the theorem was proved.

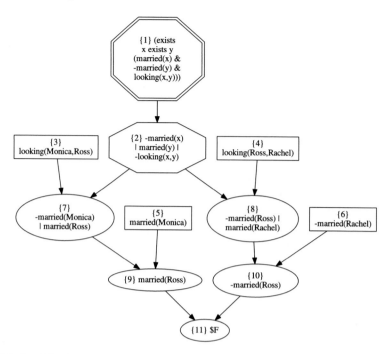

Fig. 8.1 Proof for theorem {1}

Married people always tell the truth, while unmarried people always lie. There are four friends staying on the couch in Central Perk: Ross, Monica, Rachel, and Chandler. They made the following statements:

Ross: "Rachel and I have different marital status".
Rachel: "Chandler is not married".
Monica: "I am married to Chandler".
Chandler: "Monica is lying".

How many are married, and how many are unmarried?

Listing 8.2 Finding models for "married people do not lie"

```
1   assign(domain_size,4).
2   assign(max_models,−1).
3
4   list(distinct).
5     [ross, chandler, monica, rachel].
6   end_of_list.
7
8   formulas(married).
9     all x (married(x)  <-> m(x)).
10    m(ross)     <-> (married(rachel) -> −married(ross)).
11    m(rachel)   <-> −married(chandler).
12    m(monica)   <-> (married(monica) & married(chandler)).
13    m(chandler) <-> −m(monica).
14  end_of_list.
```

Solution

With four individuals we set the domain size to 4. Let the predicate $m(x)$ formalising the message of person x. The predicate is true if x is married, and false otherwise: $\forall x\ (married(x) \leftrightarrow m(x))$. Ross says that he has a different marital status than Rachel:

$$m(ross) \leftrightarrow (married(rachel) \rightarrow \neg married(ross)) \qquad (8.2)$$

Rachel says that Chandler is not married: $m(rachel) \leftrightarrow \neg married(chandler)$. Monica says that she is married to Chandler:

$$m(monica) \leftrightarrow (married(monica) \wedge married(chandler)) \qquad (8.3)$$

Lastly, Chandler says that Monica is lying: $m(chandler) \leftrightarrow \neg m(monica)$.
Mace4 finds a single model where $chandler = 0, monica = 1, rachel = 2, ross = 3$:

$$relation(married(\_), [\,1, 0, 0, 1\,])$$

Here only Chandler and Ross are married (see Fig. 8.2).

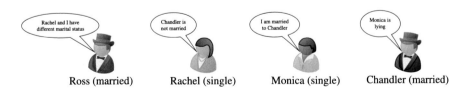

Ross (married)   Rachel (single)   Monica (single)   Chandler (married)

Rachel and I have different marital status · Chandler is not married · I am married to Chandler · Monica is lying

Fig. 8.2 The single model found by Mace4

Puzzle 75. Minos and aminos: we are both married

News has reached me, Auspicious King, of a curious town in which every inhabitant is either a Mino or an Amino. "Oh my goodness, what are they?" asked the king. "The Minos are worshipers of a good god; whereas the Aminos worship an evil god. The Minos always tell the truth-they never lie. The Aminos never tell the truth-they always lie. All members of one family are of the same type. Thus given any pair of brothers, they are either both Minos or both Aminos. Now, I heard a story of two brothers, Bahman and Perviz, who were once asked if they were married. They gave the following replies:

Bahman: We are both married.
Perviz: I am not married.
Is Bahman married or not? And what about Perviz? (taken from Smullyan (1996))

Listing 8.3 Finding models for minos and aminos

```
1   assign(max_models, −1).
2   assign(domain_size,2).
3
4   formulas(minosaminos).
5     all x (inhabitant(x) −> mino(x) | amino(x)).
6     all x (mino(x) −> −amino(x)).
7     mino(x) −> m(x).
8     amino(x) −> −m(x).
9     same_family(x,y) −> mino(x) & mino(y) | amino(x) & amino(y).
10    brother(x,y) −> same_family(x,y).
11    −brother(x,x).
12    same_family(x,x).
13  end_of_list.
14
15  formulas(puzzle).
16    inhabitant(Bahman) & inhabitant(Perviz) & (Bahman != Perviz).
17    brother(Bahman,Perviz) & brother(Perviz,Bahman).
18    m(Bahman) <−> married(Bahman) & married(Perviz).
19    m(Perviz) <−> −married(Perviz).
20  end_of_list.
```

Solution

A human agent might reason as follow: The two statements are either both true or both false, since the Perviz and Bahman are in the same family and therefore of the same religion: $m(Bahman) \leftrightarrow m(Perviz)$. The statements contradict each other: one states $married(Bahman) \land married(Perviz)$ and the other $\neg married(Perviz)$. Hence they are both false, which makes the brothers *amino*. It means that Perviz is married. Bahman's statement is also false: the only situation that falsifies $m(Bahman)$ is when Bahman is not married.

Solution

The puzzle is formalised in Listing 8.3. For finding all the models with Mace4, a domain size of 2 is enough, since we will use only predicates and not functions. Every inhabitant is a Mino or an Amino: $\forall x \ (inhabitant(x) \rightarrow mino(x) \lor amino(x))$. These sets are disjoint: $\forall x \ (mino(x) \rightarrow \neg amino(x))$. The minos statements are true (i.e. $mino(x) \rightarrow m(x)$), and aminos statements are false (i.e. $amino(x) \rightarrow \neg m(x)$). All members of one family are of the same type: $same\_family(x, y) \rightarrow mino(x) \land mino(y) \lor amino(x) \land amino(y)$. We need background knowledge about family relations: First, brothers are from the same family: $brother(x, y) \rightarrow same\_family(x, y)$. Second, brother is an irreflexive relation: $\neg brother(x, x)$. Third, $same\_family$ is a reflexive relation: $same\_family(x, x)$. The last two pieces of knowledge help Mace4 to avoid generating isomorphic models.

Given the above theory about minos and aminos, the puzzle states that "there are two inhabitants":

$$inhabitant(Bahman) \land inhabitant(Perviz) \land (Bahman! = Perviz) \qquad (8.4)$$

They are brothers: $brother(Bahman, Perviz) \land brother(Perviz, Bahman)$. Note that the brother relation is not symmetric (i.e $brother(x, y) \rightarrow brother(y, x)$) because y may be a girl. The messages are formalised as follows:

$$m(Bahman) \leftrightarrow married(Bahman) \land married(Perviz) \qquad (8.5)$$

$$m(Perviz) \leftrightarrow \neg married(Perviz) \qquad (8.6)$$

A single model is found by Mace4 (see Fig. 8.3). Both brothers are amino, both statements m are false, married relation is false for domain value 0 (i.e. Bahman) and true for domain value 1 (i.e. Perviz).

Bahman: 0   Perviz: 1   amino: | 0 1   inhabitant: | 0 1   m: | 0 1
                                      | --- | --- | ---
                                      | 1 1 | 1 1 | 0 0

married: | 0 1   mino: | 0 1   brother: | 0 1   same_family: | 0 1
         | --- | --- | --- | ---
         | 0 1 | 0 0   0 | 0 1   0 | 1 1
                                   1 | 1 0   1 | 1 1

Note that background knowledge keeps Mace4 away from generating more models for relations $brother$ and $same\_family$.

The logical agent (i.e. Prover9) has the proof in Fig. 8.4 for the goal $married(Perviz)$: Here the goal {8} is negated and based on {15} it infers that the statement of Bahman is false (i.e. {20}). Applying hyper-resolution on {14}, {18}, {20}, the agent deduces that Perviz is amino (i.e. {21}). On a different line of reasoning, the agent infers the statement of Perviz is true (i.e. {19}). This further leads to a conflict. Based on the detected inconsistency, the prover validates the initial goal {8}: Perviz is married.

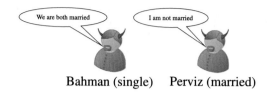

Fig. 8.3 The single model found by Mace4: both brothers are aminos

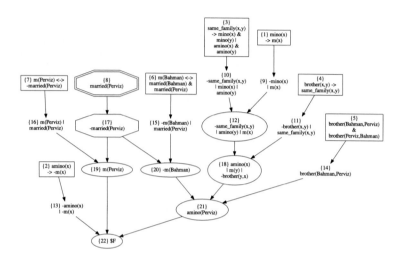

Fig. 8.4 Proof for "Perviz is married" ({8}) based on two messages: ({6}) and ({7})

Puzzle 76. Minos and aminos: we are both married or both unmarried

According to another version of the story, Oh, Auspicious King, Bahman didn't say
that they were both married; instead he said, "We are both married or both unmarried".
If that version is correct, then what can be deduced about Bahman and what can be
deduced about Perviz? (taken from Smullyan (1996))

Listing 8.4 Finding models "We are both married or both unmarried"

```
1   assign(max_models, −1).
2   assign(domain_size,2).
3
4   formulas(minosaminos).
5     all x (inhabitant(x) −> mino(x) | amino(x)).
6     all x (mino(x) −> −amino(x)).
7     mino(x) −> m(x).
8     amino(x) −> −m(x).
9     same_family(x,y) −> mino(x) & mino(y) | amino(x) & amino(y).
10    brother(x,y) −> same_family(x,y).
11    −brother(x,x).
12    same_family(x,x).
13  end_of_list.
14
15  formulas(puzzle).
16    inhabitant(Bahman) & inhabitant(Perviz) & (Bahman != Perviz).
17    brother(Bahman,Perviz) & brother(Perviz,Bahman).
18    m(Bahman) <−> (married(Bahman) & married(Perviz) |
19                   −married(Bahman) & −married(Perviz)).
20    m(Perviz) <−> −married(Perviz).
21  end_of_list.
```

Solution

A human agent would consider two cases:

1. Both sentences are true: this case leads to unmarried brothers.
2. Both sentences are false: this case leads to $married(Perviz)$ and $\neg married(Bahman)$.

In either case, Bahman is not married.
Similarly, the logical agent (Listing 8.4) finds two models:

Bahman: 0   Perviz: 1

amino: | 0 1 |
|---|
| 0 0 |

inhabitant: | 0 1 |
|---|
| 1 1 |

m: | 0 1 |
|---|
| 1 1 |

married: | 0 1 |
|---|
| 0 0 |

mino: | 0 1 |
|---|
| 1 1 |

brother: | 0 1 |
|---|
| 0 | 0 1 |
| 1 | 1 0 |

same_family: | 0 1 |
|---|
| 0 | 1 1 |
| 1 | 1 1 |

In the first model, both sentences are true (relation m). The $married$ relation is false for Bahman (value 0 in the domain).

Bahman: 0   Perviz: 1

amino: | 0 1 |
|---|
| 1 1 |

inhabitant: | 0 1 |
|---|
| 1 1 |

m: | 0 1 |
|---|
| 0 0 |

married: | 0 1 |
|---|
| 0 1 |

mino: | 0 1 |
|---|
| 0 0 |

brother: | 0 1 |
|---|
| 0 | 0 1 |
| 1 | 1 0 |

same_family: | 0 1 |
|---|
| 0 | 1 1 |
| 1 | 1 1 |

In the second model, both sentences are false (relation m). The $married$ relation is false for Bahman (value 0 in the domain). Note that in both models, the relation married is false for the individual Bahman (see Fig. 8.5). Hence, we can use Prover9 to actually prove that $\neg married(Bahman)$. The logical agent (i.e. Prover9) finds the proof in Fig. 8.6 for the goal $\neg married(Bahman)$.

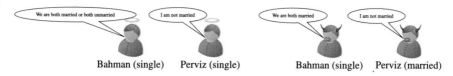

Fig. 8.5 Two models found by Mace4: In the first model (left) both brothers are minos and unmarried. In the second model (right) both brothers are aminos, Bahman is not married, and Perviz is married

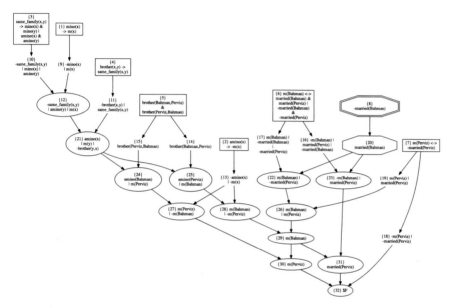

Fig. 8.6 Proof for ¬*married*(*Bahman*) (clause {8}) based on messages {6} and {7}

Puzzle 77. Marriage in company

In a certain company, Bob, Janet, and Shirley hold the positions of director, engineer, and accountant, but not necessarily in that order. The accountant, who is an only child, earns the least. Shirley, who is married to Bob's brother, earns more than the engineer. What position does each person fill? (taken from Danesi (2018))

Listing 8.5 Finding models for three jobs and three employees

```
1   assign(domain_size,3).
2   assign(max_models,-1).
3
4   formulas(assumptions).
5      bob = 0 & janet = 1 & shirley = 2.
6      accountant = 0 & engineer = 1 & director = 2.
7      all x all y (job(x) = job(y) -> x=y).
8
9      job(shirley) != engineer.          % Clue 1
10     job(shirley) != accountant.        % Clue 2
11     job(bob) != accountant.            % Clue 3
12  end_of_list.
```

Solution

We need a domain size of three elements. To avoid isomorphic models, we fix the values for each person and for each job. The function $job : person \rightarrow job$ will match the value of persons (0,1,2) with the values of jobs (0,1,2).

We assume two persons cannot have the same job: $\forall x \, \forall y \, job(x) = job(y) \rightarrow x = y$. There are three clues within the text: First, Shirley earns more than the engineer, hence she is not an engineer (line 9 in Listing 8.5). Second, since the accountant earns the least, Shirley cannot have that job (line 10). Third, we learn that Bob has a brother, but the accountant is an only child (line 11).

Mace4 finds a single model: $function(job(\_), [\ 1,\ 0,\ 2\ ])$. Here $job(0) = 1$ means $job(bob) = engineer$, $job(1) = 0$ means $job(janet) = accountant$, and $job(2) = 2$ means $job(shirley) = director$ (see Fig. 8.7).

Fig. 8.7 The single model
found by Mace4

Bob (engineer)   Janet (accountant)   Shirley (director)

Puzzle 78. What is my relationship to Teresa?

Teresa's daughter is my daughter's mother. What is my relationship to Teresa? Are you:

1. Teresa's grandmother.
2. her mother.
3. her daughter.
4. her granddaughter.
5. Teresa.

(taken from Brainzilla—www.brainzilla.com)

Listing 8.6 Adding domain knowledge to solve the quiz

```
1    assign(domain_size,4).
2    assign(max_models,−1).
3
4    formulas(background_knowledge).
5       hasDaughter(x) = y −> x != y.
6       hasDaughter(x) = y −> hasDaughter(y) != x.
7       mother(x) = y & mother(z) = y −> x = z.
8       mother(x) = y −> x != y.
9       mother(x)= y −> mother(y) != x.
10      hasDaughter(x) = y −> mother(x) != y.
11      hasDaughter(x) = y −> mother(y) = x.
12      (hasDaughter(x) = y & hasDaughter(y) = z) −> hasDaughter(z) != x.
13   end_of_list.
14
15   formulas(teresa_family).
16      hasDaughter(Teresa) = TeresaDaughter.
17      TeresaDaughter = mother(MyDaughter).
18      hasDaughter(Me)= MyDaughter.
19   end_of_list.
20
21   formulas(quiz).
22      % Me=mother(mother(Teresa)).          % 0 models  domain size = 4,5,6
23      % Me=mother(Teresa).                   % 0 models, domain size = 4,5
24        Me=TeresaDaughter.                   % 1 model,  domain size = 4
25      % Me=hasDaughter(hasDaughter(Teresa)). % 0 models, domain size = 4,5
26      % Me=Teresa.                           % 0 models, domain size = 4
27   end_of_list.
```

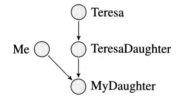

Fig. 8.8 Teresa's daughter is my daughter's mother

Solution

Two questions are: "How many individuals are assumed in this quiz?" and "What family-related knowledge should be added to solve the quiz?" First, there are four named individuals: $Teresa$, $TeresaDaughter$, $MyDaughter$ and I (Me). Hence we start with a domain size of 4. However, we do not know whether these individuals are distinct or not. Second, we formalise the following background knowledge related to family relationships (lines 4–13 in Listing 8.6). If someone x has the daughter y, then these individuals are distinct: $hasDaughter(x) = y \rightarrow x \neq y$. The relation $hasDaughter$ is antisymmetric: $hasDaughter(x) = y \rightarrow hasDaughter(y) \neq x$. The same two properties above are also valid for the $mother$ relation:

$$mother(x) = y \rightarrow x \neq y \tag{8.7}$$
$$mother(x) = y \rightarrow mother(y) \neq x \tag{8.8}$$

"Mother is unique" is formalised with: $mother(x) = y \land mother(z) = y \rightarrow x = z$. Someone's daughter cannot be someone's mother: $hasDaughter(x) = y \rightarrow mother(x) \neq y$. The relations $hasDaughter$ and $mother$ are inverse: $hasDaughter(x) = y \rightarrow mother(y) = x$.

Next, we formalise the puzzle with three equations that correspond to Fig. 8.8:

$$hasDaughter(Teresa) = TeresaDaughter \qquad (8.9)$$

$$TeresaDaughter = mother(MyDaughter) \qquad (8.10)$$

$$hasDaughter(Me) = MyDaughter \qquad (8.11)$$

Finally, we check the models found by Mace4 for each of the six questions from the quiz. First, for Teresa's grandmother $me = mother(mother(Teresa))$, we extend the domain with two more individuals for Teresa's mother and Teresa's grandmother. Given the background knowledge in Listing 8.6, Mace4 finds no model. Second, for Teresa's mother (i.e. $me = mother(Teresa)$), we extend the domain size with one more individual (i.e mother of Teresa), but Mace4 finds no model.

For $Me = TeresaDaughter$, Mace4 finds the model depicted in Fig. 8.9, where the arrow represents the $hasDaughter$ relation. In this model, Me is the same with Teresa's daughter, since both variables are linked to value 0. Note that $Me = 0$ and $TeresaDaughter = 0$, hence these variables represent the same individual. Thus, this is the answer to the quiz.

Some observations follow. Since the domain size has four individuals, Mace4 assigns the value 3 to the no-named individual in the domain. Note that Mace4 assumes $hasDaughter(myDaughter) = 3$ and $hasDaughter(3) = Teresa$. This wrong assumption can be avoided by refining the domain knowledge on family relationships. Note also that we do not use the arithmetic module. By importing this module, Mace4 will distinguish among values in the domain and it will compute 24 models instead. However, these models are not relevant for the given quiz, since only the values attached to the variables are changed. The last observation is that the distinct solution $me = son-in-law(teresa)$ does not appear in the quiz, hence supporting background knowledge for this solution was not included in Listing 8.6.

Mace4 finds no model for the last two options of the quiz, even if the domain size is increased accordingly (Fig. 8.10).

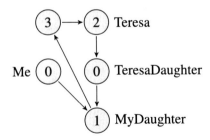

Fig. 8.9 The single model found by Mace4 for Teresa's daughter is my daughter's mother

Puzzle 79. Who is Helen's husband?

Yesterday evening, Helen and her husband invited their neighbours (two couples) for a dinner at home. The six of them sat at a round table. Helen tells you the following: "Victor sat on the left of the woman who sat on the left of the man who sat on the left of Anna. Esther sat on the left of the man who sat on the left of the woman who sat on the left of the man who sat on the left of the woman who sat on the left of my husband. Jim sat on the left of the woman who sat on the left of Roger. I did not sit beside my husband". The question: What is the name of Helen's husband? (taken from http://www.puzzlesite.nl)

Listing 8.7 Finding models for a round table

```
1    assign(domain_size,6).                                  % six places at the table
2    assign(max_models,-1).
3    set(arithmetic).
4
5    list(distinct).
6      [Anna, Esther, Helen, Jim, Roger, Victor].
7    end_of_list.
8
9    formulas(utils).
10     x > 0  -> left(x) = x + -1.
11     left(0) = domain_size + -1.                           % round table
12     beside(x,y) <-> left(x) = y | left(y) = x.
13     even(x) <-> x = 0 | x = 2 | x = 4.
14     man(x) <-> -woman(x).
15   end_of_list.
16
17   formulas(assumptions).
18     man(Roger) & man(Jim) & man(Victor).
19     woman(Helen) & woman(Anna) & woman(Esther).
20     man(HH).                                              % HH is a man
21     HH = 0.                                               % avoid isomorphisms
22     man(x) <-> even(x).                                   % men on even places
23   end_of_list.
24
25   formulas(round_table).
26     left(left(left(Anna))) = Victor.                      % Clue1
27     left(left(left(left(left(HH))))) = Esther.            % Clue2
28     left(left(Roger)) = Jim.                              % Clue3
29    -beside(HH,Helen).                                     % Clue4
30   end_of_list.
```

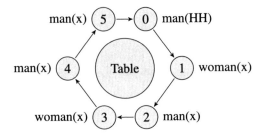

Fig. 8.10 Alternating men and women at the dinner table

Since there are six places at the table (Fig. 8.10), we set the domain size to 6. We assume that the individuals are distinct (line 6 in Listing 8.7). We define the function $left(x) = x - 1$ for $x \geq 1$ and $left(0) = 5$ (i.e. the maximum value in the domain size) in order to model a round table. In this way, the person sitting on place 0 will be on the left of place 5.

We define also the auxiliary predicate $beside(x, y)$ for x sitting besides y. That is, y is on the left of x, or x on the left of y: $beside(x, y) \leftrightarrow left(x) = y \vee left(y) = x$.

We assume that Roger, Jim, and Victor are men, while Anna, Helen, and Esther are women (lines 18–19). We note Helen's husband with HH and we assume he is a man (line 20). Since Helen's husband is one of the six people at the table, we do not need to extend the domain size. To avoid some isomorphic models, we fix the position of one person: $HH = 0$ (line 21). Since the clues mention that on the left of a woman there is a man, all the men sit on the even places (line 22).

There are four clues conveyed by Helen, which are straightforward formalised:

$$left(left(left(Anna))) = Victor \tag{8.12}$$
$$left(left(left(left(left(HH))))) = Esther \tag{8.13}$$
$$left(left(Roger)) = Jim \tag{8.14}$$
$$-beside(HH, Helen) \tag{8.15}$$

Mace4 finds a single model (see Fig. 8.11):

Model	Anna	Esther	HH	Helen	Jim	Roger	Victor
1	5	1	0	3	4	0	2

Since HH and $Roger$ have the same value 0, Roger is Helen's husband.

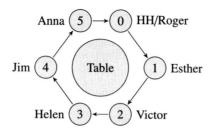

Fig. 8.11 The model found by Mace4: Roger is Helen's Husband

Puzzle 80. Arranged royal marrage

You are a young prince in situation to choose marrying one of the three princesses:
one from the north neigbouring country, one from the south, or one from the east.
Each princess secretly sent you some messages. The first one says: "If you marry the
princess from the south, there will be war". The second one says: "Marrying any of us
will not bring peace". You learn from the third princess that: "Marry me and it will be
peace, marry the south princess and there will be war". Given that you know that all
the princesses are liars, can you choose a princess that brings peace for sure?

North princess   South princess   East princess

Listing 8.8 Finding models for the arranged marriage

```
1    assign(domain_size,3).
2    assign(max_models,−1).
3
4    list(distinct).
5      [north,south,east].
6    end_of_list.
7
8    formulas(married).
9      m(north) <−> (married(south) −> −peace).
10     m(south) <−> (all x (married(x) −> −peace)).
11     m(east) <−> ((married(east) −> peace) & (married(south) −> −peace)).
12
13     −m(north) & −m(south) & −m(east).
14     married(north) | married(south) | married(east).
15     all x all y (married(x) & married(y) −> x = y).
16     peace.
17   end_of_list.
```

Solution

We assume that the three princesses are distinct (line 5 in Listing 8.8). Let the predicate $m(x)$ formalising the message of princess x. The north princess warns the prince that by marrying the princess from the south there will be war:

$$m(north) \leftrightarrow (married(south) \rightarrow \neg peace) \qquad (8.16)$$

The south princess claims that peace is not possible:

$$m(south) \leftrightarrow (\forall x \, (married(x) \rightarrow \neg peace)) \qquad (8.17)$$

The east princess guarantees for peace if the prince will marry her, but also warns against a marriage with the princess from the south:

$$m(east) \leftrightarrow ((married(east) \rightarrow peace) \wedge (married(south) \rightarrow \neg peace)) \quad (8.18)$$

The prince is aware that all princesses are lying. Hence, their messages are false: $\neg m(north) \wedge \neg m(south) \wedge \neg m(east)$. The puzzle asks for the model in which the prince marries a princess and there will be peace. We state that a marriage will take place (line 14), that the prince can marry only one princess (line 15), and the peace is a must (line 16).
Mace4 finds a single model in which $east = 0$, $north = 1$, $south = 2$, and the corresponding married relation is $married(\_) = [0, 0, 1]$.
The prince can ask his main counsellor, known as Prover9, for an explanation. In Fig. 8.12, Prover9 builds a proof based only on the message from the north princess.

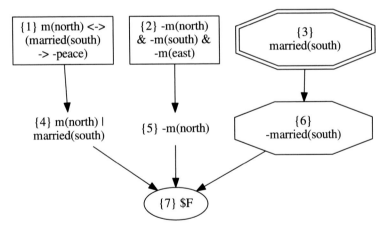

Fig. 8.12 Proof for marrying the south princess {3}

Four married men and three unmarried men are seated in a row at random. What are the chances that the two men at the ends of the row will be single? (adapted from puzzle 470 from Dudeney (2016))

$m_0 \quad m_1 \quad m_2 \quad m_3 \quad s_4 \quad s_5 \quad s_6$

Listing 8.9 Computing all models

```
1  set(arithmetic).
2  assign(domain_size,7).
3  assign(max_models,-1).
4
5  list(distinct).
6    [m0,ml,m2,m3,s4,s5,s6].
7  end_of_list.
```

Listing 8.10 Computing the favorable models

```
1  formulas(favorable_models).
2    ((s4 = 5) & (s5 = 6)) | ((s4 = 6) & (s5 = 5)) |
3    ((s4 = 5) & (s6 = 6)) | ((s4 = 6) & (s6 = 5)) |
4    ((s5 = 5) & (s6 = 6)) | ((s5 = 6) & (s6 = 5)).
5  end_of_list.
```

Solution

Let the four married persons m_1, m_2, m_3, m_4 and the three unmarried s_4, s_5, s_6. Recall the definition of probability: ratio of the number of favourable models to the total number of possible models.

To obtain all the possible models in which the seven persons can be arranged, we simply run the file in Listing 8.9. Mace4 returns 5,040 models.

To obtain all the favourable models in which two single persons are at the end of the row, one simply has to include in Listing 8.9 the constraints in Listing 8.10.

For this, we run Mace4 with two input files: `mace4 -f singleperson1.in singlepersons2.in`. Now, Mace4 finds 720 models.

Hence, the probability is 720/5040. This is indeed a useful way to use model finders to compute probabilities!

> **Puzzle 82. Five couples**
>
> My wife and I recently attended a party at which there were four other married cou-
> ples. Various handshakes took place. No one shook hands with oneself, nor with one's
> spouse, and no one shook hands with the same person more than once. After all the
> handshakes were over, I asked each person, including my wife, how many hands he (or
> she) had shaken. To my surprise each gave a different answer. How many hands did my
> wife shake? (puzzle 1.18 from Gardner (2005))

Listing 8.11 Five couples are shacking hands

```
1    set(arithmetic).
2    assign(domain_size,9).
3    assign(max_models,-1).
4
5    list(distinct).
6      [H2, H3, H4, H5, my_wife, W2, W3, W4, W5].
7    end_of_list.
8
9    formulas(five_couples).
10     me + my_wife = 8.              % first couple
11     H2 + W2      = 8.              % second couple
12     H3 + W3      = 8.              % third  couple
13     H4 + W4      = 8.              % fourth couple
14     H5 + W5      = 8.              % fifth  couple
15     W2 < W3 & W3 < W4 & W4 < W5.   % remove isomorphisms
16     4 < W2.                       % remove isomorphisms
17   end_of_list.
```

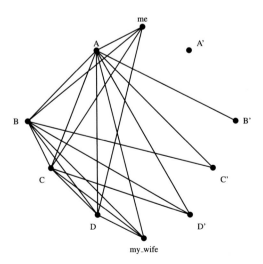

Fig. 8.13   One model found by Mace4: only *me* and *wife* shook the same number of hands

Solution

Let the no-named individuals in the couples H_i (for the husbands) and W_i (for the wives). Let the storyteller be noted *me*, and his wife *my_wife*. Since no one shook hands with oneself, nor with one's spouse, each couple shook maximum 8 hands. Thereby, we use a domain size of 9 (i.e. [0..8]). The information that each person shook a different number of hands is handled by the list of distinct elements in line 6. Since nine people each shake a different number of hands, the numbers must be 0, 1, 2, 3, 4, 5, 6, 7, and 8. Note here that *me* is not included in the list, as we don't have any information about the storyteller.

If all possible handshakes took place, then each couple should have exactly 8 handshakes. However, the statement "after all the handshakes were over" does not necessarily imply that all possible handshakes took place. Therefore, we assume that each couple has at most 8 handshakes:

$$(me + my\_wife \leq 8) \wedge (H_2 + W_2 \leq 8) \wedge (H_3 + W_3 \leq 8) \wedge (H_4 + W_4 \leq 8) \wedge (H_5 + W_5 \leq 8)$$

To avoid some isomorphic models, we set an order relation among all unnamed wives: $(W_2 < W_3) \wedge (W_3 < W_4) \wedge (W_4 < W_5)$. To avoid isomorphisms, we assume that women shook more hands than their husbands. This is simply modelled with $4 < W_2$. Mace4 finds 5 models, the only difference among them is that *me* shook 0, 1, 2, 3, or 4 hands. In all models $my\_wife = 4$. If we assume that "all the handshakes were over" means that each couple saluted each other, we have:

$$(me + my\_wife = 8) \wedge (H_2 + W_2 = 8) \wedge (H_3 + W_3 = 8) \wedge (H_4 + W_4 = 8) \wedge (H_5 + W_5 = 8)$$

In this case, the single model (depicted in Fig. 8.13) would be:

Model	me	my_wife	H_2	H_3	H_4	H_5	W_2	W_3	W_4	W_5
1	4	4	3	2	1	0	5	6	7	8

In case of hand-shaking, the principle states there is always a pair of people who will shake hands with the same number of people. With ten persons shaking at most eight hands (nine values in [0..8]), at least two persons shake the same number of hands. More general, a graph with no loops and multiple edges must contain at least two points that have the same number of lines attached to them. In this case, the graph has only two such points (i.e. *me* and *my_wife*).

Puzzle 83. A family tree

I was going through some old family photos in the attic when I stumbled upon our family tree. I studied it for a couple minutes then went back downstairs to tell my mom about the family tree. The problem is I didn't study it long enough to remember the whole thing. I only remembered a couple of things about it, and recent memories. Can you help me figure out my family tree? There are two grandparents, who had two children, who both got married and had 2 more children each. In total, there are 10 people: Alex, Bob, Caty, David, Elton, John, Lincoln, Mary, Sonia, and Tina.

Clue$_1$:   One of Elton's ancestors was David.
Clue$_2$:   John's sister gave birth to Tina.
Clue$_3$:   Mary went bowling with her nephew last Saturday.
Clue$_4$:   Alex is cousin with one of the girls.
Clue$_5$:   Bob married Mary.
Clue$_6$:   Caty is not an ancestor, nor cousin of Tina.
Clue$_7$:   Lincoln's brother showed Bob's son his baseball cards.

(adapted from www.braingle.com/brainteasers)

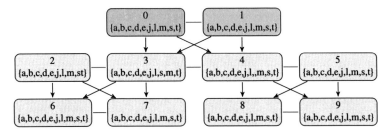

Fig. 8.14 A family of ten members over three generations. Red arrows represent the *marry* relation, blue arrows represent the *sibling* relation, and black arrows represent the *ancestor* relation. Blue nodes are males, while pink nodes are females. Unknown gender is depicted with grey nodes

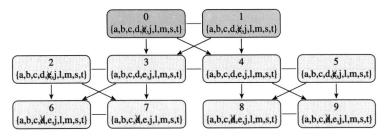

Fig. 8.15 After Clue$_1$: "One of Elton's ancestors was David"—Elton is not a grandparent and David is not a child. Since Elton has ancestors, $Elton \neq 2$ and $Elton \neq 5$

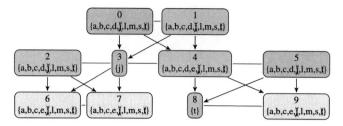

Fig. 8.16 After Clue$_2$: "John's sister gave birth to Tina"—Only the values 3 and 4 are siblings that have a child (let $j = 3$). Tina is 8 or 9 (let $t = 8$)

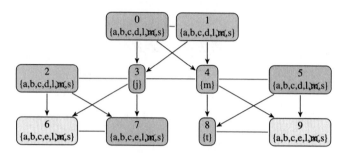

Fig. 8.17 After Clue$_3$: "Mary went bowling with her nephew last Saturday"—Mary is the aunt (i.e. $m = 4$). 6 or 7 is a boy (let the boy be 7)

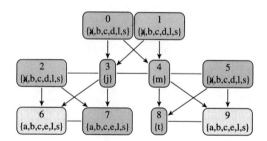

Fig. 8.18 After Clue$_4$: "Alex is cousin with one of the girls"—Alex is a child (i.e. $a \neq$ 0, 1, 2, 3, 4, 5). If 6 is boy then $a \neq 9$

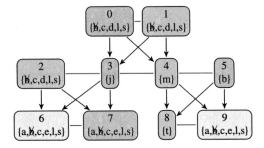

Fig. 8.19   After Clue₅: "Bob married Mary"—(i.e. $b = 5$)

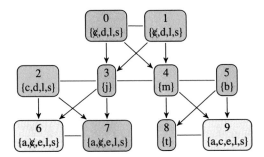

Fig. 8.20   After Clue₆: "Caty is not an ancestor, nor cousin of Tina"—Caty could be Tina's sister or Tina's aunt

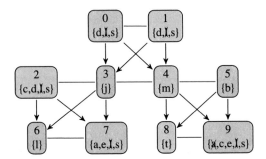

Fig. 8.21   After Clue₇: "Lincoln's brother showed Bob's son his baseball cards"—Since Lincoln has a brother $l = 6$. Bob has a male child (i.e. 9 is a boy). Since Alex is cousin with one of the girls $a \neq 9$

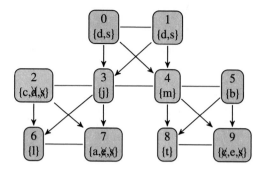

Fig. 8.22 After constraints propagation—s and d are the grandparents; then $c = 2$; then $e = 9$; then $a = 7$

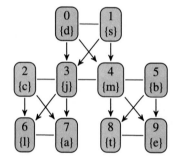

Fig. 8.23 One model found by Mace4 (where $d = 0$)

Listing 8.12 Finding models for a family tree

```
1    assign(domain_size,10).                    %ten distinct family members
2    assign(max_models,-1).
3
4    list(distinct).
5      [Alex,Bob,Caty,David,Ella,John,Lincoln,Sonia,Mary,Tina].
6    end_of_list.
7
8    formulas(utils).
9      gp(x) <-> x = 0 | x = 1.
10     p(x)  <-> x = 2 | x = 3 | x = 4 | x = 5.
```

```
5      c(x)  <-> x = 6 | x = 7 | x = 8 | x = 9.
6    end_of_list.
7
8    formulas(family_tree).
9      -gp(Ella) & -c(David) & Ella != 2 & Ella != 5.              %Clue 1
10
11     p(John) & c(Tina) & John != 2 & John != 5.                  %Clue 2
12     John = 3 -> (Tina = 8 | Tina = 9).
13     John = 4 -> (Tina = 6 | Tina = 7).
14     John = 3 -> (Mary = 4 | Ella = 4 | Caty = 4 | Sonia = 4).
15     John = 4 -> (Mary = 3 | Ella = 3 | Caty = 3 | Sonia = 3).
16
17     Mary = 3 | Mary = 4.                                        %Clue 3
18
19     c(Alex).                                                    %Clue 4
20
21     Mary = 3 -> Bob = 2.                                        %Clue 5
22     Mary = 4 -> Bob = 5.
23
24     Tina = 6 -> (Caty = 4 | Caty = 5 | Caty = 7).               %Clue 6
25     Tina = 7 -> (Caty = 4 | Caty = 5 | Caty = 6).
26     Tina = 8 -> (Caty = 2 | Caty = 3 | Caty = 9).
27     Tina = 9 -> (Caty = 2 | Caty = 3 | Caty = 8).
28
29     -gp(Lincoln) & Lincoln != 2 & Lincoln != 5.                 %Clue 7
30     -c(Bob) & c(Lincoln).
31   end_of_list.
```

Listing 8.13 Testing the family tree

```
1    formulas(test).
2    John = 3. David = 0. Tina = 8. % assumptions to avoid isomorphisms
3    % Sonia != 1. % Sonia is not the grandmather
4    % Mary != 4. % Mary is not the daughter of Sonia and David
5    % Caty != 2. % Caty is not married to John
6    % Lincoln != 6 & Lincoln != 7. % Lincoln is not the son of John
7    end_of_list.
```

Solution

Since there are ten distinct family members, we set the domain size to 10. There are three generations: two grandparents, four parents and four children. To avoid some isomorphic models, we fix some values, as follows:

$$gp(x) \leftrightarrow x = 0 \vee x = 1 \tag{8.19}$$

$$p(x) \leftrightarrow x = 2 \vee x = 3 \vee x = 4 \vee x = 5 \tag{8.20}$$

$$c(x) \leftrightarrow x = 6 \vee x = 7 \vee x = 8 \vee x = 9 \tag{8.21}$$

We assume that the two grandparents are married (red line in Fig. 8.14). One of them is the grandfather (blue node), while the other is the grandmother (pink node).

We assume that 2 is married to 3 and 4 to 5. There are three sibling relations (blue link) between: 3 and 4, 6 and 7, respectively, 8 an 9. Each node can have 10 values {a,b,c,d,e,j,l,s,m,t} that correspond to the initial letter of each family member. Given these assumptions, we start formalising the clues.

From Clue$_1$: "One of Elton's ancestors was David", we found that: Elton has an ancestor, hence she is not a grandparent. Since David is an ancestor, he cannot be a child. Since 2 and 5 do not have ancestors, Elton has neither 2 nor 5.

$$\neg gp(Elton) \wedge \neg c(David) \wedge Elton \neq 2 \wedge Elton \neq 5 \qquad (8.22)$$

The set of possible values for each node are updated in Fig. 8.15.

From Clue$_2$: "John's sister gave birth to Tina", we found that Bob has a sibling relationship.

Among the possible three sibling relationships (i.e. (3,4), (6,7), (8,9)), the nodes 6, 7, 8, and 9 do not have children. Hence Bob is either 3 or 4. If John is 3, then his pair 4 is a woman and his niece, Tina, is 8 or 9:

$$John = 3 \rightarrow (Mary = 4 \vee Elton = 4 \vee Caty = 4 \vee Sonia = 4) \qquad (8.23)$$
$$John = 3 \rightarrow (Tina = 8 \vee Tina = 9) \qquad (8.24)$$

Similarly, if John is 4, then his pair 3 is a woman and his niece, Tina, is 6 or 7:

$$John = 4 \rightarrow (Mary = 3 \vee Elton = 3 \vee Caty = 3 \vee Sonia = 3) \qquad (8.25)$$
$$John = 4 \rightarrow (Tina = 6 \vee Tina = 7) \qquad (8.26)$$

Note that these are isomorphic models, since we do not care if John is 3 or 4.

Figure 8.16 illustrates the model with $John = 3$. Note that 4 is pink coloured to indicate a woman. Here $Tina = 8$, while $Tina = 9$ would be an isomorphic model. Since $John = 3$ and $Tina = 8$ and the values are distinct, these two values are removed from the set of possible values for the other nodes.

From Clue$_3$: "Mary went bowling with her nephew last Saturday.", we found that Mary is an aunt. Since only nodes 3 and 5 have nephews or nieces, then $Mary = 3 \vee Mary = 4$. Fig. 8.17 shows the model with $Mary = 4$ (since the previous illustration has $John = 3$). We learn also that either node 6 or node 7 is a boy (i.e. Mary's nephew). Figure 8.17 assumes node 7 is a boy.

From Clue$_4$: "Alex is cousin with one of the girls" we learn that Alex is a child, since it is the only situation in which he has cousins. Hence, Alex is removed from the parents and grandparents nodes (see Fig. 8.18).

Solution

From Clue$_5$, "Bob married Mary" we know that $(Mary = 3 \rightarrow Bob = 2) \wedge (Mary = 4 \rightarrow Bob = 5)$. Figure 8.19 shows the model with $Mary = 4$ and $Bob = 5$.

From Clue$_6$, "Caty is not an ancestor, nor cousin of Tina" we learn that Caty can be Tina's sister ($Caty = 9$) or Tina's aunt ($Caty = 2$) (see Fig. 8.20). Finally, from Clue$_7$: "Lincoln's brother showed Bob's son his baseball cards", we learn that: Lincoln is not a grandparent or 2 or 5 (since he has a brother), and also Bob is not a child (since he himself has a child):

$$\neg gp(Lincoln) \wedge Lincoln \neq 2 \wedge Lincoln \neq 5 \wedge \neg c(Bob) \wedge c(Lincoln) \quad (8.27)$$

"Lincoln has a brother" is formalised with Lincoln is not the brother of Tina:

$$Tina = 8 \vee Tina = 9 \rightarrow Lincoln \neq 8 \wedge Lincoln \neq 9 \quad (8.28)$$
$$Tina = 6 \vee Tina = 7 \rightarrow Lincoln \neq 6 \wedge Lincoln \neq 7 \quad (8.29)$$

Since Lincoln has a brother, the only possibility in our depicted model is $Lincoln = 6$ (see Fig. 8.21). After constraints propagation—s and d are the grandparents; then $c = 2$; then $e = 9$; then $a = 7$ (see Fig. 8.22). Moreover, Bob has a male child (node 9 is blue in Fig. 8.23).

Based on Listing 8.12, there are 32 isomorphic models. For instance, David can be zero or one, John can be 3 or 4, Tina can be 8 or 9, Lincoln can be 6 or 7, etc. However, in all models Sonia and David were the grandparents. They gave birth to John and Mary. John married Caty, who gave birth to Lincoln and Alex. Mary married Bob and gave birth to Tina and Elton. These fixpoints can be checked by adding the test cases from Listing 8.13. If we fix $John = 3$, $David = 0$, and $Tina = 8$, 6 isomorphic models remain. One such model is depicted in Fig. 8.24.

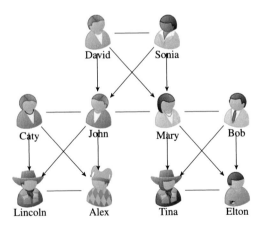

Fig. 8.24 A family tree

Puzzle 84. Uncles and aunts

The family tree consists of two grandparents, who had three children, each of whom got married and had two children. Males are Cole, Cristian, Jason, Neil, and Steve. Females are Amanda, Ashley, Beth, Erin, Kaitlyn, Katherine, Makayla, Payton, and Tammy. The clues are:

Clue$_1$ One of Makayla's cousins is Jason's son.
Clue$_2$ One of Ashley's aunts is Tammy.
Clue$_3$ Tammy's brother-in-law is Neil's son.
Clue$_4$ Kaitlyn's sister is Ashley's cousin.
Clue$_5$ Ashley's uncle, Steve, is Erin's brother-in-law.
Clue$_6$ The three uncles are Payton's dad, Cristian, and Katherine's son.
Clue$_7$ The three aunts are Kaitlyn's mom, Ashley's mom, and Cristian's sister-in-law.
Clue$_8$ Jason's brother is Ashley's dad.
Clue$_9$ Amanda's sister is Steve's niece.
Clue$_{10}$ Beth is not Cole's aunt.

(taken from www.braingle.com/brainteasers)

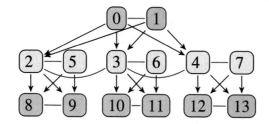

Fig. 8.25 A family of 14 members over three generations. Red arrows represent the *marry* relation, blue arrows represent the *sibling* relation, and black arrows represent the *ancestor* relation. There are 5 men (blue) and 9 women (pink). We do not know if the grandparents 0 and 1 have daughters or sons

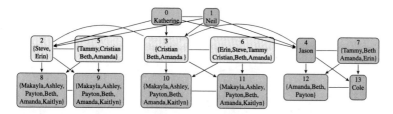

Fig. 8.26 Knowledge after Clue$_1$ to Clue$_8$

Listing 8.14 Formalising background knowledge for the family tree

```
1   assign(domain_size ,14).                    %fourteen  distinct  family  members
2   assign(max_models , -1).
3
4   list(distinct).
5     [Cole , Cristian , Jason , Neil , Steve ,
6       Amanda , Ashley , Beth , Erin , Kaitlyn , Katherine , Makayla , Payton , Tammy].
7   end_of_list .
8
9   formulas(utils ).
10    gp(x) <-> x  = 0 I x  = 1.                        %grandparents  are  on  the  1st  layer
11    (gm(x) <-> x = 0) & (gf (x) <-> x = 1).               %avoid  isomorphisms
12    p(x)  <-> x = 2 I x = 3 I x = 4 I x = 5  I x = 6  I x = 7.        %p on 2nd
13    c(x)  <-> x = 8 I x = 9 I x= 10 I x = 11 I x = 12 I x = 13.       %c on 3rd
14    family(x) <->  x = 2 I x = 3 I x = 4.               %children  of  0 and 1
15    inLaw(x)  <->  x = 5 I x = 6 I x = 7.               %spouses
16    man(x) -> -woman(x).
17    uncle(x,y) <->  man(y) & p(y) & c(x) &
18              ((x = 8  I x = 9)   -> (y = 3 I y = 6 I y = 4 I y = 7)) &
19              ((x = 10 I x = 11)  -> (y = 2 I y = 5 I y = 4 I y = 7)) &
20              ((x = 12 I x = 13)  -> (y = 2 I y = 5 I y = 3 I y = 6)).
21
22    aunt(x,y) <->  woman(y) & p(y) & c(x) &
23              ((x = 8  I x = 9)   -> (y = 3 I y = 6 I y = 4 I y = 7)) &
24              ((x = 10 I x = 11)  -> (y = 2 I y = 5 I y = 4 I y = 7)) &
25              ((x = 12 I x = 13)  -> (y = 2 I y = 5 I y = 3 I y = 6)).
26
27    cousin(x,y) <-> c(x) & c(y) &
28              ((x = 8  I x = 9)   -> (y = 10 I y = 11 I y = 12 I y = 13)) &
29              ((x = 10 I x = 11)  -> (y = 8  I y = 9  I y = 12 I y = 13)) &
30              ((x = 12 I x = 13)  -> (y = 8  I y = 9  I y = 10 I y = 11)).
31    cousin(x,y) <-> cousin(y,x).
32
33    boy(x) <-> (c(x) & man(x)).
34    all x all y ((boy(x) & boy(y)) -> x = y).
35    exists x boy(x).                                 %exactly  one  boy
36  end_of_list .
37
38  formulas(assumptions ).
39    man(Cole) & man(Cristian) & man(Jason) & man(Neil) & man(Steve).
40    woman(Amanda)   & woman(Ashley)  & woman(Beth)     &
41    woman(Erin)     & woman(Kaitlyn) & woman(Katherine) &
42    woman(Makayla) & woman(Payton)  & woman(Tammy).
43  end_of_list .
```

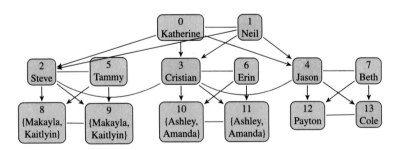

Fig. 8.27 Mace4 finds four models, given by the combinations in nodes 8, 9, 10, and 11

Listing 8.15 Finding models for a family tree

```
1    formulas ( family_tree ).
2      uncle ( Makayla , Jason ).        %Clue1 :  Makayla 's  cousins  is  Jason 's  son
3      aunt ( Ashley , Tammy ).          %Clue2 :  One  of  Ashley 's  aunts  is  Tammy
4      inLaw ( Tammy ) & gf ( Neil ).    %Clue3 :  Tammy 's  brother −in−law  is  Neil 's  son
5      c ( Kaitlyn ) & c ( Ashley ).     %Clue4 :  Kaitlyn 's  sister  is  Ashley 's  cousin
6      uncle ( Kaitlyn , Jason ).   %Kaitlyn  has  a  sister  −>  Jason  is  not  her  father
7
8      %Clue5 :  Ashley 's  uncle , Steve , is  Erin 's  brother −in−law
9      uncle ( Ashley , Steve ).
10     family ( Steve ) & inLaw ( Erin ) | inLaw ( Steve ) & family ( Erin ).
11     Steve = 2 | Erin = 2.                          %avoid  isomorphisms
12
13     %Clue6 :  The  three  uncles  are  Payton 's  dad , Cristian  and  Katherine 's  son
14     c ( Payton ) & p ( Cristian ) & gm ( Katherine ).
15     uncle ( Payton , Cristian ).
16
17     %Clue7 :  Aunts  are  Kaitlyn 's  mom , Ashley 's  mom , Cristian 's  sister −in−law
18     c ( Kaitlyn ) & c ( Ashley ) & p ( Cristian ) & cousin ( Kaitlyn , Ashley ).
19
20     %Clue 8.  Jason 's  brother  is  Ashley 's  dad
21     uncle ( Ashley , Jason ) & family ( Jason ).
22     Jason = 4 & Cole = 13.                         %avoid  isomorphisms
23
24     c ( Amanda ) & uncle ( Amanda , Steve ).% Clue9 :  Amanda 's  sister  is  Steve 's  niece
25     Amanda != 12.                      %Amanda  is  not  the  daughter  of  Jason
26
27     −aunt ( Cole , Beth ).             %Clue 10.  Beth  is  not  Cole 's  aunt
28   end_of_list .
```

Solution

Since there are fourteen distinct family members, we set the domain size to 14. There are three generations: two grandparents (predicate $gp(x)$), six parents (predicate $p(x)$), and six children (predicate $c(x)$). To avoid some isomorphisms, we fix some values, as follows:

$$gp(x) \leftrightarrow x = 0 \vee x = 1 \qquad (8.30)$$

$$p(x) \leftrightarrow x = 2 \vee x = 3 \vee x = 4 \vee x = 5 \vee x = 6 \vee x = 7 \qquad (8.31)$$

$$c(x) \leftrightarrow x = 8 \vee x = 9 \vee x = 10 \vee x = 11 \vee x = 12 \vee x = 13 \qquad (8.32)$$

These three generations are depicted in Fig. 8.25). Since there are 5 men and 9 women, there is a single grandson. We do not know if the grandparents 0 and 1 have daughters or sons. We assume 2, 3, and 4 are siblings. Since we do not care which of the two nodes in the first layer is occupied by the grandfather and which by the grandmother, we allocate the value 0 for the grandmother ($gm(x) \leftrightarrow x = 0$) and the value 1 for the grandfather ($gf(x) \leftrightarrow x = 1$) (line 11 in Listing 8.14). This helps Mace4 not to generate isomorphic models. In Fig. 8.25), the marriage relations are red, the sibling relation is blue, while hasChild relation is black.

We also need to distinguish between blood-family members and their spouses. Let the values 2, 3, and 4 for the three children of the nodes 0 and 1: $family(x) \leftrightarrow x = 2 \vee x = 3 \vee x = 4$ (line 14 in Listing 8.14). The remaining values 5, 6, and 7 are allocated for their spouses: $inLaw(x) \leftrightarrow x = 5 \vee x = 6 \vee x = 7$ (line 15). This splitting of the six parents into $family$ and $inLaw$ avoids further isomorphisms.

Then we define three auxiliary predicates: $uncle(x, y)$, $aunt(x, y)$, and $cousin(x, y)$. These predicates are not generic, but they are defined based on the values available in the domain (lines 17–31 in Listing 8.14). For instance, $uncle(x, y)$ says that x has uncle y if x is a child (i.e. $c(x)$, 3rd level in the family tree), y is a man (i.e. $man(y)$ in line 17), and also y is a parent (i.e. $p(x)$, 2nd level in the family tree). There are also three cases. First, if the child is 8 or 9, his or her uncle is 3 or 6 or 4 or 7 (line 18). Second, if the child is 10 or 11, his or her uncle is 2 or 5 or 4 or 7 (line 18). Third, if the child is 12 or 13, his or her uncle is 2 or 5 or 4 or 7 (line 18).

One relevant piece of information for solving the puzzle is the number of men and women. Note there are 5 men and 9 women (formalised in lines 39–42). One of the men is the grandfather and three men are married and they have children. It means there is only one male child in the third layer of the tree. To handle this, we define the predicate $boy(x) \leftrightarrow c(x) \wedge man(x)$, and we state that there is exactly one boy (lines 34–35).

Solution

The clues are formalised in Listing 8.15. From Clue$_1$: "One of Makayla's cousins is Jason's son" we learn that Jason is Makayla's uncle. Based on $uncle(Makayla, Jason)$ and the definition of the predicate $uncle$ the system will deduce that Makayla is a child and Jason is a parent. We learn also that the only boy is Jason's son. This piece of information will be further used when formalising the puzzle.

Clue$_2$: "One of Ashley's aunts is Tammy" is formalised with $aunt(Ashley, Tammy)$.

Clue$_3$: "Tammy's brother-in-law is Neil's son". We know that brothers-in-law are on the second layer, i.e. they are the parents. It means that Neil is the first layer, that is he is the grandfather. Since Neil's son is from the family, Tammy is a spouse of one of the three of Neil's children. Hence, this clue is formalised with: $inLaw(Tammy) \wedge gf(Neil)$.

From Clue$_4$: "Kaitlyn's sister is Ashley's cousin" we find that Kaitlyin is a child and Ashley is a child (line 5 in Listing 8.15). Since Kaitlyn has a sister, Jason is not her father (recall that Jason has a son). Thus, Jason is Kaitlyn uncle (line 6).

Clue$_5$: "Ashley's uncle, Steve, is Erin's brother-in-law" states $uncle(Ashley, Steve)$, but also that either Steve or Erin are a blood relative:

$$(family(Steve) \wedge inLaw(Erin)) \vee (inLaw(Steve) \wedge family(Erin)) \qquad (8.33)$$

Since we do not care which family node (2, 3, or 4) is occupied by Steve or Erin, we fix it to 2: $Steve = 2 \vee Erin = 2$ (line 11).

From Clue$_6$: "The three uncles are Payton's dad, Cristian and Katherine's son" we found that: (i) Payton is a child, (ii) Cristian is a parent and (iii) Katherin is the grandmother (line 14). Moreover, since Payton's dad is a distinct uncle from Cristian, then Cristian is Payton's uncle (line 15).

From Clue$_7$: "The three aunts are Kaitlyn's mom, Ashley's mom, Cristian's sister-in-law" we found that: (i) Kaitlynn and Ashley are children, (ii) Cristian is a parent and (iii) since Kaitlynn and Ashley have different mothers, they are cousins (line 18). From Clue$_8$: "Jason's brother is Ashley's dad" we found that Jason is Ashley's uncle. Since Jason is a parent and has a brother, then he is a blood relative (i.e. $family(Jason)$. Since we do not care which node is from 3 or 4 (recall that node 2 is already taken by Steve or Erin), we fix Jason = 4. Now, the system knows about the men that: Neil is the grandfather, Steve, Cristian, and Jason are parents. Hence, Cole remains to be the only grandchild. Since we know from Clue$_1$ that he is the son of Jason, we can fix it either to 12 or 13. Let $Cole = 13$ in line 21.

An illustration of the current knowledge state appears in Fig. 8.26. We know that $Steve = 2$ or $Erin = 2$. If $Steve = 2$ and given that Steve is Erin brother-in-law, then $Erin = 6$ or $Erin = 7$. If $Erin = 2$ and given that Steve is Erin brother-in-law, then $Steve = 6$ (node 7 is occupied by a woman). Note that this information is deduced by the human agent—Mace4 does not have knowledge about the marriage relations. At this moment Mace4 returns 386 models.

From Clue$_9$: "Amanda's sister is Steve's niece" says that Steve is Amanda's uncle. Since Amanda has a sister not a brother, she is not the daughter of Jason (line 25).

Finally, Clue$_{10}$ Beth is not Cole's aunt is formalised with $\neg aunt(Cole, Beth)$.

Mace4 found 4 isomorphic models. (Fig. 8.27), The four models are given by Makayla and Kaitlyn can be 8 or 9, and Ashley and Amanda can be 10 or 11. The grandparents are Katherine and Neil, who had 3 sons: Cristian, Jason, and Steve. Cristian married Erin and had Ashley and Amanda. Jason married Beth and had Payton and Cole. Steve married Tammy and had Makayla and Kaitlyn.

References

Danesi, M. (2018). *An anthropology of puzzles: The role of puzzles in the origins and evolution of mind and culture*. Bloomsbury Publishing.

Dudeney, H. E. (2016). *536 Puzzles and curious problems*. Courier Dover Publications.

Gardner, M. (2005). *The colossal book of short puzzles and problems*. W.W: Norton.

Smullyan, R. M. (1996). *The riddle of Scheherazade and other amazing puzzles ancient and modern*. New York: Knopf.

Chapter 9
Grid Puzzles

Abstract In which we enter in the world of grids. We will model grid puzzles starting with Latin or Magic squares, and continuing with Fancy queens on a chessboard, Battle puzzles, or Minesweeper. Latin squares are $n \times n$ grids containing n symbols so that each symbol appears exactly once in each row and each column. Chessboard puzzles have a long history that can be traced back to the *wheat and chessboard* puzzle posed in 1256 by Ibn Kallikan (Darling 2004). To give a flavour on modelling classical games in FOL, this chapter exemplifies the Minesweeper game. Once again, this chapter illustrates the power of FOL as a modelling tool for a plethora of grid-based types of puzzles.

$$\forall x \; (place(x) \land worth(x) \rightarrow \neg \exists y \; shortcut(y, x))$$

Beverly Sills

We enter here in the world of grids. We will model grid puzzles starting with Latin or Magic squares, and continuing with Fancy queens on a chessboard, Battle puzzles, or Minesweeper. Mace4 will be the tool of choice throughout the chapter.

Latin squares are $n \times n$ grids containing n symbols so that each symbol appears exactly once in each row and each column. Latin squares have a long history, being used on amulets in the medieval islam (c. 1200) (Darling 2004). Latin squares have been portrayed in art (e.g. famous etching *Melancholia* of Albrecht Dürer), or mathematics (Leonhard Euler's *thirty-six officers problem* posed on 1779 and solved only at the beginning of the twentieth century). Applications area for Latin squares was opened by R. A. Fischer that uses them in the design of statistical experiments. The "Latin square" puzzle in this chapter illustrates how such Latin squares can be used to design statistical experiments in the medical domain.

Chessboard puzzles have a long history that can be traced back to the *wheat and chessboard* puzzle posed in 1256 by Ibn Kallikan (Darling 2004). Recall that in this myth the chess inventor kindly refused the reward in gold offered for inventing the game. Instead, he asked for one grain for the first square, two for the second square, four for the third one. All these $2^{64} - 1$ grains would fill a granary of 40 km long, 40 km wide and 300 m tall (Darling 2004). The standard puzzles known as *kings problem*, *queen problem*, *bishops problem*, and the *knights problem* asks for the greatest number of each of these pieces that can be placed on $n \times n$ chess-like boards without attacking each other. Versions of these problems ask for the smallest number of pieces needed to occupy or attack every square. The *knight tour* is arguable to most encountered algorithmic puzzle. For more puzzles that have algorithmic touch, the eager student can consult the collection of Anany Levitin and Maria Levitin (2011). Instead of designing algorithms to solve these puzzles, we focus here on modelling them. We formalise in this chapter a variant of the queens problem, in which our queens are too fancy to stay on the main diagonals.

As stressed by Baker, the reality is that most of the current computer science students grew up with computer and video games, with many of them saying that game playing is what triggered their interest in computers (Becker 2001). The follow-up teaching strategy is to tap into this energy and use it to learn various topics in computer science. Indeed, games have the potential to integrate almost all of the concepts taught in a computer science undergraduate programme. In this line, puzzles are reasoning games adequate for teaching logic and modelling skills. To give a flavour on modelling classical games in FOL, this chapter exemplifies the Minesweeper game.

Once again, this chapter illustrates the power of FOL as a modelling tool for a plethora of grid-based types of puzzles.

Puzzle 85. A five in the middle of a grid

Here is an advertising trick that appeared in America many years ago. Place in the empty squares such figures (different in every case, and no two squares containing the same figure) so that they shall add up 15 in as many straight directions as possible. A large prize was offered, but no correct solution was received. Can the reader guess the trick? (puzzle 389 from Dudeney 2016)

---

				c_1	c_2	c_3
	5			c_4	c_5	c_6
				c_7	c_8	c_9

Listing 9.1 Finding models with $c_5 = 5$ and $sum = 15$ on each direction

```
1  assign(domain_size,10).
2  assign(max_models,-1).
3  set(arithmetic).
4
5  list(distinct).
6   [0,c1,c2,c3,c4,5,c6,c7,c8,c9].
7  end_of_list.
8
9  formulas(five).
10  c5 = 5.
11  c1 + c2 + c3 = 15 & c4 + c5 + c6 = 15 & c7 + c8 + c9 = 15.   %sum(line)
12  c1 + c4 + c7 = 15 & c2 + c5 + c8 = 15 & c3 + c6 + c9 = 15.   %sum(column)
13  c1 + c5 + c9 = 15 & c3 + c5 + c7=15.                          %sum(diagonals)
14 end_of_list.
```

Solution

Since the grid is small, we used here propositional logic. Let the grid in the illustration above. Here $c_1, \ldots, c_9$ are the propositional variables that we need to find. These values are distinct and they belong to the interval [1...9] (line 6 in Listing 9.1). The value for c_5 is given: $c_5 = 5$. We find models in which the sum on all three lines, all three columns, and the two main diagonals is 15. Mace4 finds eight solutions, depicted in Fig. 9.1.

2	7	6
9	5	1
4	3	8

2	9	4
7	5	3
6	1	8

4	3	8
9	5	1
2	7	6

4	9	2
3	5	7
8	1	6

6	1	8
7	5	3
2	9	4

6	7	2
1	5	9
8	3	4

8	1	6
3	5	7
4	9	2

8	3	4
1	5	9
6	7	2

Fig. 9.1 Eight models found by Mace4

Place the numbers 1 to 12 (one number in every design) so that they shall add up to the
same sum in the following seven different ways: each of the two center columns, each
of the two central rows, the four roses together, the four shamrocks together, and the
four thistles together. (puzzle 400 from Dudeney 2016)

	s_1	s_2	
t_1	r_1	r_2	t_2
t_3	r_3	r_4	t_4
	s_3	s_4	

Listing 9.2 Finding models for placing roses, shamrocks, and thistles in a grid

```
1   assign(domain_size,13).
2   assign(max_models,−1).
3   set(arithmetic).
4
5   list(distinct).
6     [0,s1,s2,s3,s4,r1,r2,r3,r4,t1,t2,t3,t4].
7   end_of_list.
8
9   formulas(roses).
10    s1 + s3 + r1 + r3 = s2 + s4 + r2 + r4.  %sum(column1) = sum(column2)
11    s1 + s3 + r1 + r3 = t1 + r1 + r2 + t2.  %sum(row1)    = sum(row2)
12    s1 + s3 + r1 + r3 = t3 + r3 + r4 + t4.  %sum(column1) = sum(row2)
13    s1 + s3 + r1 + r3 = r1 + r2 + r3 + r4.  %sum(column1) = sum(roses)
14    s1 + s3 + r1 + r3 = s1 + s2 + s3 + s4.  %sum(column1) = sum(shanrocks)
15    s1 + s3 + r1 + r3 = t1 + t2 + t3 + t4.  %sum(column1) = sum(thiristles)
16    %r1 < r2 & r2 < r3 & r3 < r4.           %remove isomorphisms
17  end_of_list.
```

Solution

Let s_i for shamrocks, r_i for roses, t_i for thistles. We use a domain size of 13 with all
values distinct in interval [1...12] (line 6 in Listing 9.2). We need six equations to formalise the puzzle's constraints (lines 10–15).
Mace4 finds 24,960 models. The following solution has the sum 26 on each row, column, diagonal: $r_1 = 1, r_2 = 4, r_3 = 10, r_4 = 11, s_1 = 7, s_2 = 5, s_3 = 8, s_4 = 6, t_1 =$
9, $t_2 = 12, t_3 = 2, t_4 = 3$. The grid generated by this solution is illustrated in Fig. 9.2.
Note that if one uses a variable to store the value of the sum (let's say $sum = 26$), the
domain needs to be increased up to 27 in order to accommodate this value. This unnec-
essarily increases the search space for the model finder. Note also that many models
among those 24,960 are isomorphic. To avoid some of the isomorphisms, one can add
constraints such as $r_1 < r_2 < r_3 < r_4$ (line 16). With line 16 uncommented, Mace4
returns 464 models. Four such models are depicted in Fig. 9.2.

	7	5	
9	1	4	12
2	10	11	3
	8	6	

	6	8	
4	12	9	1
11	3	2	10
	5	7	

	3	6	
11	1	5	9
4	12	8	2
	10	7	

	12	9	
11	5	6	4
10	7	8	1
	2	3	

Fig. 9.2 Four models among 24,960 found. Mace4 found also the value of the sum (i.e. 26)

Puzzle 87. Nine squares

Is it possible to have the numbers 1 to 9 positioned in a 9 square grid in a formation so that the numbers are placed so no adjacent number is next to it horizontally, vertically or diagonally?

$g1$	$g2$	$g3$
$g4$	$g5$	$g6$
$g7$	$g8$	$g9$

Listing 9.3 Finding models for no consecutive values allowed in adjacent cells

```
1   set(arithmetic).
2   assign(domain_size,10).
3   assign(max_models,-1).
4
5   list(distinct).
6     [0,g1,g2,g3,g4,g5,g6,g7,g8,g9].
7   end_of_list.
8
9   formulas(assumptions).
10    adj(x,y) <-> adj(y,x).
11    adj(x,y) -> abs(x + -y) != 1.
12
13    %Stating the adjacent cells (diagonally, vertically and horizontally)
14      adj(g1,g2) & adj(g1,g4) & adj(g1,g5).               %cell g1
15      adj(g2,g3) & adj(g2,g4) & adj(g2,g5) & adj(g2,g6).  %cell g2
16      adj(g3,g5) & adj(g3,g6).                            %cell g3
17      adj(g4,g5) & adj(g4,g7) & adj(g4,g8).               %cell g4
18      adj(g5,g6) & adj(g5,g7) & adj(g5,g8) & adj(g5,g9).  %cell g5
19      adj(g6,g8) & adj(g6,g9).                            %cell g6
20      adj(g7,g8).                                         %cell g7
21      adj(g8,g9).                                         %cell g8
22   end_of_list.
```

Solution

There are nine values available. Assume that you put a value in the cell g_5 in the middle of the grid. Among the remaining 8 values, at most seven are non-consecutive with the value used in g_5. But, g_5 is adjacent to eight other cells. It means that the remaining seven non-consecutive values do not suffice to fill all the neighbours of g_5.

Given the formalisation in Listing 9.3, Mace4 also fails to find a model. Here, we state the values are distinct and they are from the interval $[1...10]$ (lines 2 and 6). Adjacent cells x and y do not contain consecutive numbers:

$$adj(x, y) -> |x - y| \neq 1 \tag{9.1}$$

where the adjacent predicate is symmetric: $adj(x, y) \leftrightarrow adj(y, x)$. Since the grid is small, we can explicitly state which pair of cells are adjacent (lines 14–21). Since Mace4 fails to find a model, our answer to the question in the puzzle is no.

Puzzle 88. Latin square

Latin square is an $n \times n$ array filled with n different values. Each value occurs exactly once in each row and exactly once in each column. How many Latin squares are there for $n = 3$? A Latin square is normalised or in standard form if both its first row and its first column are in their natural order (i.e. 0, 1, 2). How many normalised Latin squares are for $n = 3$? What about $n = 4$?

After you solved the above puzzle, think also at the following one. Albert is a scientist that wants to test four different drugs (called A, B, C, and D) on four volunteers. He decides that every volunteer has to be tested with a different drug each week, but no two volunteers are allowed the same drug at the same time.

Listing 9.4 Finding models of a Latin square: values are distinct on each line and each column

```
1   assign(domain_size,3).
2   assign(max_models,-1).
3   set(arithmetic).
4
5   formulas(latin_square).
6       all x  all y1 all y2 (f(x, y1) = f(x, y2) -> y1 = y2).
7       all x1 all x2 all y  (f(x1, y) = f(x2, y) -> x1 = x2).
8   end_of_list.
```

Listing 9.5 To normalise a Latin square, two constraints are added

```
1   formulas(normalised_latin_square).
2       all y f(0,y)=y.
3       all x f(x,0)=x.
4   end_of_list.
```

Listing 9.6 Albert's experiment is just a normalised Latin square

```
1   formulas(albert_experiment).
2       A=0 & B=1 & C=2 & D=3.
3   end_of_list.
```

Solution

There are two constraints in a Latin square. First, the values should be distinct on each line y_i:

$$\forall x \; \forall y_1 \; \forall y_2 \; (f(x, y_1) = f(x, y_2) \rightarrow y_1 = y_2) \qquad (9.2)$$

Second, the values should be distinct on each column x_i:

$$\forall x_1 \; \forall x_2 \; \forall y \; (f(x_1, y) = f(x_2, y) \rightarrow x_1 = x_2) \qquad (9.3)$$

For $n = 3$ we set the domain size to value 3. Given the formalisation in Listing 9.4, Mace4 finds twelve solutions. Since some of them are isomorphic, one can remove isomorphisms with: `mace4 -f latinsquare.in |get_interps | isofilter`. The following five models remain:

0	1	2
1	2	0
2	0	1

0	1	2
2	0	1
1	2	0

0	2	1
1	0	2
2	1	0

0	2	1
2	1	0
1	0	2

1	0	2
0	2	1
2	1	0

For normalised Latin squares, we add two constraints (see Listing 9.5). For the first column (i.e. $x = 0$), for each line y the value in the square should be equal with y: $\forall y \; f(0, y) = y$. Similarly, for the first line (i.e. $y = 0$), for each column x the value in the square should be equal with x: $\forall x \; f(0, x) = x$. To obtain the normalised Latin square, we feed Mace4 with both formalisations in Listings 9.4 and 9.5 with: `mace4 -f latin_square.in normalised_latin_square.in`. For $n = 3$, Mace4 computes a single normalised Latin square:

f	0	1	2
0	0	1	2
1	1	2	0
2	2	0	1

For $n = 4$, we need to call Mace4 with the domain size set to 4: `mace4 -n 4 -f latinsquare.in`. The option n 4 will overwrite the `assign(domain_size,3)` in the input file. Mace4 computes 4 normalised Latin squares:

f	0 1 2 3
0	0 1 2 3
1	1 3 0 2
2	2 0 3 1
3	3 2 1 0

f	0 1 2 3
0	0 1 2 3
1	1 3 0 2
2	2 3 0 1
3	3 2 1 0

f	0 1 2 3
0	0 1 2 3
1	1 0 3 2
2	2 3 1 0
3	3 2 0 1

f	0 1 2 3
0	0 1 2 3
1	1 2 3 0
2	2 3 0 1
3	3 0 1 2

For $n = 5$, there are 56 normalised Latin squares, while for $n = 6$, Mace4 computes 9,408 models.

Consider that each row represents a different volunteer and each column represents a different week, Albert can plan the whole experiment using a Latin square. To solve the puzzle one can use the normalised Latin square with a domain of 4 and the notations in Listing 9.6: `mace4 -n 4 -f latin_square.in normalised_latin_square.in albert.in`. Mace4 finds 4 models for this task.

Puzzle 89. Magic square

A magic square is $n \times n$ square grid filled with distinct values from $1, 2, \ldots, n^2$ so that

1. each cell contains a different integer.
2. the sum of the integers in each row, column, and diagonal is equal.

Listing 9.7 Finding models for 3×3 magic squares: the domain is 16 to handle the magic sum $M = 15$

```
1    assign(domain_size,16).
2    assign(max_models,-1).
3    set(arithmetic).
4
5    formulas(magicsquare3x3).
6      gridsize = 3.
7      max = gridsize * gridsize.        %max value appearing in the grid
8      M = 15.                           %magic sum for 4x4 grid
9
10     %Sum on each row and each column is M
11     all x ((x < gridsize) -> (f(x,0) + f(x,1) + f(x,2) = M)).
12     all x ((x < gridsize) -> (f(0,x) + f(1,x) + f(2,x) = M)).
13     f(0,0) + f(1,1) + f(2,2) = M.     %sum on the main diagonal
14     f(0,2) + f(1,1) + f(2,0) = M.     %sum on the secundary diagonal
15
16     all x1 all x2 all y1 all y2        %Distinct values in each cell
17       (((x1 < gridsize) & (x2 < gridsize) &
18        (y1 < gridsize) & (y2 < gridsize) &
19        ((x1 != x2) | ( y1 != y2))) -> (f(x1,y1) != f(x2,y2))).
20
21     f(x,y) != 0.
22     ((x < gridsize)  & (y < gridsize)) -> f(x,y) <= max.
23     all x all y (((x >= gridsize) | (y >= gridsize)) -> (f(x,y) = M)).
24   end_of_list.
```

Solution

Let a grid size of 3×3, with maximum value of 9 in each cell. Let the function $f(x, y)$ containing the value in the cell on row x and column y. The sum is called the magic constant. For a magic square of size n, the magic constant is $M(n) = n(n^2 + 1)/2$. For our grid, that is $M(3) = 15$. This value forces us to set the domain size at 16 (line 1 in Listing 9.7). In a magic square, there are five constraints.

First, the values in each row sum up to the magic sum M:

$$\forall x \ ((x < gridsize) \rightarrow f(x, 0) + f(x, 1) + f(x, 2) = M) \tag{9.4}$$

Second, the sum of the values in each column is also M:

$$\forall y \ ((y < gridsize) \rightarrow f(0, y) + f(1, y) + f(2, y) = M) \tag{9.5}$$

Similarly for the main and secondary diagonal:

$$f(0, 0) + f(1, 1) + f(2, 2) = M \tag{9.6}$$
$$f(0, 2) + f(1, 1) + f(2, 0) = M \tag{9.7}$$

Fifth, the values in a magic square should be distinct:

$$\forall x_1 \ \forall x_2 \ \forall y_1 \ \forall y_2 \ (((x_1 < gridsize) \wedge (x_2 < gridsize) \wedge \tag{9.8}$$
$$(y_1 < gridsize) \wedge (y_2 < gridsize) \wedge$$
$$((x1 \neq x2) \vee (y1 \neq y2))) \rightarrow (f(x_1, y_1) \neq f(x_2, y_2)))$$

For a 3×3 grid, the values should belong to the interval [1...9]:

$$\forall x \ \forall y \ (0 < f(x, y) \wedge f(x, y) \leq max) \tag{9.9}$$

To remove irrelevant models, we set all the values outside the 4×4 grid to M:

$$\forall x \ \forall y \ ((x \geq gridsize) \vee (y \geq gridsize)) \rightarrow f(x, y) = M) \tag{9.10}$$

Mace4 finds eight magic squares (although all of them are rotations or reflections of a single one):

2	7	6
9	5	1
4	3	8

2	9	4
7	5	3
6	1	8

4	3	8
9	5	1
2	7	6

4	9	2
3	5	7
8	1	6

6	1	8
7	5	3
2	9	4

6	7	2
1	5	9
8	3	4

8	1	6
3	5	7
4	9	2

8	3	4
1	5	9
6	7	2

For 4×4 grid the magic sum is $M(4) = 34$. To explicitly handle it, we need a domain size of 35 which extends the search space for Mace4. Three 4×4 magic squares are:

1	12	15	6
14	9	4	7
3	8	13	10
16	5	2	11

1	13	4	16
8	12	5	9
14	2	15	3
11	7	10	6

1	2	15	16
13	14	3	4
12	7	10	5
8	11	6	9

Puzzle 90. Magic five-pointed star

It is required to place a different number in every circle so that the four circles in a line shall add up to 24 in all the five directions. First, show that no solution is possible with values from 1 to 10. Second, can you show that no solution exists for any 10 consecutive numbers? Third, find a solution assuming that you can use any whole numbers you like. How many solutions can you find with values up to 12? (adapted from puzzle 395 from Dudeney 2016)

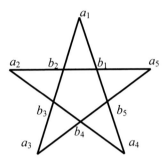

Listing 9.8 No models for values in [1...10]

```
1   set(arithmetic).
2   assign(domain_size,11).
3   assign(max_models,−1).
4
5   list(distinct).
6       [0,a1,a2,a3,a4,a5,b1,b2,b3,b4,b5].
7   end_of_list.
8
9   formulas(five_pointed_star).
10      a1 + b2 + b3 + a3 = 24.     a1 + b1 + b5 + a4 = 24.
11      a2 + b2 + b1 + a5 = 24.     a2 + b3 + b4 + a4 = 24.
12      a3 + b4 + b5 + a5 = 24.
13  end_of_list.
```

Listing 9.9 No models for 10 consecutive numbers

```
1   set(arithmetic).
2   assign(start_size,11).
3   assign(end_size,13).
4   assign(max_models,−1).
5
6   list(distinct).
```

```
7        [0,i,i+1,i+2,i+3,i+4,i+5,i+6,i+7,i+8,i+9,
8             a1,a2,a3,a4,a5,b1,b2,b3,b4,b5].
9    end_of_list.
10
11   formulas(five_pointed_star).
12       i = domain_size + −11.        a1 + b2 + b3 + a3 = 24.
13       a1 + b1 + b5 + a4 = 24.       a2 + b2 + b1 + a5 = 24.
14       a2 + b3 + b4 + a4 = 24.       a3 + b4 + b5 + a5 = 24.
15   end_of_list.
```

Solution

For the first task, to show that no solution exists in [1...10], one simply sets the domain size to 11 (line 2 in Listing 9.8). Mace4 fails to find a model for this domain.

For the second task, to show that no solution exists for any 10 consecutive numbers, one option is to iterate over the domain size: [1...10], [2...11], [3...12],...,[9...18]. We stop to 18 because 18 is the maximum value for which a solution is possible for the sum 24. This maximum value appears only in the case of $1 + 2 + 3 + 18 = 24$. We can constrain the domain even more because we know that the difference between each number in the sum should be less than 10 (since the 10 numbers are consecutive). For the domain [9...18], the maximum value added with the minimum one is already greater than the magic sum: $18 + 9 + x + y > 24$. Hence there is no solution in this interval. Note that the length of the interval equals the number of consecutive values, hence all values should be used, including the maxim one. Similarly:

For the domain [8...17]	$18 + 9 + x + y > 24$	no solution either
For the domain [7...16]	$16 + 7 + 8 + y > 24$	no solution either
For the domain [6...15]	$15 + 6 + 7 + y > 24$	no solution either
For the domain [5...14]	$14 + 5 + 6 + y > 24$	no solution either
For the domain [4...13]	$13 + 4 + 5 + 6 > 24$	no solution either
For the domain [3...12]	$12 + 3 + 4 + 5 = 24$	a solution may exist

Hence, if you really want to help Mace4, then you need to iterate over three domains only: [1...10], [2...11], [3...12] (lines 2 and 3 in Listing 9.9). One can do this with *assign(start_size, 11)* for the interval [1...10]. For the interval [3...12] one can use *assign(end_size, 13)*, and remember to add the values 2 and 3 in the list of distinct elements. Indeed, given the formalisation in Listing 9.9, Mace4 does not find any solution in these intervals. That shows there is no solution with any ten consecutive numbers.

For the third task, we set the domain size to 13 in line 2 of Listing 9.8. In this case Mace4 finds 120 models. Most of these solutions are isomorphic. One such model is: $a_1 = 1, a_2 = 2, a_3 = 3, a_4 = 4, a_5 = 5, b_1 = 9, b_2 = 8, b_3 = 12, b_4 = 6, b_5 = 10$, illustrated in Fig. 9.3.

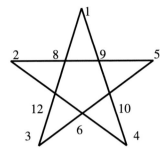

Fig. 9.3 One model among 120 found by Mace4 for values in [1...12]

(Puzzle 91. Fort Garrisons)

Here we have a system of fortifications. It will be seen that there are ten forts, connected by lines of outworks, and the numbers represent the strength of the small garrisons. The General wants to dispose these garrisons afresh so that there shall be 100 men in each of the five lines of the four forts. Can you show how it can be done? The garrisons must be moved bodily—that is to say, you are not allowed to break them up into other numbers. (puzzle 397 from Dudeney 2016)

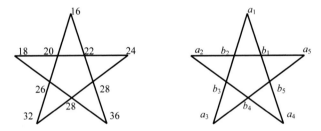

Listing 9.10 Finding models for protecting the Fort Garrisons with 100 men on each line

```
1   set(arithmetic).
2   assign(domain_size,37).
3   assign(max_models,−1).
4
5   list(distinct).
6     [a2,a3,a4,a5,b1,b2,b3,b4,b5].
7   end_of_list.
8
9   formulas(fort_garrisons).
10    a1 = 28.                        a1 + b2 + b3 + a3 = 100.
11    a1 + b1 + b5 + a4 = 100.        a2 + b2 + b1 + a5 = 100.
12    a2 + b3 + b4 + a4 = 100.        a3 + b4 + b5 + a5 = 100.
13
14    n(x) −> x=16 | x=18 | x=20 | x=22 | x=24 | x=26 | x=28 | x=32 | x=36.
15    n(a2). n(a3).  n(a4).  n(a5). n(b1). n(b2).  n(b3).  n(b4). n(b5).
16  end_of_list.
```

Solution

Observe that among the ten forts, the value 28 appears twice, and all other nine values are distinct. Hence we need nine distinct variables to model the forts. Let this nine distinct numbers $a_2, a_3, a_4, a_5, b_1, b_2, b_3, b_4, b_5$ (line 6 in Listing 9.10), with the tenth variable $a_1 = 28$. For each variable, the permitted values are from [18...36], hence the domain size is set to 37 (line 2). These permitted values [18...36] are specified by the $n(x)$ function:

$$n(x) \rightarrow x = 16 \lor x = 18 \lor x = 20 \lor x = 22 \lor x = 24 \lor x = 26 \lor x = 28 \lor x = 32 \lor x = 36$$

$$n(a_2) \land n(a_3) \land n(a_4) \land n(a_5) \land n(b_1) \land n(b_2) \land n(b_3) \land n(b_4) \land n(b_5)$$

Let the tenth node $a_1 = 28$, i.e. the value that can repeat in the fort. Next, we add the constraint that each side should be protected by 100 soldiers (lines 10–12). Mace4 finds 24 models.

Three models are depicted in Fig. 9.4. For instance, the model on the left part, we have $a_1 = 28$, $a_2 = 20$, $a_3 = 18$, $a_4 = 16$, $a_5 = 22$, $b_1 = 32$, $b_2 = 26$, $b_3 = 28$, $b_4 = 36$, $b_5 = 24$.

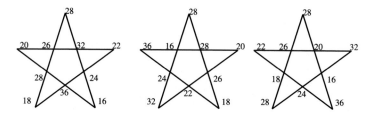

Fig. 9.4 Three solutions, among 24 models, found by Mace4: each line is protected by 100 men

Puzzle 92. Three in a row

Fill a 6 × 6 grid with Blue and White squares. When the puzzle is complete each row and column has an equal number of Blue and White squares. There will be no sequence of 3-In-A-Row squares of the same color (Fig. 9.5).

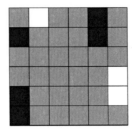

Fig. 9.5 Fill the grey cells with Blue and White

Listing 9.11 Formalising the 3-in-a-row puzzle with two functions: *b* for blue and *w* for white

```
1   set(arithmetic).
2   assign(domain_size,6).
3   assign(max_models,-1).
4
5   formulas(utils).
6     con = domain_size/2.
7     ds1 = domain_size + -1.
8     all x (x < ds1 -> s(x) = x+1) & s(ds1) = 5.          %succesor function
9   end_of_list.
10
11  formulas(3_in_a_row_with_two_functions).
12    all x all y (w(x,y)<= 1) & (b(x,y)<= 1).  %white/blue are true/false only
13    all x all y (w(x,y) = 1 -> b(x,y) = 0).  %cells can only have a colour
14    all x all y (w(x,y) = 1 | b(x,y) = 1).  %cells need to have a colour
15
16    %number(white cells) = number(blue cells) = 3 on each row/column (con=3)
17    all x w(x,0) + w(x,1) + w(x,2) + w(x,3) + w(x,4) + w(x,5) = con.
18    all y w(0,y) + w(1,y) + w(2,y) + w(3,y) + w(4,y) + w(5,y) = con.
19
20    %cannot have 3 of a kind in a row (con=3)
21    all x all y (y <= con -> -(w(x,y)=1 & w(x,s(y))=1 & w(x,s(s(y)))=1) &
22                             -(b(x,y)=1 & b(x,s(y))=1 & b(x,s(s(y)))=1)).
23
24    %cannot have 3 of a kind in a column (con=3)
25    all x all y (x <= con -> -(w(x,y)=1 & w(s(x),y)=1 & w(s(s(x)),y)=1) &
26                             -(b(x,y)=1 & b(s(x),y)=1 & b(s(s(x)),y)=1)).
27  end_of_list.
28
29  formulas(sample_puzzle).
30    b(0,4)=1 & b(1,0)=1 & b(1,4)=1 & b(4,0)=1 & b(5,0)=1.
31    w(0,1)=1 & w(3,5)=1 & w(4,5)=1.
32  end_of_list.
```

Fig. 9.6 Model found by Mace4 based on Listing 9.11

Listing 9.12 Formalising the 3-in-a-row puzzle: $f(x,y)=1$ is used for blue and $f(x,y)=0$ for white

```
1    set(arithmetic).
2    assign(domain_size,6).
3    assign(max_models,-1).
4
5    formulas(utils).
6      ds1 = domain_size + -1.
7      con = domain_size/2.                    %con=3 for a 6x6 grid
8      all x (x < ds1 -> s(x) = x+1).   s(ds1) = 5.   %successor function
9    end_of_list.
10
11   formulas(3_in_a_row_with_one_function).
12     all x all y (f(x,y) <= 1).          %f(x,y)=1 is blue, f(x,y)=0 is white
13
14     %number(white cells) = number(blue cells) = 3 on each row/column
15     all x f(x,0) + f(x,1) + f(x,2) + f(x,3) + f(x,4) + f(x,5) = con.
16     all y f(0,y) + f(1,y) + f(2,y) + f(3,y) + f(4,y) + f(5,y) = con.
17
18     %Cannot have 3-in-a-row consecutive cells of the same color
19     all x all y (y <= con -> -(f(x,y)=1 & f(x,s(y))=1 & f(x,s(s(y)))=1) &
20                              -(f(x,y)=0 & f(x,s(y))=0 & f(x,s(s(y)))=0)).
21
22     %Cannot have 3-in-a-column consecutive cells of the same color
23     all x all y (x <= con -> -(f(x,y)=1 & f(s(x),y)=1 & f(s(s(x)),y)=1) &
24                              -(f(x,y)=0 & f(s(x),y)=0 & f(s(s(x)),y)=0)).
25   end_of_list.
26
27   formulas(sample_puzzle).
28     f(0,4)=1 & f(1,0)=1 & f(1,4)=1 & f(4,0)=1 & f(5,0)=1.
29     f(0,1)=0 & f(3,5)=0 & f(4,5)=0.
30   end_of_list.
```

Solution

We provide two solutions: First, let the board 6×6 (line 1 in Listing 9.11). To have a generic implementation for $n \times n$ grids, we use the `domain_size` constant available in Mace4. For instance, the number of consecutive blue cells in a row is half of the domain size (line 6 in Listing 9.11, where $con = domain\_size/2$). We will also use the successor function $s(x) = x + 1$ (lines 7–8). We will use two functions:

$$w(x, y) = \begin{cases} 1 & \text{if } (x, y) \text{ is white} \\ 0 & \text{if } (x, y) \text{ is not white} \end{cases} \quad b(x, y) = \begin{cases} 1 & \text{if } (x, y) \text{ is blue} \\ 0 & \text{if } (x, y) \text{ is not blue} \end{cases}$$

Each cell needs to be either blue or white:

$$\forall x \; \forall y \; (w(x, y) = 1 \vee b(x, y) = 1) \tag{9.11}$$

The advantage of using functions instead of predicates is that we can use their values in computations. Thus, the constraint that the number of white cells should be 3 in each row is formalised with:

$$\forall x\ w(x, 0) + w(x, 1) + w(x, 2) + w(x, 3) + w(x, 4) + w(x, 5) = con \quad (9.12)$$

Similarly, the number of white cells should be 3 in each column is formalised with:

$$\forall y\ w(0, y) + w(1, y) + w(2, y) + w(3, y) + w(4, y) + w(5, y) = con \quad (9.13)$$

Note there is no need to write the similar constraints for the blue function, as the functions are complementary. Note also that Eqs. (9.12) and (9.13) break the generality of the implementation, because the sum is computed exactly for six terms.
There can't be consecutive three of a kind in one row:

$$\forall x\ \forall y\ (y \leq 3 \rightarrow \neg(w(x, y) = 1 \land w(x, y + 1)) = 1 \land w(x, y + 2) = 1)$$
$$\land\neg(b(x, y) = 1 \land b(x, y + 1)) = 1 \land b(x, y + 2) = 1)) \quad (9.14)$$

Also, there can't be consecutive three of a kind in one column:

$$\forall x\ \forall y\ (x \leq 3 \rightarrow \neg(w(x, y) = 1 \land w(x + 1, y) = 1 \land w(x + 2, y) = 1)$$
$$\land\neg(b(x, y) = 1 \land b(x + 1, y) = 1 \land b(x + 2, y) = 1)) \quad (9.15)$$

In line 21, we used the constant con instead of value 3 and the successor function s. With two functions for blue and white cells, the formalisation of init state is straightforward (lines 30–31). Given the formalisation in Listing 9.11, Mace4 computes a single model:

b	0 1 2 3 4 5		w	0 1 2 3 4 5
0	0 0 1 0 1 1		0	1 1 0 1 0 0
1	1 1 0 0 1 0		1	0 0 1 1 0 1
2	0 0 1 1 0 1		2	1 1 0 0 1 0
3	0 1 1 0 1 0		3	1 0 0 1 0 1
4	1 1 0 1 0 0		4	0 0 1 0 1 1
5	1 0 0 1 0 1		5	0 1 1 0 1 0

This model is depicted in Fig. 9.6.

Solution

The second formalisation (Listing 9.12) uses only one function $f(x, y) = 1$ for blue, and $f(x, y) = 0$ for white, instead of two functions $blue$ and $white$. Because the domain size is set to 6, we need to explicitly set the range of f to $\{0, 1\}$:
$\forall x\ \forall y\ (f(x, y) \leq 1)$. Each row and column contains exactly three white cells:

$$\forall x\ f(x, 0) + f(x, 1) + f(x, 2) + f(x, 3) + f(x, 4) + f(x, 5) = 3 \quad (9.16)$$
$$\forall y\ f(0, y) + f(1, y) + f(2, y) + f(3, y) + f(4, y) + f(5, y) = 3 \quad (9.17)$$

In each row, one cannot have 3 consecutive cells of the same colour, either blue or white:

$$\forall x\, \forall y\, (y \leq 3 \rightarrow \neg(f(x, y) = 1 \land f(x, s(y)) = 1 \land f(x, s(s(y))) = 1)$$
$$\land\neg(f(x, y) = 0 \land f(x, s(y)) = 0 \land f(x, s(s(y))) = 0)) \quad (9.18)$$

Here the successor function s is used to increase the column y. Note that the constraint is already true starting with the fourth column (i.e. $x > 3$). Also, in each column, one cannot have 3 consecutive cells of the same colour:

$$\forall x\, \forall y\, (x \leq 3 \rightarrow \neg(f(x, y) = 1 \land f(s(x), y) = 1 \land f(s(s(x)), y) = 1)$$
$$\land\neg(f(x, y) = 0 \land f(s(x), y) = 0 \land f(s(s(x)), y) = 0)) \quad (9.19)$$

Here s is used to increase the row x. Based on the implementation in Listing 9.12, Mace4 outputs the same single model:

f	0	1	2	3	4	5
0	0	0	1	0	1	1
1	1	1	0	0	1	0
2	0	0	1	1	0	1
3	0	1	1	0	1	0
4	1	1	0	1	0	0
5	1	0	0	1	0	1

Puzzle 93. Star battle

Each puzzle is divided into a fixed number of cages. Place stars in an $n \times n$ grid so that:

1. Each cage, row, and column must contain the same number of stars (1 or 2 stars, depending on the puzzle variety).
2. Stars may not reside in adjacent cells (not even diagonally).

The 5×5 puzzle in Fig. 9.7 has one star in each cage. (taken from Kleber 2013)

Listing 9.13 Finding models for the 5×5 star battle puzzle:

```
1   set(arithmetic).
2   assign(domain_size,5).
3   assign(max_models,−1).
4
5   formulas(utils).
6       ds1 = domain_size + −1.
7       all x (x < ds1 −> s(x) = x+1).   %s(ds1) = ds1. %successor function
```

```
 8      ds1 = domain_size + −1.
 9      all x (x < ds1 −> s(x) = x+1).    %s(ds1) = ds1. %successor function
10      all x (x > 0 −> p(x) = x + −1).   %p(0) = 0.      %predecessor function
11      stars = 1.                        %only one star per block
12    end_of_list.
13
14    formulas( star_battle ).
15      all x all y (f(x,y) < 2).              %cells can be 0 (no star) or 1 (star)
16      f(x,y1) = 1 & f(x,y2) = 1 −> y1 = y2.       %only one star per row
17      f(x1,y) = 1 & f(x2,y) = 1 −> x1 = x2.       %only one star per column
18
19      %no two stars can be adjacent horizontally, vertically or diagonally
20      %the sum of the values adjacent to a star must be 1
21      f(x,y) = 1 −> f(p(x),p(y)) + f(p(x),y) + f(p(x),s(y)) +
22                       f(x,p(y))      +                   f(x,s(y))      +
23                       f(s(x),p(y)) + f(s(x),y) + f(s(x),s(y)) = 0.
24    end_of_list.
25
26    formulas( sample_puzzle ).
27      f(0,0) + f(0,1) = stars.
28      f(0,2) + f(0,3) + f(0,4) + f(1,4) + f(1,3) + f(1,2) = stars.
29      f(1,0) + f(1,1) + f(2,0) + f(2,1) = stars.
30      f(3,0) + f(3,1) + f(3,2) + f(2,2) + f(2,3) + f(2,4) = stars.
31      f(4,0) + f(4,1) + f(4,2) + f(4,3) + f(4,4) + f(3,4) + f(3,3) = stars.
32    end_of_list.
```

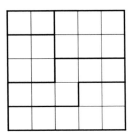

Fig. 9.7   One star in each line, column, cage. No stars in adjacent cells are allowed

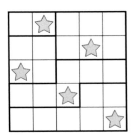

Fig. 9.8   One model found by Mace4

The current puzzle assumes one star in each cage. Hence we fix the number of stars with $star = 1$. Let $f(x, y) = 1$ for a star in cell (x, y) and $f(x, y) = 0$ for no star. We limit the range of the function f only to these two values with $\forall x\ \forall y\ f(x, y) < 2$. Next, only one star is allowed on row or column:

$$f(x, y_1) = 1 \wedge f(x, y_2) = 1 \rightarrow y_1 = y_2 \tag{9.20}$$
$$f(x_1, y) = 1 \wedge f(x_2, y) = 1 \rightarrow x_1 = x_2 \tag{9.21}$$

The constraint that no two stars are adjacent horizontally, vertically or diagonally means that a cell with a star has all its eight neighbours zero:

$$f(x, y) = 1 \rightarrow f(p(x), p(y)) + f(p(x), y) + f(p(x), s(y)) +$$
$$f(x, p(y)) + \qquad\qquad f(x, s(y)) + \tag{9.22}$$
$$f(s(x), p(y)) + f(s(x), y) + f(s(x), s(y)) = 0$$

Here $s(x)$ is the successor function, while $p(x)$ is the predecessor function, defined in the utils formulas (lines 7–8 in Listing 9.13). Note that we did not define the $s(x)$ for $x = domain\_size$ and $p(x)$ for $x = 0$. For this reason Mace4 outputs four isomorphic models, all of them having the unique solution in Fig. 9.8.

Puzzle 94. Star battle reloaded

Enter exactly one star in each row, each column, and each area of the grid. Cells with stars must not touch each other orthogonally or diagonally. (taken from Kleber 2013).

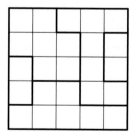

Listing 9.14 Using predicate for finding models of the star battle puzzle

```
1   set(arithmetic).
2   assign(domain_size,5).
3   assign(max_models,-1).
4
5   formulas(star_battle).
6       all x exists y s(x,y).                    %at least a star on each line
```

```
 7    s(x,y1) & s(x,y2) -> y1 = y2.              %at most a star on each line
 8    s(x1,y) & s(x2,y) -> x1 = x2.              %at most one star on each column
 9    s(x1,y1) & s(x2,y2) &                      %at most one star on diagonal + no adj
10       (abs(x2 + -x1) = abs(y2 + -y1) & abs(x2 + -x1) = 1)
11              -> (x1 = x2) & (y1 = y2).
12    end_of_list.
13
14    formulas(sample_puzzle).
15       s(0,0) | s(0,1) | s(1,0) | s(1,1) | s(1,2) | s(2,1) | s(2,2).       %1 zone
16       s(2,0) | s(3,0).                                                    %2 zone
17       s(0,2) | s(0,3) | s(0,4) | s(1,3) | s(2,3) | s(3,3) | s(3,4).       %3 zone
18       s(1,4) | s(2,4).                                                    %4 zone
19       s(4,0) | s(4,1) | s(4,2) | s(4,3) | s(4,4) | s(3,1) | s(3,2).       %5 zone
20    end_of_list.
```

Solution

We need a domain size of 5 for the given 5×5 puzzle. Let the predicate $s(x, y)$ being true when a star is in cell (x, y) and false otherwise. There are four constraints: First, there is at least one star on each line x: $\forall x \, \exists y \, s(x, y)$. Second, there is at most one star on each line: $s(x_1, y) \wedge s(x_2, y) \rightarrow x_1 = x_2$ (i.e. if there is a star on line x_1 and a star on x_2, then x_1 is the same with x_2). Third, there is at most one star on each column: $s(x, y_1) \wedge s(x, y_2) \rightarrow y_1 = y_2$. Fourth, there is at most one star on each diagonal, but also no adjacent stars are allowed:

$$s(x_1, y_1) \wedge s(x_2, y_2) \wedge |x_2 - x_1| = |y_2 - y_1| \wedge |x_2 - x_1| = 1 \rightarrow (x_1 = x_2) \wedge (y_1 = y_2) \tag{9.23}$$

Mace4 found one model:

```
                                    s:| 0 1 2 3 4
                                    --+----------
                                    0 | 0 0 0 1 0
                    f1:| 0 1 2 3 4  1 | 0 1 0 0 0
                    ---+----------  2 | 0 0 0 0 1
                       | 3 1 4 0 2  3 | 1 0 0 0 0
                                    4 | 0 0 1 0 0
```

Here the Skolem function f_1 occurs due to the existential quantifier in line 6 in Listing 9.14, and it indicates the columns on which the stars appear. The model of function s is illustrated in Fig. 9.9.

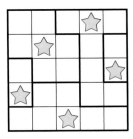

Fig. 9.9 The single model found by Mace4

Puzzle 95. Fancy queens

I have placed a queen in one of the white squares of the board shown. Place 7 more queens in white squares so that no 2 of the 8 queens are in line horizontally, vertically, or diagonally (adapted from puzzle 113 from Kordemsky 1992).

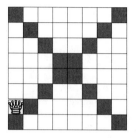

Listing 9.15 Fancy queens don't like main diagonals

```
1    set(arithmetic).
2    assign(max_models,−1).
3    assign(domain_size,8).    %queens can be placed in one column from 0 to 7
4
5    formulas(classic_queens).
6      all x exists y Q(x, y).              %each row has at least one queen
7      Q(x,y1) & Q(x,y2) −> y1 = y2.        %each row has at most one queen
8      Q(x1,y) & Q(x2,y) −> x1 = x2.        %each column has at most one queen
9
10     %each / diagonal has at most one queen
11     Q(x1,y1) & Q(x2,y2) & (x2 + −x1 = y2 + −y1) −> x1 = x2 & y1 = y2.
12
13     %each \ diagonal has at most one queen
14     Q(x1,y1 ) & Q(x2,y2) & (x1 + −x2 = y2 + −y1) −> x1 = x2 & y1 = y2.
15   end_of_list.
16
17   formulas(fancy_queens).
18     Q(x,y) −> x != y.                    %no queen on the main diag
19     Q(x,y) −> x != domain_size + −y + −1.  %no queen on the sec. diag
20     Q(6,0).                              %there is a queen at (6,0)
21   end_of_list.
```

Solution

The problem resembles the queens problem (lines 5–15 in Listing 9.15). The queens should not attack each other diagonally, along the line and on the column. Assume the queens can be placed in one column from 0 to 7 (line 3). Assume the predicate $Q(x, y)$ is true when a queen is on the position (x, y). Each row has at most one queen: $Q(x, y_1) \wedge Q(x, y_2) \rightarrow y_1 = y_2$. Each column has at most one queen: $Q(x_1, y) \wedge Q(x_2, y) \rightarrow x_1 = x_2$. We also need to specify that each row has at least one queen: $\forall x \, \exists y \, Q(x, y)$. Moreover, each diagonal has at most one queen:

$$Q(x_1, y_1) \wedge Q(x_2, y_2) \wedge (x_2 - x_1 = y_2 - y_1) \rightarrow x_1 = x_2 \wedge y_1 = y_2 \quad (9.24)$$
$$Q(x_1, y_1) \wedge Q(x_2, y_2) \wedge (x_1 - x_2 = y_2 - y_1) \rightarrow x_1 = x_2 \wedge y_1 = y_2 \quad (9.25)$$

Here Eq. (9.24) states the constraint for the main diagonal, and Eq. (9.25) for the secondary diagonal. If we use only these formulas, we have the classical 8 queens puzzle. Note that the formalisation is generic for the n-queens problem. One has only to change the domain size. Mace4 can find the models for several domain sizes, as listed below:

Domain size	4	5	6	7	8	9	10
Models found	2	10	4	40	92	352	724
Remove isomorphisms	2	5	4	16	26	41	60

The isomorphisms are removed with: mace4 -f fancyqueens.in | interps | isofilter. For instance, for the domain size of 8, Mace4 identifies 92 models, from which 26 are not isomorphisms.

In addition to the classical queens puzzle, the fancy queens cannot be placed on the two diagonals. The module $formulas(fancy\_queens)$ (lines 18–19) handles these constraints. First, no queen on the main diagonal is formalised with: $Q(x, y) \rightarrow x \neq y$. For the second diagonal, we have: $Q(x, y) \rightarrow x \neq domain\_size - y - 1$. Note here the usage of the variable $domain\_size$ that makes the implementation general for $n \times n$ boards. Since there is a queen in 6 and 0 (i.e. Q(6, 0), line 20), Mace4 finds a single model, depicted also in Fig. 9.10:

Fig. 9.10 Unique model found by Mace4 for the 8 × 8 fancy queens puzzle

---

Puzzle 96. Playing minesweeper

A brave soldier has to find the mines in the given enemy map. The number in a cell indicates how many of the eight neighbouring cells contain a mine. A numbered cell does not contain a mine. No clue says that there are zero mines around.

1	1	1		2		3		3
1	1							
1	1	1		4	4	3		
								2
				7				
5		6				3		1
				8				1
						3		
		5	4			3		2

c	n	n	n	n	n	n	n	c
e	i	i	i	i	i	i	i	w
e	i	i	i	i	i	i	i	w
e	i	i	i	i	i	i	i	w
e	i	i	i	i	i	i	i	w
e	i	i	i	i	i	i	i	w
e	i	i	i	i	i	i	i	w
e	i	i	i	i	i	i	i	w
c	s	s	s	s	s	s	s	c

Fig. 9.11 Minesweeper: 9×9 map with initial clues (left); the modelling approach (right)

Solution

The number in a cell indicates how many of the eight neighbouring cells contain a mine. Since there are maximum eight cells, the domain is minim 9. The domain is increased if the grid is larger than 8. For a 8×8 grid, the domain of 9 values suffices. Let $mine(x, y) : [0...X_{max}] \times [0...Y_{max}] \rightarrow \{0, 1\}$ the function indicating whether a mine exists or not in location (x, y). Mine is a function and not a predicate because we need to perform computations (e.g. addition) on it. Here, we define the successor function s and the predecessor p:

$$s(x) = \begin{cases} x+1 & x \leq 7 \\ 7 & x = 8 \end{cases} \qquad\qquad p(x) = \begin{cases} x-1 & x \geq 1 \\ 1 & x = 0 \end{cases}$$

Let $f(x, y)$ the function representing the clues that signal how many mines are nearby location (x, y). We know that a numbered cell does not contain a mine: $f(x, y) \neq 0 \rightarrow mine(x, y) = 0$.

Listing 9.16 Finding models for a 9×9 minesweeper

```
1    set(arithmetic).
2    assign(max_models,-1).
3    assign(domain_size,9).
4
5    formulas(rules_minesweeper).
6      s(0)=1. s(1)=2. s(2)=3. s(3)=4. s(4)=5. s(5)=6. s(6)=7. s(7)=8. s(8)=7.
7      p(0)=1. p(1)=0. p(2)=1. p(3)=2. p(4)=3. p(5)=4. p(6)=5. p(7)=6. p(8)=7.
8      mine(x,y)=1 | mine(x,y)=0.   %mine is a function to support computations
9      f(x,y)!=0  -> mine(x,y)=0.   %cells with numbers do not contain mines
10
11     %corner has only 3 neighbours
12     f(x,y)!=0 & ((x=0 & y=0) | (x=8 & y=0) | (x=0 & y=8) | (x=8 & y=8))
13        -> mine(x,s(y)) + mine(s(x),y) + mine(s(x),s(y)) = f(x,y).
14
15     f(x,y)!=0  & ((x=0 | x=8) & (y>0 & y<8))              %margin up-down
```

```
16      -> mine(x,p(y)) + mine(p(x),p(y)) + mine(s(x),y) +
17         mine(s(x),s(y)) + mine(x,s(y)) = f(x,y).
18
19      f(x,y)!=0 & ((y=0 | y=8) & (x>0 & x<8))          %margin left-right
20      -> mine(p(x),y) + mine(p(x),s(y)) + mine(x,s(y)) +
21         mine(s(x),s(y)) + mine(s(x),y) = f(x,y).
22
23      f(x,y)!=0 & (y>0 & x>0 & x<8 & y<8)              %middle
24      -> mine(p(x),p(y)) + mine(p(x),y) + mine(p(x),s(y)) + mine(x,p(y)) +
25         mine(x,s(y)) + mine(s(x),p(y)) + mine(s(x),y) + mine(s(x),s(y))=f(x,y).
26   end_of_list.
27
28   formulas(map_minesweeper).
29      f(0,0)=1. f(0,1)=1. f(0,2)=1. f(0,3)=0. f(0,4)=2.   %line 1
30      f(0,5)=0. f(0,6)=3. f(0,7)=0. f(0,8)=3.
31      f(1,0)=1. f(1,1)=0. f(1,2)=1. f(1,3)=0. f(1,4)=0.   %line 2
32      f(1,5)=0. f(1,6)=0. f(1,7)=0. f(1,8)=0.
33      f(2,0)=1. f(2,1)=1. f(2,2)=1. f(2,3)=0. f(2,4)=4.   %line 3
34      f(2,5)=4. f(2,6)=3. f(2,7)=0. f(2,8)=0.
35      f(3,0)=0. f(3,1)=0. f(3,2)=0. f(3,3)=0. f(3,4)=0.   %line 4
36      f(3,5)=0. f(3,6)=0. f(3,7)=0. f(3,8)=2.
37      f(4,0)=0. f(4,1)=0. f(4,2)=0. f(4,3)=0. f(4,4)=7.   %line 5
38      f(4,5)=0. f(4,6)=0. f(4,7)=0. f(4,8)=0.
39      f(5,0)=5. f(5,1)=0. f(5,2)=6. f(5,3)=0. f(5,4)=0.   %line 6
40      f(5,5)=0. f(5,6)=3. f(5,7)=0. f(5,8)=1.
41      f(6,0)=0. f(6,1)=0. f(6,2)=0. f(6,3)=0. f(6,4)=8.   %line 7
42      f(6,5)=0. f(6,6)=0. f(6,7)=0. f(6,8)=1.
43      f(7,0)=0. f(7,1)=0. f(7,2)=0. f(7,3)=0. f(7,4)=0.   %line 8
44      f(7,5)=0. f(7,6)=0. f(7,7)=3. f(7,8)=0.
45      f(8,0)=0. f(8,1)=5. f(8,2)=0. f(8,3)=4. f(8,4)=0.   %line 9
46      f(8,5)=0. f(8,6)=3. f(8,7)=0. f(8,8)=2.
47   end_of_list.
```

Solution

There are four cases: (1) corners; (2) north and south border; (3) east and west border; (4) inner cells. In Fig. 9.11 (right), these cases are: (1) corners c have 3 neighbours, (2) north n and south s margins have 5 neighbours; (3) e and w have 5 neighbours, and (4) inner cells i have 8 neighbours. First, if there is a clue in a corner, then the sum of its three neighbours should equal the value of the clue:

$$f(x, y) \neq 0 \land (x = 0 \land y = 0) \lor (x = XMax \land y = 0)$$
$$\lor (x = 0 \land y = YMax) \lor (x = XMax \land y = YMax) \qquad (9.26)$$
$$\rightarrow mine(x, s(y)) + mine(s(x), y) + mine(s(x), s(y)) = f(x, y).$$

Second, if the clue is on the south border ($x = 0$) or on the north border ($x = 8$), but not on corners ($y \neq 0$ or $y \neq 8$), then the sum of its five neighbours equals the value of the clue:

$$f(x, y) \neq 0 \land (x = 0 \lor x = 8) \land (y > 0 \lor y < 8)$$
$$\rightarrow mine(x, p(y)) + mine(s(x), p(y)) + mine(s(x), y) + mine(s(x), s(y)) + mine(x, s(y)) = f(x, y)$$

Third, if the clue is on the west border ($y = 0$) or east border ($y = 8$), but not on corners ($x \neq 0$ or $x \neq 8$), then the sum of its five neighbours should equal the value of the clue:

$$f(x, y) \neq 0 \wedge (y = 0 \vee y = 8) \wedge (x > 0 \vee x < 8)$$
$$\rightarrow mine(p(x), y) + mine(p(x), s(y)) + mine(x, s(y)) + mine(s(x), s(y)) + mine(s(x), y) = f(x, y)$$

Fourth, if the clue is not on any border, then the sum of its eight neighbours should equal the value of the clue:

$$f(x, y) \neq 0 \wedge (x \neq 0 \wedge x \neq 8 \wedge y \neq 0 \wedge y \neq 8)$$
$$\rightarrow mine(p(x), p(y)) + mine(p(x), y) + mine(p(x), s(y)) + mine(x, p(y)) +$$
$$mine(x, s(y))mine(s(x), p(y)) + mine(s(x), y) + mine(s(x), s(y)) = f(x, y).$$
$$(9.27)$$

For the puzzle in Fig. 9.11 Mace4 computes a single model, given by the functions f and $mine$. The solution is depicted in Fig. 9.12.

f:	0	1	2	3	4	5	6	7	8
0	1	1	1	0	2	0	3	0	3
1	1	0	1	0	0	0	0	0	0
2	1	1	1	0	4	4	3	0	0
3	0	0	0	0	0	0	0	0	2
4	0	0	0	0	7	0	0	0	0
5	5	0	6	0	0	0	3	0	1
6	0	0	0	0	8	0	0	0	1
7	0	0	0	0	0	0	0	3	0
8	0	5	0	4	0	0	3	0	2

mine:	0	1	2	3	4	5	6	7	8
0	0	0	0	0	0	0	0	1	0
1	0	1	0	0	1	1	0	1	1
2	0	0	0	0	0	0	0	0	1
3	0	0	0	0	1	1	0	0	0
4	1	1	0	1	0	1	0	0	1
5	0	1	0	1	1	1	0	0	0
6	1	1	0	1	0	1	0	0	0
7	1	1	1	1	1	1	1	0	1
8	1	0	1	0	0	0	0	1	0

1	1	1		2		3	◆	3
1	◆	1					◆	◆
1	1	1		4	4	3		◆
				◆	◆			2
◆	◆		◆	7	◆			◆
5	◆	6	◆	◆	◆	3		1
◆	◆		◆	8	◆			1
◆	◆	◆	◆	◆	◆	◆	3	◆
◆	5	◆	4			3	◆	2

Fig. 9.12 Unique solution found by Mace4: the functions f and $mine$ are plotted on the same grid

References

Becker, K. (2001). Teaching with games: The minesweeper and asteroids experience. *Journal of Computing Sciences in Colleges, 17*(2), 23–33.

Darling, D. (2004). *The universal book of mathematics from Abracadabra to Zeno's paradoxes.* New York: Wiley.

Dudeney, H. E. (2016). *536 puzzles and curious problems.* Courier Dover Publications.

Kleber, J. (2013). A style for typesetting logic puzzles.

Kordemsky, B. A. (1992). *The Moscow puzzles: 359 mathematical recreations.* Courier Corporation.

Levitin, A., & Levitin, M. (2011). *Algorithmic puzzles.* Oxford: Oxford University Press.

Chapter 10
Japanese Puzzles

Abstract In which we take the reader on a trip to Japan, by solving Sudoku-like puzzles: Killer Sudoku, Futoshiki, Kakurasu, Takuzo, Kakuro, or Kendoku. Based on the examples modelled in this chapter, students can build their own formalisations on many variants of Sudoku.

$Mind(undisciplined) \land \forall x \, (Mind(x) \land x \neq undisciplined \rightarrow$
$moreDisobedient(undisciplined, x))$
$Mind(disciplined) \land \forall x \, (Mind(x) \land x \neq disciplined \rightarrow$
$moreObedient(disciplined, x))$
$moreObedient(x, y) \leftrightarrow moreDisobedient(y, x).$

Gautama Buddha

This chapter takes you on a trip to Japan, by solving Sudoku-like puzzles: Killer Sudoku, Futoshiki, Kakurasu, Takuzo, Kakuro, or Kendoku. These are special grid-based puzzles that you have been training with the previous chapter, including Latin squares. Sudoku is indeed a Latin square with one additional constraint: the values should be distinct in some given regions. Note also that solving a Sudoku grid is the same as solving a graph colouring problem. Each Sudoku cell is a vertex, while edges represent the links between cells in a line, a column, or a block. A digit (i.e. distinct colour among nine colours available) is assigned to each vertex so that connected vertices have different values.

The game has European origins that can be traced back to 1779 to Euler's *officer problem*. The problem asks you to arrange 36 officers in 6 × 6 grid, so that one officer from each of the six regiments appears in each row and column. This is a Latin square of order 6, whose search for solution led to developments in combinatorics Darling (2004). In 1979, Howard Garns, a retired architect and puzzles fan, created the current version of the game under the name *Number Place*. The game was introduced in Japan in 1984 under the name Sudoku Jussien (2007). In 1989, the first piece of software able to produce Sudoku grids appeared under the name DigiHunt.

Sudoku has been the root of countless variants by adding new constraints, including uniqueness value on diagonal, common cells between the regions, the digits
within a region sum up to a given value (i.e. Killer Sudoku in this chapter), and the
sum of digits on columns and lines are restricted to some value (i.e. Kakuro in this
chapter). Based on the examples modelled in this chapter, students can build their
own formalisations on many variants of Sudoku.

Puzzle 97. Killer Sudoku

Fill the cell with the numbers from 0 to size-1 of the puzzle. The numbers may occur
only once in each row, column, and coloured area if specified. In addition to Sudoku, a
Killer Sudoku grid is divided into cages, shown with dashed lines (Fig. 10.1). The sum
of the numbers in a cage must equal the small number in its top-left corner. The same
number cannot appear in a cage more than once. (taken from Kleber (2013))

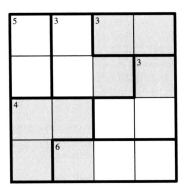

Fig. 10.1 Killer Sudoku has cages with numbers that sum up to a given value

Listing 10.1 Finding models for Killer Sudoku by extending a Latin square

```
1   assign(domain_size,4).
2   assign(max_models,−1).
3   set(arithmetic).
4
5   formulas(latin_square).
6       all x   all y1 all y2 (f(x,y1) = f(x,y2) −> y1 = y2).
7       all x1 all x2 all y  (f(x1,y) = f(x2,y) −> x1 = x2).
8   end_of_list.
9
10  formulas(killer_sudoku).
```

```
11     same_color(2,3) & same_color(0,1).  % let two sets {0,1} and {2,3}.
12     —same_color(0,2).                    % 0 and 2 are not from the same set
13     all x same_color(x,x).                                    % reflexive
14     all x all y (same_color(x,y) -> same_color(y,x)).         % symmetric
15     all x all y all z (same_color(x,y) &                      % transitive
16        same_color(y,z)  -> same_color(x,z)).
17
18     all x1 all y1 all x2 all y2 (same_color(x1,x2) &    % Zone 2: up right
19        same_color(y1,y2) & f(x1,y1) = f(x2,y2) &
20        x1 + x2 >= 4 & y1 + y2 < 2 -> x1 = x2 & y1 = y2).
21
22     all x1 all y1 all x2 all y2 (same_color(x1,x2) & % Zone 3: bottom left
23        same_color(y1,y2) & f(x1,y1) = f(x2,y2) &
24        x1 + x2 < 2 & y1 + y2 >= 4 -> x1 = x2 & y1 = y2).
25  end_of_list.
26
27  formulas(sample_puzzle_with_six_cages).
28    f(0,2) + f(0,3)          = 3.            % this cage has 2 cells
29    f(0,0) + f(1,0)          = 5.            % this cage has 2 cells
30    f(0,1) + f(1,1) + f(1,2) = 3.            % this cage has 3 cells
31    f(1,3) + f(2,3) + f(2,2) = 3.            % this cage has 3 cells
32    f(2,0) + f(2,1) + f(3,0) = 4.            % this cage has 3 cells
33    f(3,1) + f(3,2) + f(3,3) = 6.            % this cage has 3 cells
34  end_of_list.
35
36  list(distinct).  % The same number cannot appear in a cage more than once
37    [f(0,2),f(0,3)].        [f(0,0),f(1,0)].        [f(0,1),f(1,1),f(1,2)].
38    [f(1,3),f(2,3),f(2,2)]. [f(2,0),f(2,1),f(3,0)]. [f(3,1),f(3,2),f(3,3)].
39  end_of_list.
```

Solution

First, we introduce the constraints for the Latin square, i.e. the values should be distinct on rows and columns:

$$\forall x \,\forall y_1 \,\forall y_2 \,(f(x, y_1) = f(x, y_2) \to y_1 = y_2) \tag{10.1}$$

$$\forall x_1 \,\forall x_2 \,\forall y \,(f(x_1, y) = f(x_2, y) \to x_1 = x_2) \tag{10.2}$$

Second, we define the coloured regions. There are four zones:

$Zone_1 : x < 2 \land y < 2$ $Zone_2 : x < 2 \land y \geq 2$

$Zone_3 : x \geq 2 \land y < 2$ $Zone_4 : x \geq 2 \land y < 2$

To formalise these zones, we define two sets {0,1} and {2,3}:

$$same\_color(2, 3) \land same\_color(0, 1) \tag{10.3}$$

where 0 and 2 do not belong to the same set: $\neg same\_color(0, 2)$. The predicate *same_color* is reflexive, symmetric, and transitive:

$$\forall x \; same\_color(x, x) \qquad (10.4)$$
$$\forall x \; \forall y \; (same\_color(x, y) \rightarrow same\_color(y, x)) \qquad (10.5)$$
$$\forall x \; \forall y \; \forall z \; (same\_color(x, y) \wedge same\_color(y, z) \rightarrow same\_color(x, z)) \qquad (10.6)$$

A coloured zone should have unique numbers within it. For the upper right $Zone_2$, the columns $x_i \geq 2$ and rows $y_i < 2$. To have a unique number in $Zone_2$, the function $f(x, y)$ should be injective within this zone:

$$\forall x_1 \; \forall y_1 \; \forall x_2 \; \forall y_2 \; (same\_color(x_1, x_2) \wedge same\_color(y_1, y_2)$$
$$\wedge x_1 + x_2 \geq 4 \wedge y_1 + y_2 < 2 \wedge f(x_1, y_1) = f(x_2, y_2) \rightarrow x_1 = x_2 \wedge y_1 = y_2)$$
$$(10.7)$$

Similarly, for the bottom left $Zone_3$ the columns $x_i < 2$ and rows $y_i \geq 2$. To have a unique number in $Zone_3$, the function $f(x, y)$ should be injective within this zone:

$$\forall x_1 \; \forall y_1 \; \forall x_2 \; \forall y_2 \; (same\_color(x_1, x_2) \wedge same\_color(y_1, y_2)$$
$$\wedge x_1 + x_2 < 2 \wedge y_1 + y_2 \geq 4 \wedge f(x_1, y_1) = f(x_2, y_2) \rightarrow x_1 = x_2 \wedge y_1 = y_2) \qquad (10.8)$$

Third, we formalise the sum constraints for each cage. There are six cages formalised in lines 28–33. For instance, the cage from top-right corner that sums to 5 has two cells. Hence $f(0, 2) + f(0, 3) = 3$. The cage from the bottom corner that sums to 6 has 3 cells. Hence $f(3, 1) + f(3, 2) + f(3, 3) = 6$.
Mace4 found a single model for the function $f(x, y)$ (depicted in Fig. 10.2).

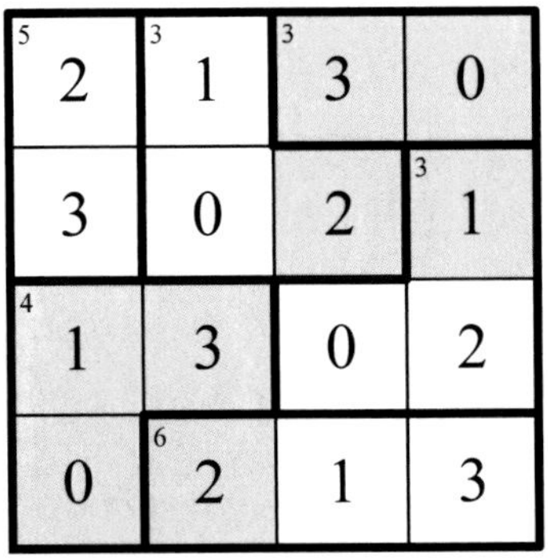

Fig. 10.2 The single model found by Mace4

You have to fill the grid with numbers so that:

1. The numbers are from 0 to the size-1 of the grid (i.e. 0 to 3 for a 4 × 4 puzzle).
2. Each row and column must contain only one instance of each number.
3. The numbers should satisfy the comparison signs—less than or greater than (Fig. 10.3).

(taken from Kleber (2013))

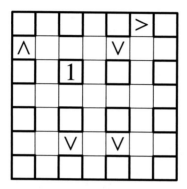

Fig. 10.3 A 4 × 4 Futoshiki puzzle with value 1 as the starting value

Listing 10.2 Finding models for Futoshiki puzzles by extending a Latin square

```
1    set(arithmetic).
2    assign(domain_size,4).
3    assign(max_models,-1).
4
5    formulas(latin_square).
6      f(x,y1) = f(x,y2)  -> y1 = y2.
7      f(x1,y) = f(x2,y)  -> x1 = x2.
8    end_of_list.
9
10   formulas(futoshiki).
11     f(0,2) > f(0,3).      f(0,0) < f(1,0).      f(0,3) > f(1,3).
12     f(2,1) > f(3,1).      f(2,2) > f(3,2).
13     f(1,1) = 1.           %initial value
14   end_of_list.
```

Futoshiki is just a Latin square with additional order constraints on some cells. We start by introducing the usual constraints for Latin square, i.e. the numbers should be distinct on lines and columns:

$$\forall x \ \forall y_1 \ \forall y_2 \ (f(x, y_1) = f(x, y_2) \rightarrow y_1 = y_2) \tag{10.9}$$

$$\forall x_1 \ \forall x_2 \ \forall y \ (f(x_1, y) = f(x_2, y) \rightarrow x_1 = x_2) \tag{10.10}$$

Then, we just need to formalise the order relation stated in the initial problem. There are two types of constraints: the order relations and the starting value:

$f(0, 2) > f(0, 3)$ $f(0, 0) < f(1, 0).$ $f(0, 3) > f(1, 3)$
$f(2, 1) > f(3, 1)$ $f(2, 2) > f(3, 2)$ $f(1, 1) = 1$

The single model found by Mace4 is depicted in Fig. 10.4.

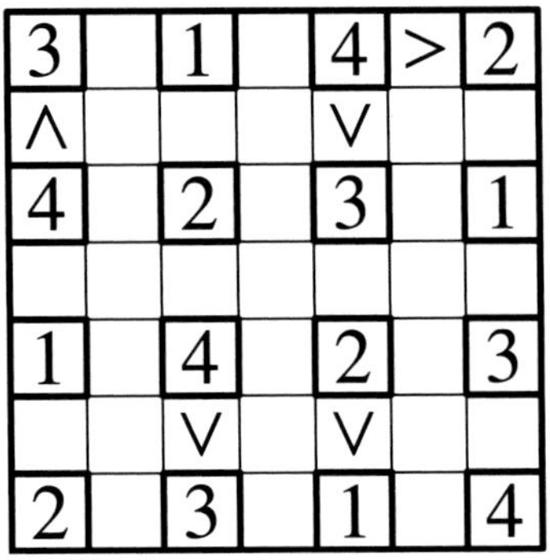

Fig. 10.4 Single model found by Mace4 for the Futoshiki puzzle

Puzzle 99. Kaos Sudoku

Fill the cell with the numbers from 0 to size-1 of the puzzle. Each number can appear only once—in each cage, column, and row (Fig. 10.5). (taken from Kleber (2013))

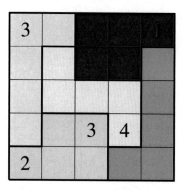

Fig. 10.5 Kaos Sudoku: values are unique in each row, column and cage

Listing 10.3 Finding models for Kaos Sudoku by extending a latin square

```
1    assign(domain_size,5).
2    assign(max_models,-1).
3    set(arithmetic).
4
5    list(distinct).
6       [f(0,0),f(0,1),f(1,0),f(2,0),f(3,0)].      % pink cage
7       [f(0,2),f(0,3),f(0,4),f(1,2),f(1,3)].      % blue cage
8       [f(1,1),f(2,1),f(2,2),f(2,3),f(3,3)].      % gray cage
9       [f(1,4),f(2,4),f(3,4),f(4,4),f(4,3)].      % orange cage
10      [f(4,0),f(4,1),f(4,2),f(3,1),f(3,2)].      % yellow cage
11   end_of_list.
12
13   formulas(latin_square).
14      all x  all y1 all y2 (f(x,y1) = f(x,y2) -> y1 = y2).
15      all x1 all x2 all y  (f(x1,y) = f(x2,y) -> x1 = x2).
16   end_of_list.
17
18   formulas(kaos_puzzle).
19        f(0,0)=3.    f(0,4)=1.    f(3,2)=3.    f(3,3)=4.  f(4,0)=2.
20      % f(0,0)=4.    f(0,4)=2.    f(3,2)=4.    f(3,3)=5.  f(4,0)=3.
21   end_of_list.
```

Solution

First, we introduce the constraints for the Latin square, i.e. the values should be distinct
on rows and columns:

$$\forall x \; \forall y_1 \; \forall y_2 \; (f(x, y_1) = f(x, y_2) \rightarrow y_1 = y_2) \tag{10.11}$$

$$\forall x_1 \; \forall x_2 \; \forall y \; (f(x_1, y) = f(x_2, y) \rightarrow x_1 = x_2) \tag{10.12}$$

Second, we state that each cage contains distinct values. This is handled by using a distinct list for each cage. Since there are five cages we need five distinct lists (lines 6–
10 in Listing 10.3). Last, we state the initial constraints of the current puzzle (line 19).
Mace4 finds a single model, illustrated in Fig. 10.6.

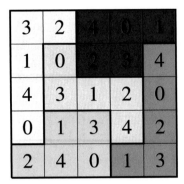

Fig. 10.6 The unique solution found by Mace4

Puzzle 100. Kakurasu

The goal is to make some of the cells black in such a way that:

1. The black cells on each row sum up to the number on the right.
2. The black cells on each column sum up to the number on the bottom.
3. If a black cell is first on its row/column, its value is 1. If it is second, its value is 2, etc. (taken from Kleber (2013)) (Fig. 10.7)

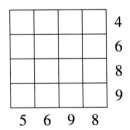

Fig. 10.7 A 4 × 4 Kakurasu puzzle

Listing 10.4 Formalising the Kakurasu puzzle with: $f(x,y)=0$ for white and $f(x,y)=1$ for black

```
1   set(arithmetic).
2   assign(domain_size,4).
3   assign(max_models,-1).
```

```
 5   formulas(kakurasu).
 6     all x all y f(x,y) < 2.
 7
 8     f(0,0) + f(0,1) * 2 + f(0,2) * 3 + f(0,3) * 4 = 4.      % first line
 9     f(1,0) + f(1,1) * 2 + f(1,2) * 3 + f(1,3) * 4 = 6.      % second line
10     f(2,0) + f(2,1) * 2 + f(2,2) * 3 + f(2,3) * 4 = 8.      % third line
11     f(3,0) + f(3,1) * 2 + f(3,2) * 3 + f(3,3) * 4 = 9.      % fourth line
12     f(0,0) + f(1,0) * 2 + f(2,0) * 3 + f(3,0) * 4 = 5.      % first column
13     f(0,1) + f(1,1) * 2 + f(2,1) * 3 + f(3,1) * 4 = 6.      % second column
14     f(0,2) + f(1,2) * 2 + f(2,2) * 3 + f(3,2) * 4 = 9.      % third column
15     f(0,3) + f(1,3) * 2 + f(2,3) * 3 + f(3,3) * 4 = 8.      % fourthcolumn
16   end_of_list.
```

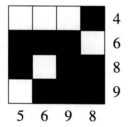

Fig. 10.8 The single solution found by Mace4 for the Kakurasu puzzle

Listing 10.5 Formalising the Kakurasu puzzle with two functions: *right* and *bottom* store the initial constraints.

```
 1   set(arithmetic).
 2   assign(domain_size,11).              % max n(n+1)/2 for nxn grid
 3   assign(max_models,-1).
 4
 5   formulas(kakurasu2).
 6     all x all y f(x,y) < 2.
 7     all x  (f(x,0) + f(x,1)*2 + f(x,2)*3 + f(x,3)*4 = right(x)).
 8     all y  (f(0,y) + f(1,y)*2 + f(2,y)*3 + f(3,y)*4 = bottom(y)).
 9     all x (x > 3 -> right(x) = 0 & bottom(x)= 0).
10     all x all y (x > 3 | y > 3 -> f(x,y)=0).
11   end_of_list.
12
13   formulas(puzzle).
14     right(0)  = 4 & right(1)  = 6 & right(2)  = 8 & right(3)  = 9.
15     bottom(0) = 5 & bottom(1) = 6  & bottom(2) = 9 & bottom(3) = 8.
16   end_of_list.
```

Listing 10.6 Avoiding the use the additional functions *right* and *bottom*. The values on the right and bottom can be fixed in the result function *f*

```
1   set(arithmetic).
2   assign(domain_size,11).        % max n(n+1)/2 for grid size n
3   assign(max_models,-1).
4
5   formulas(kakurasu3).
6     all x all y (x < 4 & y < 4 ->
7        (f(x,y) < 2) &
8        (f(x,0) + f(x,1)*2 + f(x,2)*3 + f(x,3)*4 = f(x,4)) &
9        (f(0,y) + f(1,y)*2 + f(2,y)*3 + f(3,y)*4 = f(4,y))).
10
11    all x all y (x > 4 | y > 4 -> f(x,y)=0).
12    f(4,4)=0.
13  end_of_list.
14
15  formulas(puzzle).
16    f(0,4) = 4 & f(1,4) = 6  & f(2,4) = 8 & f(3,4) = 9.
17    f(4,0) = 5 & f(4,1) = 6  & f(4,2) = 9 & f(4,3) = 8.
18  end_of_list.
```

Solution

We provide three solutions.

First, in Listing 10.4, let the `domain_size` equal to the size of the grid. Since any cell in the grid can either be black or white, $\forall x \, \forall y \, f(x, y) < 2$, that is, $f(x, y) = 0$ denotes a white cell, and $f(x, y) = 1$ denotes a black cell. Next, to model the rules for the sum, the following template is used:

$$f(x, y_1) * 1 + f(x, y_2) * 2 + \cdots + f(x, y_n) * n = value \qquad (10.13)$$

For a 4×4 grid, there are 8 such equations (lines 8–15 in Listing 10.4). The solution found by Mace4 is (Fig. 10.8)

f	0	1	2	3
0	0	0	0	1
1	1	1	1	0
2	1	0	1	1
3	0	1	1	1

Second, in Listing 10.5, a more general formalisation is provided. "The black cells on each row x sum up to the number on the right" is modelled with function $right(x)$:

$$\forall x \, (f(x, 0) + f(x, 1) * 2 + f(x, 2) * 3 + f(x, 3) * 4 = right(x)) \qquad (10.14)$$

"The black cells on each column y sum up to the number on the bottom" is represented with the function $bottom(x)$:

$$\forall y\ (f(0, y) + f(1, y) * 2 + f(2, y) * 3 + f(3, y) * 4 = bottom(y)) \qquad (10.15)$$

Here, the functions right and bottom store the sum values:

$$right(0) = 4 \wedge right(1) = 6 \wedge right(2) = 8 \wedge right(3) = 9$$
$$bottom(0) = 5 \wedge bottom(1) = 6 \wedge bottom(2) = 6 \wedge bottom(3) = 8 \qquad (10.16)$$

The maximum value for both $right(x)$ and $bottom(x)$ is when there are four black cells, that is, 1+2+3+4=10. To cover this maximum value we need to set the domain size to 11. That is, for a grid of $n \times n$ the domain size is $n(n + 1)/2 + 1$.

However, the domain of all functions is from 0 to 3. For input values larger than 3, we fix the functions to 0 in order to remove some isomorphic models:

$$\forall x\ (x > 3 \rightarrow right(x) = 0 \wedge bottom(x) = 0) \qquad (10.17)$$
$$\forall x\ \forall y\ (x > 3 \vee y > 3 \rightarrow f(x, y) = 0) \qquad (10.18)$$

Third, in Listing 10.6, we store the values of $right(x)$ and $bottom(x)$ in function f:

$$(\forall x\ (bottom(x) = f(4, x))) \wedge (\forall x\ (right(x) = f(x, 4))) \qquad (10.19)$$

With this formalisation we need to set to 0 values of f only for input larger than 4:

$$\forall x\ \forall y\ (x > 4 \vee y > 4 \rightarrow f(x, y) = 0) \qquad (10.20)$$

Note that there is specified value for the bottom-right corner (i.e. $f(4, 4)$). To avoid generation of distinct models for $f(4, 4)$ we fix the value with $f(4, 4) = 0$ (line 12 in Listing 10.6).

Puzzle 101. Takuzu

Some cells start out filled with black or white circles. Place circles in the empty cells so that:

1. Each row and each column must contain an equal number of white and black circles.

2. More than two circles of the same colour can't be adjacent.

3. Each row and column is unique (Fig. 10.9).

(taken from Kleber (2013))

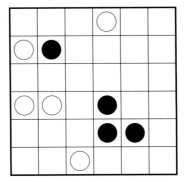

Fig. 10.9 A 6 × 6 Takuzo

Listing 10.7 Finding models for a 6 × 6 Takuzo

```
1    set(arithmetic).
2    assign(domain_size,6).
3    assign(max_models,−1).
4
5    formulas(utils).
6        s(0) = 1.  s(1) = 2.  s(2) = 3.  s(3) = 4.   s(4) = 5.   s(5) = 0.
7        p(1) = 0.  p(2) = 1.  p(3) = 2.  p(4) = 3.   p(5) = 4.   p(0) = 5.
8    end_of_list.
9
10   formulas(takuzo).
11       all x all y (f(x,y) < 2).     % only black f(x,y)=1, and whites f(x,y)=0
12       all x (f(x,0) + f(x,1) + f(x,2) + f(x,3) + f(x,4) + f(x,5)) = 3.
13       all y (f(0,y) + f(1,y) + f(2,y) + f(3,y) + f(4,y) + f(5,y)) = 3.
14       f(p(x),y) + f(x,y) + f(s(x),y) < 3. % max 3 adjacent black in a column y
15       f(p(x),y) + f(x,y) + f(s(x),y) > 0. % min 1 adjacent black in a column y
16       f(x,p(y)) + f(x,y) + f(x,s(y)) < 3. % max 3 adjacent black in a line x
17       f(x,p(y)) + f(x,y) + f(x,s(y)) > 0. % min 1 adjacent black in a line x
18   end_of_list.
19
20   formulas(sample_puzzle).
21       f(0,4) = 0.    f(1,0) = 1.   f(3,0) = 0.   f(3,1) = 0.  f(5,2) = 0.  % white
22       f(1,1) = 1.    f(3,3) = 0.   f(4,3) = 1.   f(4,4) = 1.               % black
23   end_of_list.
```

⎧Solution⎫

Since the given puzzle is a 6 × 6 grid, we need a domain size of 6. Let function
$f(x, y)$ to store the solution. There are only two values for f: $f(x, y) = 0$ for white
circles and $f(x, y) = 1$ for black circles. Since the current domain size is larger (i.e.
6), we restrict the range of f to 0 and 1: $\forall x \, \forall y \; f(x, y) < 2$ (line 11 in Listing 10.7).
To set exactly three black circles on each line x, the sum of the six values $f(x, y)$ on
the line x should be 3:

$$\forall x \; (f(x, 0) + f(x, 1) + f(x, 2) + f(x, 3) + f(x, 4) + f(x, 5)) = 3 \qquad (10.21)$$

It means that on all lines x there are three values of 0 and three values of 1. Similarly, for each column y:

$$\forall y (f(0, y) + f(1, y) + f(2, y) + f(3, y) + f(4, y) + f(5, y)) = 3 \qquad (10.22)$$

That is, there are three values of 0 and three values of 1 on each column y.

To model that there are maximum two adjacent circles of the same colour, we need the successor and predecessor functions (lines 6–7). Maximum two adjacent black circles on a column y are formalised with:

$$f(x - 1, y) + f(x, y) + f(x + 1, y) < 3 \qquad (10.23)$$

If the sum equals 2, there are exactly 2 adjacent black circles. If the sum equals 1, there are no 2 adjacent black circles. If the sum equals 0, all three adjacent cells are white, which is not allowed. To avoid this case we add:

$$f(p(x), y) + f(x, y) + f(s(x), y) > 0 \qquad (10.24)$$

We do the same for maximum 2 adjacent black circles on a row x (lines 16–17):

$$0 < f(x, y - 1) + f(x, y) + f(x, s(y)) < 3 \qquad (10.25)$$

Last, we give the positions for the initial white and black circles. There are 5 white circles (line 21) and 4 black circles (line 22). Mace4 finds a single model, which is illustrated in Fig. 10.10.

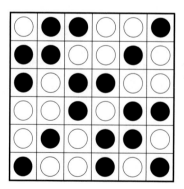

Fig. 10.10 The single solution found by Mace4

Puzzle 102. Kakuro

Kakuro is like a crossword puzzle with numbers. Each "word" must add up to the number provided in the clue above it or to the left. Words can only use the numbers 1 through 9, and a given number can only be used once in a "word". (taken from Kleber (2013)) (Fig. 10.11)

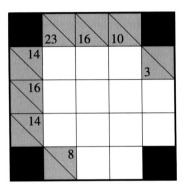

Fig. 10.11 A Kakuro puzzle: each value can be used only once in each "word"

Solution

Let function $f(x, y)$ store the solution. Since the largest value in the matrix is 23, one option is to use a domain size of 24. Hence, the matrix for f is 24×24, but we are interested only in the left upper corner of 5×5. We fix the values outside this 5×5 grid to 0 (line 17 in Listing 10.8):

$$\forall x \, \forall y \, (((x > 4) \vee (y > 4)) \rightarrow f(x, y) = 0) \tag{10.26}$$

Inside the 5×5 zone of interest there are three types of cells: black, green, and white. The black corners are fixed to value 0 (line 17). The green cells store the sums that constrain the puzzle (lines 27–29). The white cells contain distinct numbers from 1 to 9 (lines 19–20). We split the white cells into two zones: the 3×3 grid from $f(1, 1)$ to $f(4, 4)$, respectively, the cells $f(4, 2)$, $f(4, 3)$, $f(2, 4)$, $f(3, 4)$. First, the 3×3 is a Latin square (lines 6–9). Second, the values from the four cells should be distinct on the corresponding rows and columns (lines 10–12). Last, we write eight equations representing the required sum on each row and each column (lines 31–38). Mace4 found 17 models, with 3 of them listed in Fig. 10.12.

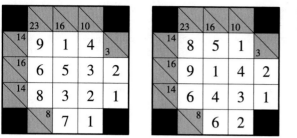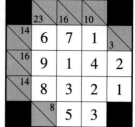

Fig. 10.12   Three models among 17 found by Mace4

Listing 10.8 Finding models for the Kakuro puzzle

```
1   set(arithmetic).
2   assign(domain_size,24).
3   assign(max_models,−1).
4
5   formulas(unique_values_in_white_cells).
6     all x all y1 all y2 (f(x,y1) = f(x,y2) & (x > 0) & (x < 4) &
7        (y1 > 0) & (y1 < 4) & (y2 > 0) & (y2 < 4) −> y1 = y2).
8     all x1 all x2 all y (f(x1,y) = f(x2,y) & (x1 > 0) & (x1 < 4) &
9        (x2 > 0) & (x2 < 4) & (y > 0) & (y < 4)−> x1 = x2).
10    all x ((x > 0 & x < 4) −> (f(4,2) != f(x,2)) & (f(4,3) != f(x,3))).
11    all y ((y > 0 & y < 4) −> (f(2,4) != f(2,y)) & (f(3,4) != f(3,y))).
12    f(4,2) != f(4,3).     f(2,4) != f(3,4).
13  end_of_list.
14
15  formulas(5x5_left_upper_corner).
16    all x all y (((x > 4) | (y > 4)) −> f(x,y) = 0).     % out 5x5 is black
17    f(0,0) = 0. f(4,0) = 0. f(0,4) = 0. f(4,4) = 0. % corners are black
18
19    all x all y (((x > 0) & (x < 4)) & ((y > 0) & (y < 4))
20        −> ((f(x,y) > 0) & (f(x,y) < 10))).
21
22    f(2,4) >= 1 & f(2,4) <= 9.      f(3,4) >= 1 & f(3,4) <= 9.
23    f(4,2) >= 1 & f(4,2) <= 9       f(4,3) >= 1 & f(4,3) <= 9.
24  end_of_list.
25
26  formulas(sample_puzzle).
27    f(0,1) = 23.  f(0,2) = 16.   f(0,3) = 10.     % first line
28    f(1,0) = 14.  f(2,0) = 16.   f(3,0) = 14.     % first column
29    f(4,1) = 8.   f(1,4) = 3.                     % last line & last column
30
31    f(1,0) = f(1,1) + f(1,2) + f(1,3).            % sum(white) on line 1
32    f(2,0) = f(2,1) + f(2,2) + f(2,3) + f(2,4).   % sum(white) on line 2
33    f(3,0) = f(3,1) + f(3,2) + f(3,3) + f(3,4).   % sum(white) on line 3
34    f(4,1) = f(4,2) + f(4,3).                     % sum(white) on line 4
35    f(0,1) = f(1,1) + f(2,1) + f(3,1).            % sum(white) on column 1
36    f(0,2) = f(1,2) + f(2,2) + f(3,2) + f(4,2).   % sum(white) on column 2
37    f(0,3) = f(1,3) + f(2,3) + f(3,3) + f(4,3).   % sum(white) on column 3
38    f(1,4) = f(2,4) + f(3,4).                     % sum(white) on column 4
39  end_of_list.
```

> **Puzzle 103. Daily neighbours**
>
> Complete an $n \times n$ grid so that every row and column contains every number from
> [0..n-1] exactly once. You also have to satisfy the neighbour constraint (⋈):
>
> 1. A symbol between two cells means the numbers are consecutive.
> 2. No symbol between cells means the numbers are not consecutive (Fig. 10.13).
>
> (taken from Kleber (2013))

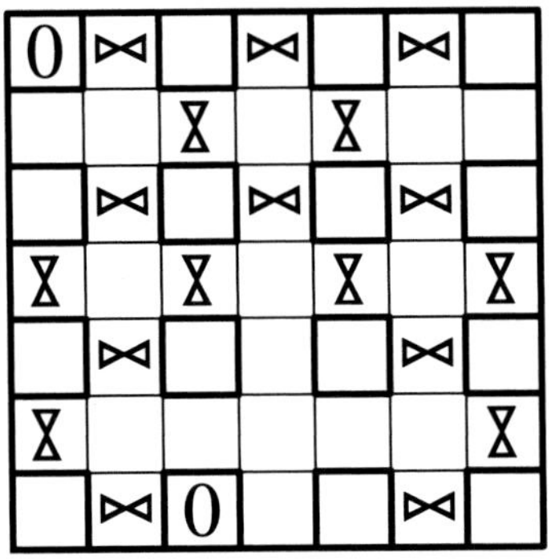

Fig. 10.13 Sample 4 × 4 neighbours puzzle

Listing 10.9 Finding models for the daily neighbours puzzle

```
1    set(arithmetic).
2    assign(domain_size, 4).
3    assign(max_models, −1).
4
5    formulas(latin_square).
6      all x all y1 all y2 (f(x,y1) = f(x,y2) -> y1 = y2).
7      all x1 all x2 all y (f(x1,y) = f(x2,y) -> x1 = x2).
8    end_of_list.
9
10   formulas(neighbours).
11     all x1 all x2 all y1 all y2 (n(x1,y1,x2,y2)
12       -> f(x1,y1) = (f(x2,y2) + 1) | f(x2,y2) = (f(x1, y1) + 1)).
13
14     all x1 all x2 all y1 all y2 (−n(x1,y1,x2,y2)
15       -> f(x1,y1) != (f(x2,y2)+1) & f(x2,y2) != (f(x1,y1) + 1)).
16   end_of_list.
17
18   formulas(sample_puzzle).
19     f(0,0) = 0.    f(3,1) = 0.
20
21     n(0,0,0,1).    n(0,1,0,2).    n(0,2,0,3).
22     −n(0,0,1,0).   n(0,1,1,1).    n(0,2,1,2).    −n(0,3,1,3).
23     n(1,0,1,1).    n(1,1,1,2).    n(1,2,1,3).
24     n(1,0,2,0).    n(1,1,2,1).    n(1,2,2,2).    n(1,3,2,3).
25     n(2,0,2,1).    −n(2,1,2,2).   n(2,2,2,3).
26     n(2,0,3,0).    −n(2,1,3,1).   −n(2,2,3,2).   n(2,3,3,3).
27     n(3,0,3,1).    −n(3,1,3,2).   n(3,2,3,3).
28   end_of_list.
```

Solution

The puzzle is a Latin square with two additional constraints on the neighbour relation $\bowtie$. We represent the $\bowtie$ relation with the predicate $n(x_1, y_1, x_2, y_2)$. For example, $n(1, 2, 1, 3)$ says there is a $\bowtie$ between the cell $(1, 2)$ and the cell $(1, 3)$. First, if two cells are neighbours then they should contain consecutive numbers:

$$\forall x_1 \, \forall x_2 \, \forall y_1 \, \forall y_2 \, n(x_1, y_1, x_2, y_2) \rightarrow f(x_1, y_1) = f(x_2, y_2) + 1 \lor f(x_2, y_2) = f(x_1, y_1) + 1$$
(10.27)

Second, if two cells are not neighbours then they do not contain consecutive numbers:

$$\forall x_1 \, \forall x_2 \, \forall y_1 \, \forall y_2 \, \neg n(x_1, y_1, x_2, y_2) \rightarrow f(x_1, y_1) \neq f(x_2, y_2) + 1 \land f(x_2, y_2) \neq f(x_1, y_1) + 1$$

The two above implications can be combined into one equivalence. Mace4 found the unique solution from Fig. 10.14.

Fig. 10.14 Single model found by Mace4

Puzzle 104. Kendoku

Fill in the $n \times n$ grid with numbers from 1 to n complying with operation. No number is repeated in any row and column. The calculation rules of each group should also be satisfied. (taken from Kleber (2013)) (Fig. 10.15)

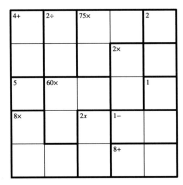

Fig. 10.15 Sample 5 × 5 Kendoku

Listing 10.10 Finding models for the 5 × 5 Kendoku puzzle

```
1   assign(domain_size,6).
2   assign(max_models,−1).
3   set(arithmetic).
4
5   formulas(kendoku).
6     all x all y1 all y2                    % distinct values on rows x
7     ((x < 5) & (y1 < 5) & (y2 < 5) & f(x,y1) = f(x,y2) −> y1 = y2).
8
9     all y all x1 all x2                    % distinct values on columns y
10    ((y < 5) & (x1 < 5) & (x2 < 5) & f(x1,y) = f(x2,y) −> x1 = x2).
11
12    all x all y ((x < 5 & y < 5) −> f(x,y) != 0).% values are from 1 to 5
13    all x all y ((x > 4 | y > 4) −> f(x,y)  = 0).% last column and line is 0
14  end_of_list.
15
16  formulas(sample_puzzle).
17    f(0,0) + f(1,0) = 4.          f(0,1) / f(1,1) = 2.          % row 0
18    f(0,2) * f(0,3) * f(1,2) = 75.   f(0,4) = 2.                % row 0
19    f(1,3) * f(2,3) = 2.                                        % row 1
20    f(2,1) * f(2,2) * f(3,1) = 60.   f(2,0) = 5.  f(2,4) = 1.   % row 2
21    f(3,0) * f(4,0) * f(4,1) = 8.    f(3,2) * f(4,2) = 2.       % row 3
22    abs(f(3,3) + −f(3,4)) = 1.                                  % row 3
23    f(4,3) + f(4,4) = 8.                                        % row 4
24  end_of_list.
```

> **Solution**
>
> Let $f(x, x)$ the function contain the values. As we need to fill values from 1 to 5, we need to set a domain size of 6. We put zero in the last line and last column of the 6 × 6 grid:

$$\forall x \, \forall y \, ((x > 4 \vee y > 4) \rightarrow f(x, y) = 0) \tag{10.28}$$

The values we are interested in are from 1 to 5:

$$\forall x \, \forall y \, ((x < 5 \wedge y < 5) \rightarrow f(x, y) > 0) \tag{10.29}$$

The values are distinct on each line in [0..4]:

$$\forall x \, \forall y_1 \, \forall y_2 \, ((x < 5) \wedge (y_1 < 5) \wedge (y_2 < 5) \wedge f(x, y_1) = f(x, y_2) \rightarrow y_1 = y_2) \tag{10.30}$$

The values are distinct in the first five columns:

$$\forall y \, \forall x_1 \forall x_2 \, ((y < 5) \wedge (x_1 < 5) \wedge (x_2 < 5) \wedge f(x_1, y) = f(x_2, y) \rightarrow x_1 = x_2) \tag{10.31}$$

Mace4 found one solution, depicted also in Fig. 10.16.

f	0	1	2	3	4	5
0	1	4	3	5	2	0
1	3	2	5	1	4	0
2	5	3	4	2	1	0
3	2	5	1	4	3	0
4	4	1	2	3	5	0
5	0	0	0	0	0	0

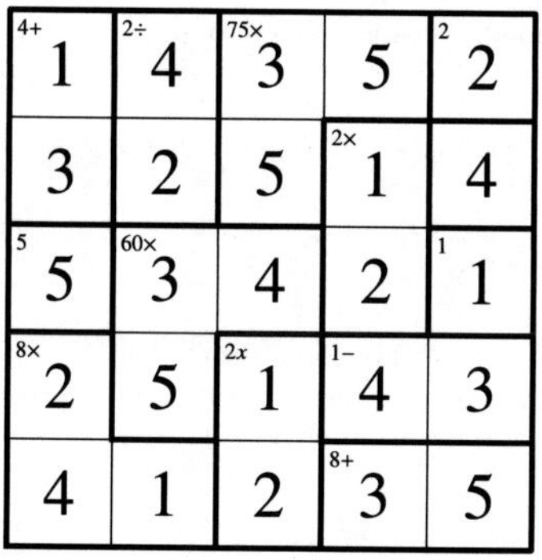

Fig. 10.16 The single model found by Mace4 for Kendoku

Enter the numbers 1 to M into an $n \times n$ grid. Each number can appear only once in
each column and row. Following the labyrinth from the outside inwards, then the
given number sequence must be repeated continuously. (taken from Kleber (2013))
(Fig. 10.17)

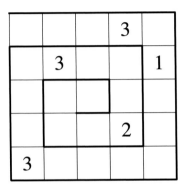

Fig. 10.17 A magic 5 labyrinth to be fill with the sequence 1,2,3

Listing 10.11 Finding models for the magic labyrinth

```
1    set(arithmetic).
2    assign(domain_size,5).
3    assign(max_models,−1).
4
5    formulas(magic_labyrinth).
6       f(x,y) != 4.
7       f(x,y1) !=0 -> (f(x,y1) = f(x,y2)  -> y1 = y2).   % distinct if not empty
8       f(x1,y) !=0 -> (f(x1,y) = f(x2,y)  -> x1 = x2).   % distinct if not empty
9       f(x,0) + f(x,1) + f(x,2) + f(x,3) + f(x,4) = 6.   % 1+2+3=6
10
11      all y1 all y2 ((y1 < y2 & f(0,y1) != 0 & f(0,y2) != 0)        % seg. 1
12         -> f(0,y1) < f(0,y2)).
```

```
13 │  all x1 all x2 ((x1 < x2 & f(x1,4) != 0 & f(x2,4) != 0        % seg. 2
14 │      & f(0,4) != 3) -> f(x1,4) < f(x2,4)).
15 │  all y1 all y2 ((y1 < y2 & f(4,y1) != 0 & f(4,y2) != 0        % seg. 3
16 │      & f(4,4) != 3) -> f(4,y1) > f(4,y2)).
17 │  all x1 all x2 ((x1 < x2 & f(x1,0) != 0 & f(x2,0) != 0        % seg. 4
18 │      & f(4,0) != 3 & x1 != 0) -> f(x1,0) > f(x2,0)).
19 │
20 │  all y1 all y2 ((y1 < y2 & f(1,y1) != 0 & f(1,y2) != 0        % seg. 5
21 │      & f(1,0) != 3 & y2 < 4) -> f(1,y1) < f(1,y2)).
22 │  all x1 all x2 ((x1 < x2 & f(x1,3) != 0 & f(x2,3) != 0        % seg. 6
23 │      & f(1,3) != 3 & x1 > 0 & x2 < 4) -> f(x1,3) < f(x2,3)).
24 │  all y1 all y2 ((y1 < y2 & f(3,y1) != 0 & f(3,y2) != 0        % seg. 7
25 │      & f(3,3) !=3 & y1 > 0 & y2 < 4) -> f(3,y1) > f(3,y2)).
26 │  end_of_list.
27 │
28 │  formulas(sample_puzzle).
29 │    f(0,3) = 3. f(1,1) = 3. f(1,4) = 1. f(3,3) = 2. f(4,0) = 3.
30 │  end_of_list.
```

Solution

Since the grid is 5×5, we set the domain size to 5. Let the function $f(x, y)$ handle the solution. The functions should contain the values from the given sequence 1, 2, 3. We note the empty cells with 0. Hence, the function may have all values from the domain, but 4 (line 6 in Listing 10.11). Each number can appear only once in each column and row. This holds only for the values 1, 2, 3 in the given sequence, but not for the empty cells:

$$f(x, y_1) \rightarrow (f(x, y_1) = f(x, y_2) \rightarrow y_1 = y_2 \qquad (10.32)$$
$$f(x, y_1) \rightarrow (f(x_1, y) = f(x_2, y) \rightarrow x_1 = x_2 \qquad (10.33)$$

Next, we note that each row or column contains exactly two empty cells. One option to formalise this is: $(\exists y\, f(x, y) = 1) \wedge (\exists y\, f(x, y) = 2) \wedge (\exists y\, f(x, y) = 3)$. This works fine, but if one wants to avoid Skolem constants, the following formalisation can be used:

$$f(x, 0) + f(x, 1) + f(x, 2) + f(x, 3) + f(x, 4) = 6 \qquad (10.34)$$

Equation 10.34 together with 10.32 and 10.33 guarantee that each line and column contain only permutations of the sequence $\langle 0, 0, 1, 2, 3 \rangle$. At this step, Mace4 computes 66,240 models.

Next, we split the labyrinth in segments. The first segment is represented by the first row ($x = 0$). In line 0, the sequence $\langle 1, 2, 3 \rangle$ is kept if the function $f(x, y)$ increases as the column y increases. This does not apply to the empty cells. Thus, excepting the empty cells (i.e. $f(0, y) = 0$), the function is monotonically increasing in row 0:

$$\forall y_1\, \forall y_2\, ((y_1 < y_2 \wedge f(0, y_1) \neq 0 \wedge f(0, y_2) \neq 0) \rightarrow f(0, y_1) < f(0, y_2)) \quad (10.35)$$

The second segment is represented by the last column ($y = 4$). Here, excepting the empty cells (i.e. $f(x, 4) = 0$), the function is monotonically increasing as the line x increases:

$$\forall x_1 \, \forall x_2 \, ((x_1 < x_2 \wedge f(x_1, 4) \neq 0 \wedge f(x_2, 4) \neq 0 \wedge f(0, 4) \neq 3) \rightarrow f(x_1, 4) < f(x_2, 4))$$
$$(10.36)$$

Observe here that value 3 can appear at the end of the first row (i.e. $f(0, 4) = 3$). That's why Eq. 10.36 states that the monotonicity of line 4 applies when $f(0, 4) \neq 3$. The third segment represents the last line ($x = 4$). On this line, the function is monotonically decreasing as the column y increases. This does not apply to empty cells or when the value 3 appears in the last column (i.e. $f(4, 4) = 3$):

$$\forall y_1 \, \forall y_2 \, ((y_1 < y_2 \wedge f(4, y_1) \neq 0 \wedge f(4, y_2) \neq 0 \wedge f(4, 4) \neq 3) \rightarrow f(4, y_1) > f(4, y_2))$$
$$(10.37)$$

The fourth segment is smaller: it represents the first column ($y = 0$), excepting the cell $(0,0)$. Here the function is monotonically decreasing as the line x increases:

$$\forall x_1 \, \forall x_2 \, ((x_1 < x_2 \wedge f(x_1, 0) \neq 0 \wedge f(x_2, 0) \neq 0 \wedge f(4, 0) \neq 3 \wedge x1 \neq 0) \rightarrow f(x_1, 0) > f(x_2, 0))$$

Observe once again that the monotonicity does not hold for empty spaces, and when the value 3 appears at the start of this segment.

Solution

The fifth segment is also shorter, as the last column is not included. Here, the values of the function increase as the column y increases:

$$\forall y_1 \, \forall y_2 \, ((y_1 < y_2 \wedge f(1, y_1) \neq 0 \wedge f(1, y_2) \neq 0 \wedge f(1, 0) \neq 3 \wedge y_2 < 4) \rightarrow f(1, y_1) < f(1, y_2))$$

The next segment (line 22–23) is even shorter: it contains three cells from the column $y = 3$. Similar to the second segment, the function increases as the rows increase:

$$\forall x_1 \, \forall x_2 \, ((x_1 < x_2 \wedge f(x_1, 3) \neq 0 \wedge f(x_2, 3) \neq 0 \wedge f(1, 3) \neq 3 \wedge x_1 > 0 \wedge x_2 < 4) \rightarrow f(x_1, 3) < f(x_2, 3))$$

The seventh segment includes three cells from the row $x = 4$. Similar to the third segment, the function decreases as the column increases:

$$\forall y_1 \, \forall y_2 \, ((y_1 < y_2 \wedge f(3, y_1) \neq 0 \wedge f(3, y_2) \neq 0 \wedge f(3, 3) \neq 3 \wedge y_1 > 0 \wedge y_2 < 4) \rightarrow f(3, y_1) > f(3, y_2))$$

Given these constraints and the initial values on line 29, Mace4 computes the single model depicted in Fig. 10.18.

	1	2	3	
2	3			1
		3	1	2
1			2	3
3	2	1		

Fig. 10.18 The solution found by Mace4 for the magic labyrinth

---

Puzzle 106. Stars and arrows

Enter a star in some empty cells of the grid. Each arrow points to at least one star.
Arrows point to a whole row, column, or diagonal, also through other stars and arrows.
The numbers on the left and top of the grid indicate how many stars are located in the
row or column. (taken from Kleber (2013)) (Fig. 10.19)

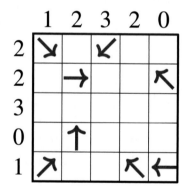

Fig. 10.19 Stars and Arrow sample puzzle of 5 × 5

Solution

We use two functions: $g(x, y)$ states the positions of the arrows, and $f(x, y)$ states if
there is a star or not on row x and column y. Since there are eight distinct arrows

and an empty cell, we need a domain size of 9. We code each arrow with a distinct value: " $\leftarrow$ " $= L = 1$, " $\rightarrow$ " $= R = 2$, " $\uparrow$ " $= U = 3$, " $\downarrow$ " $= D = 4$, " $\diagup$ " $= DL = 5$, " $\diagdown$ " $= DR = 6$, " $\diagdown$ " $= UL = 7$, " $\diagup$ " $= UR = 8$ (lines 14–15 in Listing 10.12). In the given 5×5 grid, all the arrows are within the first 5 rows and 5 columns. For the remaining domain $[5, 8] \times [5, 8]$, we fix the function g to zero: $\forall x\, \forall y\, (x > 4 \vee y > 4 \rightarrow g(x, y) = 0)$. The function $f(x, y) = 0$ for no star, and $f(x, y) = 1$ for star in (x, y) within the $[0, 4] \times [0, 4]$ grid: $\forall x\, \forall y\, (x < 5 \vee y < 5 \rightarrow f(x, y) \le 1$ (lines 6–7).

The number of stars on each row is kept on column $y = 5$, while the number of stars in each column is kept on row $x = 5$ (lines 8–9):

$$f(x, 0) + f(x, 1) + f(x, 2) + f(x, 3) + f(x, 4) = f(x, 5) \tag{10.38}$$

$$f(0, y) + f(1, y) + f(2, y) + f(3, y) + f(4, y) = f(5, y) \tag{10.39}$$

Outside this grid, we fix the function f to zero: $\forall x\, \forall y\, (x > 5 \vee y > 5 \rightarrow f(x, y) = 0)$, and also $f(5, 5) = 0$ (lines 11–12). One constraint states that there are no stars in the cells containing arrows:

$$\forall x\, \forall y\, ((x \le 4 \wedge y \le 4 \wedge g(x, y) \ne 0) \rightarrow f(x, y) = 0) \tag{10.40}$$

Listing 10.12 Finding models for the stars and arrows puzzle

```
1    set(arithmetic).
2    assign(domain_size,9).                    % 8 arrows and an empty space
3    assign(max_models,1).
4
5    formulas(stars_and_arrows).
6      all x all y (x < 5 & y < 5 ->
7        (f(x,y) < 2) &
8        (f(x,0) + f(x,1) + f(x,2) + f(x,3) + f(x,4) = f(x,5)) &
9        (f(0,y) + f(1,y) + f(2,y) + f(3,y) + f(4,y) = f(5,y))).
10
11     all x all y (x > 5 | y > 5 -> f(x,y) = 0).
12     f(5,5)=0.
13
14     N A = 0.  L  = 1.    R = 2.   U = 3.    D = 4.
15              D L  = 5.  D R = 6.  U L = 7.  U R = 8.
16
17     all x all y (x > 4 | y > 4 -> g(x,y) = 0).
18
19     % Where is arrow, there is no star
20     all x all y ((x <= 4 & y <= 4 & g(x,y) != 0) -> f(x,y) = 0).
21
22     % Each arrow points to at least one star
23     g(x,y) = R  -> exists z (z > y & f(x,z) = 1).
24     g(x,y) = L  -> exists z (z < y & f(x,z) = 1).
25     g(x,y) = U  -> exists z (z < x & f(z,y) = 1).
26     g(x,y) = D  -> exists z (z > x & f(z,y) = 1).
27     g(x,y) = D R -> (exists x1 exists y1
```

```
28        (x1 > x & y1 > y & (x1 + −x = y1 + −y) & f(x1,y1) = 1)).
29    g(x,y) = D L  −>  (exists x1 exists y1
30        (x1 > x & y1 < y & (x1 + −x = y + −y1) & f(x1,y1) = 1)).
31    all x all y ((x > 0 & y < 4 & g(x,y) = U R) −> (exists x1 exists y1
32        (x1 < x & y1 > y & (x + −x1 = y1 + −y) & f(x1,y1) = 1))).
33    all x all y ((x > 0 & y > 4 & g(x,y) = U L) −> (exists x1 exists y1
34        (x1 < x & y1 < y & (x + −x1 = y1 + −y) & f(x1,y1) = 1))).
35  end_of_list.
36
37  formulas(sample_puzzle).
38    f(0,5) = 2 & f(1,5) = 2 & f(2,5) = 3 & f(3,5) = 0 & f(4,5) = 1.
39    f(5,0) = 1 & f(5,1) = 2 & f(5,2) = 3 & f(5,3) = 2 & f(5,4) = 0.
40
41    g(0,0) = D R.  g(0,1) = N A.  g(0,2) = D L.  g(0,3) = N A.  g(0,4) = N A.
42    g(1,0) = N A.  g(1,1) = R.    g(1,2) = N A.  g(1,3) = N A.  g(1,4) = U L.
43    g(2,0) = N A.  g(2,1) = N A.  g(2,2) = N A.  g(2,3) = N A.  g(2,4) = N A.
44    g(3,0) = N A.  g(3,1) = U.    g(3,2) = N A.  g(3,3) = N A.  g(3,4) = N A.
45    g(4,0) = U R.  g(4,1) = N A.  g(4,2) = N A.  g(4,3) = U L.  g(4,4) = L.
46  end_of_list.
```

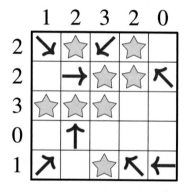

Fig. 10.20 The model found by Mace4 for the stars and arrows puzzle

---

Solution

We know that each arrow points to at least one star. We use a distinct equation for each arrow type. In case of the right arrow in $g(x, y) = $ "$\rightarrow$", there should be a star on the same row x and a right column $z > y$ (line 23):

$$(g(x, y) = \text{ "}\rightarrow\text{ "}) \rightarrow \exists z \, (z > y \wedge f(x, z) = 1) \tag{10.41}$$

In case of the left arrow in $g(x, y) = $ "$\leftarrow$", there should be a star on the same row x and a left column $z < y$ (line 24):

$$(g(x, y) = \text{ "}\leftarrow\text{ "}) \rightarrow \exists z \, (z < y \wedge f(x, z) = 1) \tag{10.42}$$

In case of up arrow $g(x, y) = $ "$\uparrow$", there should be a star on the same column y and an up column $z < x$ (line 25):

$$(g(x, y) = \text{``}\uparrow\text{''}) \rightarrow \exists z \, (z < x \wedge f(z, y) = 1) \qquad (10.43)$$

In case of down arrow $g(x, y) = \text{``}\downarrow\text{''}$, there should be a star on the same column y and a down column $z > x$ (line 25):

$$(g(x, y) = \text{``}\downarrow\text{''}) \rightarrow \exists z \, (z > x \wedge f(z, y) = 1) \qquad (10.44)$$

In case of down-left arrow $g(x, y) = \text{``}\swarrow\text{''}$, there should be a star on the right row $x_1 > x$ and down column $y_1 > y$ (line 27–28):

$$(g(x, y) = \text{``}\swarrow\text{''}) \rightarrow \exists x_1 \, \exists y_1 \, (x_1 > x \wedge y_1 < y \wedge (x_1 - x = y - y_1) \wedge f(x_1, y_1) = 1)) \qquad (10.45)$$

In case of down-right arrow $g(x, y) = \text{``}\searrow\text{''}$, there should be a star on the left row $x_1 < x$ and down column $y_1 < y$ (line 29–30):

$$(g(x, y) = \text{``}\searrow\text{''}) \rightarrow \exists x_1 \, \exists y_1 \, (x_1 < x \wedge y_1 < y \wedge (x_1 - x = y - y_1) \wedge f(x_1, y_1) = 1)) \qquad (10.46)$$

For the "$\nearrow$" and "$\nwarrow$" we avoid situations in which such arrows are on the first row (i.e. $x = 0$) or on the last column of the grid (i.e. $y = 4$). For the "$\nearrow$" on row x and column y, there should be a star on top row $x_1 < x$ and on a right column $y_1 > y$:

$$\forall x \, \forall y \, ((x > 0 \wedge y < 4 \wedge g(x, y) = \text{``}\nearrow\text{''}) \rightarrow \qquad (10.47)$$
$$(\exists x_1 \, \exists y_1 \, (x_1 < x \wedge y_1 > y \wedge (x - x_1 = y_1 - y) \wedge f(x_1, y_1) = 1)))$$

For the "$\nwarrow$" on row x and column y, there should be a star on top row $x1 < x$ and on a left column $y_1 < y$:

$$\forall x \, \forall y \, ((x > 0 \wedge y > 4 \wedge g(x, y) = \text{``}\nwarrow\text{''}) \rightarrow \qquad (10.48)$$
$$(\exists x_1 \, \exists y_1 \, (x_1 < x \wedge y_1 < y \wedge (x - x_1 = y - y_1) \wedge f(x_1, y_1) = 1)))$$

Finally, we formalise the given puzzle. The number of stars on each row x in [0..4] is handled in column $y = 5$ (line 38). The number of stars on each column y in [0..4] is handled in column $x = 5$ (line 39). The arrows positions are completely specified in lines 41–45.

The first model found by Mace4 is depicted in Fig. 10.20.

Puzzle 107. Tents and trees

Draw tents in the grid. Next to each tree, a tent must be entered in a horizontal or vertical adjacent cell, which is associated with this tree. The numbers next to the grid indicate the quantity of tents in each row or column. No tent can stand directly next to another one, not even diagonally. (taken from Kleber (2013))

Solution

We use the function $f(x, y) = 1$ for tent on row x and column y, and $f(x, y) = 0$ for empty space. By defining f as a function instead of predicate, we can use it to compute the number of tents on each row or column, simply by summing the values of f on that row or column. We store the summing values in the last row and last column of the function f. For the given 5×5 puzzle, we use a domain size of 6: the tents will be deployed in the first five rows and columns $[0..4] \times [0..4]$, while the sums are stored on row $x = 5$ and column $y = 5$, as illustrated below:

f	0	1	2	3	4	5
0	–	–	–	–	–	3
1	–	–	–	–	–	0
2	–	–	–	–	–	2
3	–	–	–	–	–	0
4	–	–	–	–	–	2
5	1	2	1	1	2	0

The number of tents on each row is kept on the last column, while the number of tents in each column is kept on the last row:

$$f(x, 0) + f(x, 1) + f(x, 2) + f(x, 3) + f(x, 4) = f(x, 5) \qquad (10.49)$$
$$f(0, y) + f(1, y) + f(2, y) + f(3, y) + f(4, y) = f(5, y) \qquad (10.50)$$

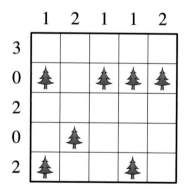

Fig. 10.21 Tents and trees puzzle of 5×5

Listing 10.13 Formalising the tents and trees puzzle

```
1    set(arithmetic).
2    assign(domain_size,6).                    % 5x5 grid
3    assign(max_models,−1).
4
5    formulas(utils).
6       ds1 = domain_size + −1.
7       ds2 = ds1 + −1.
8       x < ds2 −> s(x) = x + 1.      s(ds2) = p(ds2).      s(ds1) = ds1.
9       x > 0   −> p(x) = x + −1.     p(0)  = s(0).
10   end_of_list.
11
12   formulas(tents_and_trees).
13      all x all y (x < ds1 & y < ds1 −>
14         (f(x,y) < 2) &
15         (f(x,0) + f(x,1) + f(x,2) + f(x,3) + f(x,4) = f(x,ds1)) &
16         (f(0,y) + f(1,y) + f(2,y) + f(3,y) + f(4,y) = f(ds1,y))).
17      f(ds1,ds1) = 0.
18
19      t(x,y) −> f(x,y)=0.                     % no tent in the cell with a tree
20
21      % no tent can stand directly next to another one, not even diagonally.
22      all x all y ((x < ds1 & y < ds1 & f(x,y) = 1) −>
23         (f(x,s(y)) = 0 & f(x,p(y)) = 0 & f(s(x),y) = 0 & f(p(x),y) = 0 &
24         f(p(x),p(y)) = 0 & f(p(x),s(y)) = 0 &
25         f(s(x),p(y)) = 0 & f(s(x),s(y)) = 0)).
26   end_of_list.
27
28   formulas(trees_map).
29      all y (−t(0,y) &−t(2,y) &−t(ds1,y)).
30      all x −t(x,ds1).
31      t(1,0) &−t(1,1) &  t(1,2) &  t(1,3) &  t(1,4).
32      −t(3,0) &  t(3,1) &−t(3,2) &−t(3,3) &−t(3,4).
33      t(4,0) &−t(4,1) &−t(4,2) &  t(4,3) &−t(4,4).
34   end_of_list.
35
36   formulas(sample_puzzle).
37      f(0,ds1) = 3 & f(1,ds1) = 0 & f(2,ds1) = 2 & f(3,ds1) = 0 & f(4,ds1)=2.
38      f(ds1,0) = 1 & f(ds1,1) = 2 & f(ds1,2) = 1 & f(ds1,3) = 1 & f(ds1,4)=2.
39   end_of_list.
```

Solution

We use the predicate $t(x, y)$ to signal if there is a tree on row x and column y. One constraint states that there are no tents in the cells containing trees: $t(x, y) \rightarrow f(x, y) = 0$. We also know that no tent can stand directly next to another one, not even diagonally. That is, if there is a tent on row x and column y (i.e. $f(x, y) = 1$), no tent can be placed in any of the eight neighbours. In lines 23–25, each neighbour is modelled by the successor and predecessor functions s and p (defined in lines 8–9). Finally, we formalise the given puzzle. The tree map is modelled in lines 28–34. Line 29 states there is no tree on rows 0, 2, and 5 (i.e. the last one stores the number of tents on each column). Line 30 says there is no tree on the last column (used to store the number of tents on each row). Lines 31–33 specify the trees, as they appear in Fig. 10.21. The number of tents on each row is specified in line 37. The number of tents on each column is modelled in line 38. The single model found by Mace4 is depicted in Fig. 10.22.

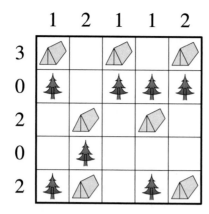

Fig. 10.22 The single model found by Mace4 for the tents and trees puzzle

Puzzle 108. Sun and moon

Enter exactly one star and one dark cloud in each row and each column of the grid, so that the planets are illuminated as specified. The stars shine horizontally or vertically arbitrarily far, but not through a planet or a dark cloud. (taken from Kleber (2013)) (Fig. 10.23)

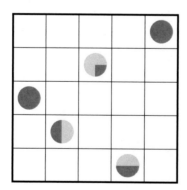

Fig. 10.23 Sun and moon sample puzzle

Solution

First, we need to set the domain size. Since there are 10 possibilities for a planet to shine, plus the case with no planet in a cell, we need a domain size of 11 elements:

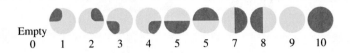

Let the values above for each case, and let the function $p(x, y)$ stating which type of the shining is row x and column y. Note that light comes from above in cases 3, 4, 6, and 9. Light comes from below in cases 1, 2, 5, and 9. Light comes from left in cases 2, 3, 8, and 9. Light comes from right in cases 1, 4, 7, and 9:

$$light\_above(x, y) \leftrightarrow (p(x, y) = 3 \vee p(x, y) = 4 \vee p(x, y) = 6 \vee p(x, y) = 9) \qquad (10.51)$$

$$light\_below(x, y) \leftrightarrow (p(x, y) = 1 \vee p(x, y) = 2 \vee p(x, y) = 5 \vee p(x, y) = 9) \qquad (10.52)$$

$$light\_left(x, y) \leftrightarrow (p(x, y) = 2 \vee p(x, y) = 3 \vee p(x, y) = 8 \vee p(x, y) = 9) \qquad (10.53)$$

$$light\_right(x, y) \leftrightarrow (p(x, y) = 1 \vee p(x, y) = 4 \vee p(x, y) = 7 \vee p(x, y) = 9) \qquad (10.54)$$

There are also four equations for each dark side:

$$dark\_above(x, y) \leftrightarrow (p(x, y) = 5 \vee p(x, y) = 10) \qquad (10.55)$$

$$dark\_below(x, y) \leftrightarrow (p(x, y) = 6 \vee p(x, y) = 10) \qquad (10.56)$$

$$dark\_left(x, y) \leftrightarrow (p(x, y) = 7 \vee p(x, y) = 10) \qquad (10.57)$$

$$dark\_right(x, y) \leftrightarrow (p(x, y) = 8 \vee p(x, y) = 10) \qquad (10.58)$$

Listing 10.14 Finding models for the sun and moon puzzle

```
1   set(arithmetic).
2   assign(domain_size,11).            % ten ways to shine a planet + no planet
3   assign(max_models,-1).
4
5   formulas(utils).
6    gs = 5.     % grid size
7    s  = 1.     % star
8    c  = 2.     % cloud
9    light_above(x,y) <-> (p(x,y) = 3 | p(x,y) = 4 | p(x,y) = 6 | p(x,y) = 9).
10   light_below(x,y) <-> (p(x,y) = 1 | p(x,y) = 2 | p(x,y) = 5 | p(x,y) = 9).
11   light_left(x,y)  <-> (p(x,y) = 2 | p(x,y) = 3 | p(x,y) = 8 | p(x,y) = 9).
12   light_right(x,y) <-> (p(x,y) = 1 | p(x,y) = 4 | p(x,y) = 7 | p(x,y) = 9).
13
14   dark_above(x,y)  <-> (p(x,y) = 5 | p(x,y) = 10).
15   dark_below(x,y)  <-> (p(x,y) = 6 | p(x,y) = 10).
16   dark_left(x,y)   <-> (p(x,y) = 7 | p(x,y) = 10).
17   dark_right(x,y)  <-> (p(x,y) = 8 | p(x,y) = 10).
18  end_of_list.
19
20  formulas(sun_and_moon).
21   p(x,y) != 0 -> f(x,y) = 0.  % where is planet, there is no star or cloud
22   f(x,y) = 0 | f(x,y) = s | f(x,y) = c.
23   all x all y ((x >= gs | y >= gs) -> f(x,y) = 0).
24
25   % at least one star and one dark cloud in each row and each column
```

```
26    all x ((x < gs) -> ((exists y1 (f(x,y1)=s)) & (exists y2 (f(x,y2)=c)))).
27    all y ((y < gs) -> ((exists x1 (f(x1,y)=s)) & (exists x2 (f(x2,y)=c)))).
28
29    %at most one star and one dark cloud in each row and each column
30    all x  all y1 all y2 ((f(x,y1) = s & f(x,y2) = s) -> y1 = y2).
31    all x1 all x2 all y  ((f(x1,y) = s & f(x2,y) = s) -> x1 = x2).
32    all x  all y1 all y2 ((f(x,y1) = c & f(x,y2) = c) -> y1 = y2).
33    all x1 all x2 all y  ((f(x1,y) = c & f(x2,y) = c) -> x1 = x2).
34
35    (light_above(x,y) & f(x1,y) = s) -> (x1 < x &
36                               (all x2 ((x2 > x1 & x2 < x) -> f(x2,y) = 0))).
37    (light_below(x,y) & f(x1,y) = s) -> (x1 > x &
38                               (all x2 ((x2 < x1 & x2 > x) -> f(x2,y) = 0))).
39    (light_left(x,y) & f(x,y1) = s) -> (y1 < y &
40                               (all y2 ((y2 > y1 & y2 < y) -> f(x,y2) = 0))).
41    (light_right(x,y) & f(x,y1) = s) -> (y1 > y &
42                               (all y2 ((y2 < y1 & y2 > y) -> f(x,y2) = 0))).
43
44    (dark_above(x,y)  & f(x1,y) = s & f(x2,y) = c) -> (x1 > x |
45                                                       (x2 < x & x1 < x2)).
46    (dark_below(x,y)  & f(x1,y) = s & f(x2,y) = c) -> (x1 < x |
47                                                       (x2 > x & x1 > x2)).
48    (dark_left(x,y)   & f(x,y1) = s & f(x,y2) = c) -> (y1 > y |
49                                                       (y2 > y1 & y2 < y)).
50    (dark_right(x,y)  & f(x,y1) = s & f(x,y2) = c) -> (y1 < y |
51                                                       (y2 > y & y1 > y2)).
52    end_of_list.
53
54    formulas(planet_latin_square).
55      all x all y ((x >= gs | y >= gs) -> p(x,y) = 0).
56      p(0,4) = 10 & p(1,2) = 3 & p(2,0) = 10 & p(3,1) = 7 & p(4,3) = 6.
57      all x  all y1 all y2 ((p(x,y1) != 0 & p(x,y2) != 0) -> y1 = y2).
58      all x1 all x2 all y  ((p(x1,y) != 0 & p(x2,y) != 0) -> x1 = x2).
59    end_of_list.
```

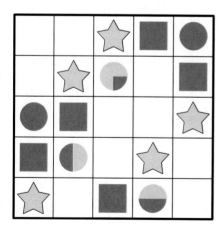 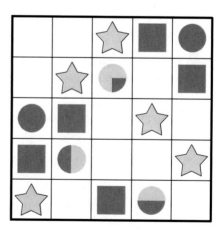

Fig. 10.24 Two models found by Mace4

Solution

We use function $f(x, y)$ to place stars and clouds. There are three values for f: $f(x, y) = 0$ for an empty cell, $f(x, y) = s$ for a star (where $s = 1$) or $f(x, y) = c$ for a cloud (where $c = 2$). Of course, where there is a planet, there is no star or cloud: $p(x, y) \neq 0 \to f(x, y) = 0$.

Since the domain size is larger (i.e. 11) than the grid size $gs = 5$, we fix $f(x, y) = 0$ outside the area of interest:

$$\forall x \, \forall y \, ((x \geq gs \lor y \geq gs) \to f(x, y) = 0) \tag{10.59}$$

We learn there are exactly one star and one dark cloud in each row and column. This is formalised with the "at least one" and "at most one" constraints. First, at least one star and one dark cloud in each row x (Eq. (10.60)) and each column y (Eq. (10.61)), within the interest area $[0..gs] \times [0..gs]$ is modelled with:

$$\forall x \, ((x < gs) \to ((\exists y_1 \, (f(x, y_1) = s)) \land (\exists y_2 \, (f(x, y_2) = c)))) \tag{10.60}$$

$$\forall y \, ((y < gs) \to ((\exists x_1 \, (f(x_1, y) = s)) \land (\exists x_2 \, (f(x_2, y) = c)))) \tag{10.61}$$

Second, at most one star in each row x and each column y:

$$\forall x \, \forall y_1 \, \forall y_2 \, ((f(x, y_1) = s \land f(x, y_2) = s) \to y_1 = y_2) \tag{10.62}$$

$$\forall x_1 \forall x_2 \, \forall y \, ((f(x_1, y) = s \land f(x_2, y) = s) \to x_1 = x_2) \tag{10.63}$$

Note, there is no need to focus on the area of interest, as outside $[0..gs] \times [0..gs]$, $f(x, y) = 0$, hence no stars here. The same at most constraint is also stated for clouds (lines 32–33).

Next, we specify the logical constraints of the puzzle. First, if light comes from above the cell (x, y) and the unique star is at $f(x_1, y) = 1$ then the star is above (x, y) (i.e. row $x_1 < x$) and all rows x_2 between x (the planet) and x_1 (the star) are empty:

$(light\_above(x, y) \land f(x_1, y) = s) \to (x_1 < x \land (\forall x_2 \, ((x_2 > x_1 \land x_2 < x) \to f(x_2, y) = 0)))$

Second, if light comes from below of the cell (x, y) and the unique star is at $f(x_1, y) = 1$ then the star is below (x, y) (i.e. row $x_1 > x$) and all rows x_2 between x (the planet) and x_1 (the star) are empty:

$light\_below(x, y) \land f(x_1, y) = s \to (x_1 > x \land (\forall x_2 \, ((x_2 < x_1 \land x_2 > x) \to f(x_2, y) = 0)))$ (10.64)

Third, if light comes from the left of the cell (x, y) and the unique star is at $f(x, y_1) = 1$ then the star is on the left of cell (x, y) (i.e. column $y_1 < y$) and all columns y_2 between y_1 (the star) and y (the planet) are empty:

$(light\_left(x, y) \land f(x, y_1) = s) \to (y_1 < y \land (\forall y_2 \, ((y_2 > y_1 \land y_2 < y) \to f(x, y_2) = 0)))$ (10.65)

Fourth, if light comes from the right of the cell (x, y) and the unique star is at $f(x, y_1) = 1$ then the star is on the right of cell (x, y) (i.e. column $y_1 > y$) and all columns y_2 between y (the planet) and y_1 (the star) are empty:

$(light\_right(x, y) \land f(x, y_1) = s) \to (y_1 > y \land (\forall y_2 \, ((y_2 < y_1 \land y_2 > y) \to f(x, y_2) = 0)))$ (10.66)

> **Solution**
>
> There are also four cases when the planet is dark. When there is dark above the cell (x, y) and the star situated at row x_1 and column y (i.e. $f(x_1, y) = s$) and the cloud at $f(x_2, y) = c$) then either the star is below the planet (i.e. row $x_1 > x$) or the cloud is between the star and the planet ($x_1 < x_2 < x$).
>
> $$(dark\_above(x, y) \wedge f(x_1, y) = s \wedge f(x_2, y) = c) \rightarrow (x_1 > x \vee (x_2 < x \wedge x_1 < x_2)) \quad (10.67)$$
>
> For the remaining three cases (i.e. $dark\_below$, $dark\_left$, $dark\_right$), the formalisation appears in lines 46–51.
>
> Finally, we formalise the initial state of the puzzle. Outside the zone of interest, there is no planet:
>
> $$\forall x \, \forall y \, ((x \geq gs \vee y \geq gs) \rightarrow p(x, y) = 0) \quad (10.68)$$
>
> One option is to explicitly state the planets and the empty spaces. A second modelling option is based on the observation that there is exactly one planet both on each row and each column:
>
> $$\forall x \, \forall y_1 \, \forall y_2 \, ((p(x, y_1) \neq 0 \wedge p(x, y_2) \neq 0) \rightarrow y_1 = y_2) \quad (10.69)$$
> $$\forall x_1 \forall x_2 \, \forall y((p(x_1, y) \neq 0 \wedge p(x_2, y) \neq 0) \rightarrow x_1 = x_2) \quad (10.70)$$
>
> For instance, Eq. (10.69) says there is a planet on row x and column y_1 (i.e. $(p(x, y_1) \neq 0)$ and there is also a planet on the same row but a different column y_2, then the columns y_1 and y_2 must be the same. Based on this modelling, it is enough to specify only the five planets and the reasoner will deduce that the other cells are empty. The given planets are
>
> $$p(0, 4) = 10 \wedge p(1, 2) = 3 \wedge p(2, 0) = 10 \wedge p(3, 1) = 7 \wedge p(4, 3) = 6 \quad (10.71)$$
>
> Two solutions found by Mace4 are depicted in Fig. 10.24 Note that both models would not be valid if the puzzle would specify that the stars shine also on the diagonals.

References

Darling, D. (2004). *The universal book of mathematics from abracadabra to Zeno's paradoxes.* Wiley.

Jussien, N. (2007). *A to Z of Sudoku.* ISTE Ltd.

Kleber, J. (2013). *A style for typesetting logic puzzles.*

Chapter 11
Russian Puzzles

Abstract In which we take the reader on a trip to Russia, by modelling the "Moscow puzzles" from Kordemsky's book Kordemsky (1992). The book is a collection of 359 diverse puzzles: cryptoarithmetics, geometrical, dominos, magic squares, and combinatorial, which makes Martin Gardner call it "marvelously varied".

$\forall x\ \forall y\ (old(x) \rightarrow belive(x, y))$
$\forall x\ \forall y\ (middleAged(x) \rightarrow suspect(x, y))$
$\forall x\ \forall y\ (young(x) \rightarrow know(x, y))$

Oscar Wilde

This chapter takes you on a trip to Russia, by modelling the "Moscow puzzles" from Kordemsky's book Kordemsky (1992). The book is a collection of 359 diverse puzzles: cryptoarithmetics, geometrical, dominos, magic squares, and combinatorial, which makes Martin Gardner call it "marvelously varied". The aim of this collection was mainly didactic, as Kordemsky states in the introduction: "All materials in this book are devoted to the educational aim—to spur creative thinking, to further perfection of mathematical knowledge".

The trip starts with a puzzle having a Soviet scenery: the young communist boys and girls have to decorate a hydroelectric powerhouse. Given 12 flags, their task is to arrange 6 flags on each side.

The next stops on the trip are even more difficult: you are asked to find all the solutions to some combinatorial problems. Finding all solutions requires a disciplined mind. Instead, this is an adequate task for Mace4, given a proper formalisation of the puzzle. However, a complete and correct formalisation is not always easy to design. But once designed, the model looks simple.

Another stop on the trip checks your abilities as military commander. Your forces are arranged so that 11 boys defend each side of a fort. After the first assault you "lost" four boys, after the second you lost another four, after the third one you lost another four, and so in the fourth assault. You also lost two boys in the fifth attempt of the enemy to conquer the fort. Interestingly, due to your military (more exactly your logical) abilities, you managed to keep in each assault exactly 11 boys to protect each side of the fort. Mace4 finds the solution based on five linear equations.

Let's start our journey to fascinating Russia.

> **Puzzle 109. Arranging flags**
>
> Komsomol youths have built a small hydroelectric powerhouse. Preparing for its open-ing, young communist boys and girls are decorating the powerhouse on all four sides with garlands, electric bulbs, and small flags. There are 12 flags. At first they arrange the flags 4 on a side, as shown, but then they see that the flags can be arranged 5 or even 6 on a side. How? (puzzle 19 from Kordemsky (1992)) (Fig. 11.1)

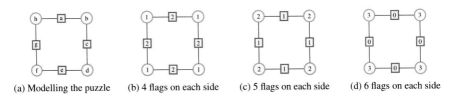

(a) Modelling the puzzle   (b) 4 flags on each side   (c) 5 flags on each side   (d) 6 flags on each side

Fig. 11.1 Arranging flags

Listing 11.1 Finding models for flags arranged by young communists

```
1    set(arithmetic).
2    assign(domain_size,7).
3    assign(max_models,-1).
4
5    formulas(arranging_flags).
6      a + b + c + d + e + f + g + h = 12.  % maximum 12 flags
7      i = 6. % i = 5. % i=4                % number of flags for each side
8
9      a + b + h = i.          b + c + d = i.
10     d + e + f = i.          f + g + h = i.
11
12     h=b & b=f & f=d.        % each corner has the same number of flags
13     a=c & c=e & e=g.        % each middle has the same number of flags
14   end_of_list.
```

(Solution)

There are eight positions in which the flags can be placed, noted by $a, b, c, d, e, f, g,$ and h. There are 12 flags available (line 6 in Listing 11.1). Let i the number of flags on each side. The variable i takes successively three values: 4, 5, 6 (line 7). When $i = 6$ the domain size should be set to 7. Note that all the other variables are less than 6, since their sum should be 6. The puzzle is modelled by the four equations in lines 9–10.

Without considering symmetry, there are 7 solutions for arranging 6 flags on each side:

Model	a	b	c	d	e	f	g	h	i
1	0	0	0	6	0	0	0	6	6
2	0	1	0	5	0	1	0	5	6
3	0	2	0	4	0	2	0	4	6
4	0	6	0	0	0	6	0	0	6
5	**0**	**3**	**0**	**3**	**0**	**3**	**0**	**3**	**6**
6	0	4	0	2	0	4	0	2	6
7	0	5	0	1	0	5	0	1	6

If we add the symmetry conditions then each corner and middle position should have the same number of flags (lines 26–28 and 31–33). In this case only the model 5 remains.

If 5 flags need to be arranged ($i = 5$), Mace4 finds 37 models without symmetry conditions, respectively, 1 model when symmetry conditions are active:

Model	a	b	c	d	e	f	g	h	i
1	1	2	1	2	1	2	1	2	5

For $i = 4$ flags on each side, Mace4 finds the following model:

Model	a	b	c	d	e	f	g	h	i
1	2	1	2	1	2	1	2	1	4

(Puzzle 110. Keep it even)

Take 16 objects (pieces of paper, coins, plums, checkers) and put them in 4 rows of 4 each. Remove 6, leaving an even number of objects in each row and each column. How many solutions are there? (puzzle 21 from Kordemsky (1992)) (Fig. 11.2)

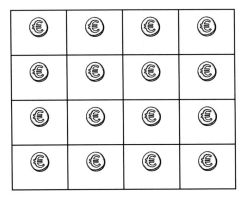

Fig. 11.2 Remove 6 and leave even number in each row and column

Listing 11.2 Finding models for removing 6 items

```
1    set(arithmetic).
2    assign(domain_size,4).
3    assign(max_models,-1).
4
5    formulas(assumptions).
6      all x (evenCol(x) <-> (f(0,x) + f(1,x) + f(2,x) + f(3,x)) mod 2 = 0).
7      all x (evenRow(x) <-> (f(x,0) + f(x,1) + f(x,2) + f(x,3)) mod 2 = 0).
8
9      evenCol(0) & evenCol(1) & evenCol(2) & evenCol(3).
10     evenRow(0) & evenRow(1) & evenRow(2) & evenRow(3).
11
12     all x all y (f(x,y) = 0 | f(x,y) = 1).
13
14     f(0,0) + f(1,0) + f(2,0) + f(3,0) +
15     f(0,1) + f(1,1) + f(2,1) + f(3,1) +
16     f(0,2) + f(1,2) + f(2,2) + f(3,2) +
17     f(0,3) + f(1,3) + f(2,3) + f(3,3) = 10.
18   end_of_list.
```

Solution

We define the predicate $evenCol(x)$ being true when the number of coins on column x is even: $\forall x \ (evenCol(x) \leftrightarrow (f(0,x) + f(1,x) + f(2,x) + f(3,x)) \bmod 2 = 0)$. Similarly, $evenRow(x)$ is true when the number of coins on line x is even:

$$\forall x \ (evenRow(x) \leftrightarrow (f(x,0) + f(x,1) + f(x,2) + f(x,3)) \bmod 2 = 0) \quad (11.1)$$

Here the function $f(x, y) = 1$ if cell (x, y) contains a coin, and 0 otherwise. Hence, we restrict its range to 0 and 1: $\forall x \ \forall y \ (f(x, y) = 0 \lor f(x, y) = 1)$. After removing 6 coins, the remaining value is 10 (lines 14–17 in Listing 11.2).
Mace4 finds 96 models, with 10 coins. Six such models are depicted in Fig. 11.3:

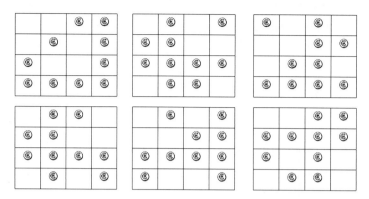

Fig. 11.3 Six models with 10 coins remaining (even number on each row and column)

Puzzle 111. A magic triangle

I have placed the numbers 1, 2, and 3 at the vertices of a triangle. Arrange 4, 5, 6, 7, 8, and 9 along the sides of the triangle so that the numbers along each side add to 17. This is harder: without being told which numbers to place at the vertices, make a similar arrangement of the numbers from 1 through 9, adding up to 20 along each side. (Several solutions are possible.) (puzzle 22 from Kordemsky (1992)) (Fig. 11.4)

Listing 11.3 Sum to 17 and vertices are $A = 1$, $B = 2$, and $C = 3$

```
1   set(arithmetic).
2   assign(domain_size,10).
3   assign(max_models,−1).
4
5   list(distinct).
6     [AB1,AB2,AC1,AC2,BC1,BC2,0,1,2,3].   %some values are taken by A, B, C
7   end_of_list.
8
9   formulas(assumptions).
10    A = 1 & B = 2 & C = 3.                %initial values
11    A + B + AB1 + AB2 = 17.               %values on AB sum to 17
12    A + C + AC1 + AC2 = 17.               %values on AC sum to 17
13    B + C + BC1 + BC2 = 17.               %values on BC sum to 17
14  end_of_list.
```

Listing 11.4 Sum to 20 and no constraints on the vertices

```
1   set(arithmetic).
2   assign(domain_size,10).
3   assign(max_models,−1).
4
5   list(distinct).
6      [A,B,C,AB1,AB2,AC1,AC2,BC1,BC2,0].
7   end_of_list.
8
9   formulas(assumptions).
10     A + B + AB1 + AB2 = 20.
11     A + C + AC1 + AC2 = 20.
12     B + C + BC1 + BC2 = 20.
13  end_of_list.
```

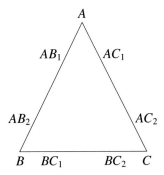

Fig. 11.4 Numbers along each side should add to 17

Solution

Since the vertices have values 1, 2, and 3, you need at least two more numbers on each edge to reach 17. Let AB_1 and AB_2 the values for the edge AB, AC_1 and AC_2 the values for the edge AC, and BC and BC_2 the values for the edge BC. The numbers are from 1 to 9 so we set the domain size at 10 (see Listing 11.3). One cannot put the same number on different edges, hence they are distinct. Also, the numbers 0, 1, and 2 are already taken by the vertices A, B, and C.

There are 16 models for the sum of 17:

Model	AB_1	AB_2	AC_1	AC_2	BC_1	BC_2
1	**5**	**9**	**6**	**7**	**4**	**8**
2	5	9	6	7	8	4
3	5	9	7	6	4	8
4	5	9	7	6	8	4
5	**6**	**8**	**4**	**9**	**5**	**7**
6	6	8	4	9	7	5
7	6	8	9	4	5	7
8	6	8	9	4	7	5
9	9	5	6	7	4	8
10	9	5	6	7	8	4
11	9	5	7	6	4	8
12	9	5	7	6	8	4
13	8	6	4	9	5	7
14	8	6	4	9	7	5
15	8	6	9	4	5	7
16	8	6	9	4	7	5

Among the above models, only two solutions are non-isomorphic: 1 and 5. These models are depicted in Fig. 11.5.

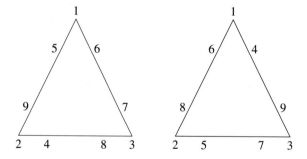

Fig. 11.5 Two models for a magic triangle with sum equal to 17

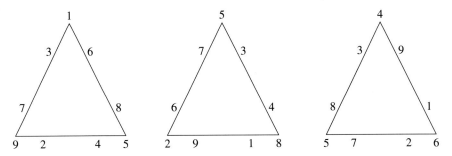

Fig. 11.6 Three models for a magic triangle with sum equal to 20

Puzzle 112. From 1 through 19

Write the numbers from 1 through 19 in the circles so that the numbers in every 3 circles on a straight line sum up to 30. (puzzle 34 from Kordemsky (1992)) (Fig. 11.7)

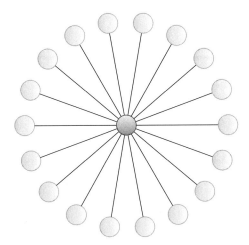

Fig. 11.7 Sum of each straight line should be 30

Listing 11.5 Modelling the "1 to 19" puzzle in propositional logic

```
1   set(arithmetic).
2   assign(domain_size,20).
3   assign(max_models,1).
4
5   list(distinct).
6   [t1, t2, t3, t4, t5, t6, t7, t8, t9,
7    t10, t11, t12, t13, t14, t15, t16, t17, t18, center, 0].
8   end_of_list.
9
10  formulas(assumptions).
11  t1 = 1.                        %avoid isomorphisms
12  t1 + t10 + center = 30.        t2 + t11 + center = 30.
13  t3 + t12 + center = 30.        t4 + t13 + center = 30.
14  t5 + t14 + center = 30.        t6 + t15 + center = 30.
15  t7 + t16 + center = 30.        t8 + t17 + center = 30.
16  t9 + t18 + center = 30.
17  end_of_list.
```

Listing 11.6 Finding the possible values for the center

```
1   set(arithmetic).
2   assign(domain_size,22).
3   assign(max_models,-1).
4
5   formulas(assumptions).
6   9 * Sum + Center = 190.
7   Center < 20.
8   end_of_list.
```

Solution

A straightforward model in propositional logic appears in Listing 11.5. Since there are 19 values available, the domain is set to 20. The numbers are distinct and they start from 1 to 19. The sum between two opposite numbers and the center is 30 (e.g. $t_1 + t_{10} + center = 30$). To avoid some isomorphic models, we assign a specific value for a random node (e.g. $t_1 = 1$). Similarly, one can specify an order relation between elements in each sum (hence we do not care about models in which, for instance, $t_1 = 11, t_{10} = 9, centre = 10$ and $t_1 = 9, t_{10} = 11, centre = 10$. The first computed model has $centre = 10$ and

t_1	t_2	t_3	t_4	t_5	t_6	t_7	t_8	t_9	t_{10}	t_{11}	t_{12}	t_{13}	t_{14}	t_{15}	t_{15}	t_{17}	t_{18}
1	18	17	16	15	14	13	12	11	19	2	3	4	5	6	7	8	9

Listing 11.7 Modelling "1 to 19" puzzle in FOL, here the center is fixed to the value 1

```
1    set(arithmetic).
2    assign(domain_size,20).
3    assign(max_models,1).
4
5    formulas(assumptions).
6      %let the position of each circle to be 0 to 18
7      %opposing circles have a difference of 9
8      all x all y (abs(x+ −y) = 9 & x<18 & y<18 <−> opus(x,y) & opus(y,x)).
9
10     %otherwise, the circles are not opposite
11     all x all y (abs(x+ −y) != 9 | x>17 | y>17 <−> −opus(x,y) & −opus(y,x)).
12
13     %sum of two opposite cicles is the same
14     %for center = 1 the sum is (190−1)/9=21.
15     all x all y (opus(x,y) −> t(x) + t(y) = 21).
16
17     %each circle contains distinct values
18     all x all y (x!=y & x <= 17 & y <= 17 −> t(x) != t(y)).
19
20     %the value are from 1 to 19, but 1 is already taken by the center
21     all x (x <= 17 −> t(x) > 1).
22
23     t(18)=0.                                      %avoid isomorphisms
24     all x all y (x <= 8 & y > 8) −> t(x) < t(y).  %avoid isomorphisms
25   end_of_list.
```

Solution

Note that once the value for centre is found, the problem is trivial. One question is which are the possible values for the centre for the problem to be satisfiable?
The centre of the circle is contained in each line. Hence, the relaxed problem is to find 9 distinct pairs of two numbers from 1 to 19 such that their sum is the same, let's say *Sum*. Hence, $9 * Sum$ plus the value in centre should equal the sum of all available 19 numbers. That is:

$$Sum = 1 + 2 + 3 + \cdots + 18 + 19 = \frac{19 * 20}{2} = 190 \qquad (11.2)$$

Thus, the constraint for the centre value is:

$$9 * Sum + Centre = 190 \qquad (11.3)$$

The implementation of this sub-problem appears in Listing 11.6. Here, the maximum value for *Sum* is obtained when $Centre = 1$:

$$Sum = \frac{190 - Centre}{9} \leq \frac{189}{9} \leq 21 \qquad (11.4)$$

Hence, the domain size is set to 22. Mace4 outputs 3 models for the centre value:

Model	Sum	Centre
1	19	19
2	20	10
3	21	1

Now we can reduce the search space by instructing Mace4 to search for models where centre is 1, 10, or 19 only. Hence, there are three cases to follow:

1. $Centre = 1 \Rightarrow t(x) \in \{1, 2, ..., 19\} \setminus \{1\}$
2. $Centre = 10 \Rightarrow t(x) \in \{1, 2, ..., 19\} \setminus \{10\}$
3. $Centre = 19 \Rightarrow t(x) \in \{1, 2, ..., 19\} \setminus \{19\}$

In the following, we will use a formalisation in first-order logic (see Listing 11.7). We note the position of each circle from 0 to 17. The difference between two opposite circles equals 9 (line 8 in Listing 11.7):

$$\forall x \, \forall y \, |x - y| = 9 \wedge x \leq 17 \wedge y \leq 17 \Rightarrow opus(x, y) \wedge opus(y, x) \qquad (11.5)$$

The function $t(x)$ returns the value at position x. For the first case (i.e. $Centre = 1$ and $Sum = 21$), the domain and range of the function circle are: $t(x) : \{0, ..., 17\} \rightarrow \{1, ..19\} \setminus \{1\}$. The range of the function are values from 1 to 19, but 1 is already taken by the centre: $\forall x \, (x \leq 17) \rightarrow (t(x) \neq 1)$.

Next we avoid some isomorphic models, with two constraints. First, we fix a starting point with $t(18) = 0$. Second, we specify that positions from 0 to 8 should contain the smaller value in the sum. Hence, positions from 9 to 17 will contain the larger value in the sum:

$$\forall x \, \forall y \, (x \leq 8 \wedge y > 8) \rightarrow t(x) < t(y) \qquad (11.6)$$

That is, we do not distinguish between, e.g. $1 + 5 + 15 = 21$, respectively, $1 + 15 + 5 = 21$.

Solution

For each case, one solution is listed below:

C	Sum	0	1	2	3	4	5	6	7	8	9	10	11	12	13	14	15	16	17
1	21	2	3	4	5	6	7	8	9	10	19	18	17	16	15	14	13	12	11
10	20	1	2	3	4	5	6	7	8	9	19	18	17	16	15	14	13	12	11
19	19	1	2	3	4	5	6	7	8	9	18	17	16	15	14	13	12	11	10

Three solutions for each possible centre value are depicted in Fig. 11.8.

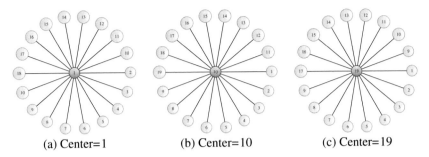

(a) Center=1   (b) Center=10   (c) Center=19

Fig. 11.8 Three solutions for each possible values of the centre

---

Puzzle 113. A duel in arithmetic

The Mathematics Circle in our school had this custom: Each applicant was given a simple problem to solve a little mathematical nut to crack, so to speak. You could become a full member only if you solved the problem. An applicant named Vitia was given this array:

$$
\begin{array}{ccc}
1 & 1 & 1 \\
3 & 3 & 3 \\
5 & 5 & 5 \\
7 & 7 & 7 \\
9 & 9 & 9 \\
\end{array}
$$

He was asked to replace 12 digits with zeros so that the sum would be 20. Vitia thought a little, then wrote rapidly two solutions:

$$
\begin{array}{ll}
\begin{array}{l}
0\ 1\ 1+ \\
0\ 0\ 0+ \\
0\ 0\ 0+ \\
0\ 0\ 0+ \\
\underline{0\ 0\ 9} \\
\ \ \ 2\ 0
\end{array}
&
\begin{array}{l}
0\ 1\ 0+ \\
0\ 0\ 3+ \\
0\ 0\ 0+ \\
0\ 0\ 7+ \\
\underline{0\ 0\ 0+} \\
\ \ \ 2\ 0
\end{array}
\end{array}
$$

He smiled and said: "If you substitute just ten zeros for digits, the sum will be 1,111. Try it!" The Circle's president was taken aback briefly. He not only solved Vitia's problem, but also he improved it: "Why not replace only 9 digits with zeros and still get 1,111?" As the debate continued, ways of getting 1,111 by replacing 8, 7, 6, and 5 digits with zeros were found. Solve the six forms of this problem. (puzzle 48 from Kordemsky (1992))

Listing 11.8 Modelling a duel in arithmetic in Mace4

```
1   set(arithmetic).
2   assign(domain_size,2).
3   assign(max_models,-1).
4
5   formulas(assumptions).
6      1 * 100 * A1 + 1 * 10 * A2 + 1 * 1 * A3 +
7      3 * 100 * B1 + 3 * 10 * B2 + 3 * 1 * B3 +
8      5 * 100 * C1 + 5 * 10 * C2 + 5 * 1 * C3 +
9      7 * 100 * D1 + 7 * 10 * D2 + 7 * 1 * D3 +
10     9 * 100 * E1 + 9 * 10 * E2 + 9 * 1 * E3 = 1111.
11
12     % substitute 10 digits with zeros: 15-10 = 5 digits remaining.
13     A1+A2+A3 + B1+B2+B3 + C1+C2+C3 + D1+D2+D3 + E1+E2+E3 = 5.
14
15     %(substitute 9 digists with zeros)
16     %A1+A2+A3+ B1+B2+B3+ C1+C2+C3 + D1+D2+D3 + E1+E2+E3 = 6.
17  end_of_list.
```

Solution

Let k be the number of replaced digits. The first solution uses the following formalisation:

$$
\begin{array}{rcccc}
1 * & A_1 & A_2 & A_3 & + \\
3 * & B_1 & B_2 & B_3 & + \\
5 * & C_1 & C_2 & C_3 & + \\
7 * & D_1 & D_2 & D_3 & + \\
9 * & E_1 & E_2 & E_3 & \\
\hline
& 1 & 1 & 1 & 1
\end{array}
$$

All the 15 variables A_i, B_i, C_i, D_i, and E_i ($i \in [1..3]$) can be 0 or 1. Therefore, we set the domain size at 2. The important observation here is that when all the variables occur their sum is 15. For $k = 10$ substituted digits, the remaining sum is $15 - k = 5$. Mace4 outputs two solutions:

Model	A_1	A_2	A_3	B_1	B_2	B_3	C_1	C_2	C_3	D_1	D_2	D_3	E_1	E_2	E_3
1	0	0	0	0	1	0	0	0	1	0	1	1	1	1	1
2	1	1	1	1	0	0	0	0	0	1	0	0	0	0	0

These models correspond to the following two solutions:

$$
\begin{array}{rccc}
1 * & 0 & 0 & 0 & + \\
3 * & 0 & 1 & 0 & + \\
5 * & 0 & 0 & 1 & + \\
7 * & 0 & 1 & 1 & + \\
9 * & 1 & 1 & 1 & \\
\hline
& 1 & 1 & 1 & 1
\end{array}
\qquad
\begin{array}{rccc}
1 * & 1 & 1 & 1 & + \\
3 * & 1 & 0 & 0 & + \\
5 * & 0 & 0 & 0 & + \\
7 * & 1 & 0 & 0 & + \\
9 * & 0 & 0 & 0 & \\
\hline
& 1 & 1 & 1 & 1
\end{array}
$$

When replacing $k = 9$ digits, the sum of the variables should be 6, for $k = 8$ digits the sum is 7, for $k = 7$ the sum is 8, for $k = 6$ the sum is 9, and for $k = 5$ the sum is 10.

Substituted digits (k)	10	9	8	7	6	5	4
Number of models	2	5	6	3	1	1	0

Note that for $k = 4$ there is no solution. Next, a solution is shown for each situation $k \in \{9, 8, 7, 6, 5\}$.

```
      k=9                          k=8                          k=7
1 *  1  1  1  +          1 *  1  1  1  +          1 *  1  1  0  +
3 *  0  1  0  +          3 *  0  0  1  +          3 *  0  1  0  +
5 *  0  0  0  +          5 *  0  0  0  +          5 *  0  1  1  +
7 *  0  1  0  +          7 *  0  0  1  +          7 *  0  0  1  +
9 *  1  0  0            9 *  1  1  0            9 *  1  0  1
    ───────────             ───────────             ───────────
     1  1  1  1             1  1  1  1             1  1  1  1

            k=6                             k=5
   1 *  1  0  0  +              1 *  1  1  1  +
   3 *  1  1  0  +              3 *  1  1  1  +
   5 *  1  0  1  +              5 *  1  0  0  +
   7 *  0  1  1  +              7 *  0  1  1  +
   9 *  0  1  1                9 *  0  1  0
       ───────────────            ───────────────
        1  1  1  1                1  1  1  1
```

Puzzle 114. Twenty

There are three ways to add four odd numbers and get 10:

$$1 + 1 + 3 + 5 = 10$$
$$1 + 1 + 1 + 7 = 10$$
$$1 + 3 + 3 + 3 = 10$$

Changes in the order of numbers do not count as new solutions. Now add eight odd numbers to get 20. To find all 11 solutions you will need to be systematic. (puzzle 49 from Kordemsky (1992))

Listing 11.9 Add eight odd numbers to get 20

```
1   set(arithmetic).
2   assign(domain_size,14).
3   assign(max_models,−1).
4
5   formulas(assumptions).
6     A + B + C + D + E + F + G + H = 20.
7
8     odd(x) <-> x mod 2 = 1.
9     odd(A) & odd(B) & odd(C) & odd(D) & odd(E) & odd(F) & odd(G) & odd(H).
10
11    A >= B & B >= C & C >= D & D >= E & E >= F & F >= G & G >= H.
12  end_of_list.
```

Solution

Let $[A..H]$ be the eight numbers which sum to 20 (line 6 in Listing 11.9). We need to define the *odd* predicate (line 8), and to constrain the values to be odd (line 9). Since the minimum odd value is 1, and there are 8 variables that should be summed to 20, then each value cannot be larger than 13. Hence, we restrict the domain size to values from the interval [0..13] (line 2). Mace4 computes the following 11 models:

Model	A	B	C	D	E	F	G	H
1	3	3	3	3	3	3	1	1
2	5	3	3	3	3	1	1	1
3	5	5	3	3	1	1	1	1
4	5	5	5	1	1	1	1	1
5	7	3	3	3	1	1	1	1
6	7	5	3	1	1	1	1	1
7	7	7	1	1	1	1	1	1
8	9	3	3	1	1	1	1	1
9	9	5	1	1	1	1	1	1
10	11	3	1	1	1	1	1	1
11	13	1	1	1	1	1	1	1

Puzzle 115. Order the numbers

The diagram shows 1 through 10 (in order) at the tips of five diameters. Only once does the sum of two adjacent numbers equal the sum of the opposite two numbers: $10 + 1 = 5 + 6$. Elsewhere, for example, $1 + 2 \neq 6 + 7$ or $2 + 3 \neq 7 + 8$. Rearrange the numbers so that all such sums are equal. You can expect more than one solution to this problem. How many basic solutions are there? How many variants (not including simple rotations of variants)? (puzzle 51 from Kordemsky (1992)) (Fig. 11.9)

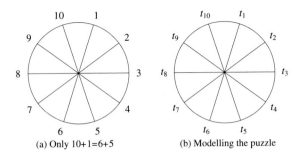

Fig. 11.9   Order the numbers

Listing 11.10 Finding models for ordering 10 numbers on a circle

```
1    set(arithmetic).
2    assign(domain_size,11).
3    assign(max_models,−1).
4
5    list(distinct).
6        [t1,t2,t3,t4,t5,t6,t7,t8,t9,t10,0].
7    end_of_list.
8
9    formulas(assumptions).
10       t1 + t2 = t6 + t7.          t2 + t3 = t7 + t8.
11       t3 + t4 = t8 + t9.          t4 + t5 = t9 + t10.
12       t5 + t6 = t10 + t1.
13       t10=10.
14   end_of_list.
```

Solution

We need ten distinct variables t_i, with i from 1 to 10. Hence, we set the domain size to 11 and we constrain the variables to be distinct one of each other, and different from zero (line 6). Next, we state that two adjacent variables should have the same sum with their corresponding variables from the end of the diameter. That is, $t_i + t_{i+1} = t_{i+5} + t_{i+5+1}$ for $i \in [1..5]$ and $t_{11} = t_1$.

For the equations in lines 10–12, Mace4 computes 480 models. To filter isomorphic models (i.e. rotated solutions), we can fix a variable (e.g. $t_{10} = 10$, line 13). In this case, Mace4 outputs 48 solutions. Two such solutions appear in Fig. 11.10.

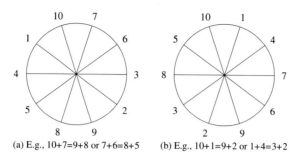

(a) E.g., 10+7=9+8 or 7+6=8+5   (b) E.g., 10+1=9+2 or 1+4=3+2

Fig. 11.10 Two solutions among 48 found by Mace4

Puzzle 116. A mysterious box

Misha brought a pretty little box for his sister Irochka from his Crimean summer camp. She was not of school age yet, but could count to 10. She liked the box because she could count 10 sea shells along each side, as shown. One day, Irochka's mother, while cleaning the box, accidentally broke 4 shells. "No great trouble", Misha said. He unstuck some of the remaining 32 shells, then pasted them on so that there were again 10 shells along each side of the cover, and the box was as symmetrical as before. How did he do it? A few days later, when the box fell on the floor and 6 more shells were crushed, Misha again redistributed the shells—though not quite so symmetrically—so Irochka could count 10 on each side. Can you find one of the many solutions? (puzzle 100 from Kordemsky (1992)) (Fig. 11.11)

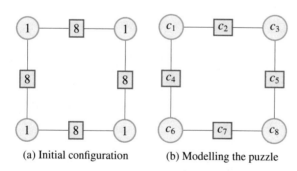

(a) Initial configuration   (b) Modelling the puzzle

Fig. 11.11 A mysterious box

Listing 11.11 Finding models for the mysterious box

```
1   assign(domain_size,9).
2   assign(max_models,-1).
3   set(arithmetic).
4
5   list(distinct).
6     [c1,0].  [c2,0].  [c3,0].  [c4,0]. [c5,0].  [c6,0].  [c7,0].  [c8,0].
7   end_of_list.
8
9   formulas(assumptions).
10    (c2 = c4) & (c2 = c5) & (c2 = c7) & ((c1 = c8) | (c3 = c6)).   %symmetry
11
12    c1 + c2 + c3 = 10.      c6 + c7 + c8 = 10.
13    c1 + c4 + c6 = 10.      c3 + c5 + c8 = 10.
14
15    c1 + c2 + c3 + c4 + c5 + c6 + c7 + c8 = 26.         %next 32, next 26
16  end_of_list.
```

Solution

Let $c_1, c_2, c_3, c_4, c_5, c_6, c_7$, and c_8 be the eight places. Since the initial configuration has a maximum of 8, by gradually braking it, this value decreases only. Hence, a domain size of 9 suffices (line 1 in Listing 11.11). We assume at least one shell in each position (line 6). The box should always be symmetric. We assume that symmetry means: (i) there is the same number of shells in the middle places (line 10), and (ii) the opposite corners have equal number of shells (line 11). The sum is gradually decreasing: 36, then 32, then 26.

For 32 shells remaining, Mace4 found 3 models (one depicted in Fig. 11.12)

Models (32 shells)	c_1	c_2	c_3	c_4	c_5	c_6	c_7	c_8
1	1	6	3	6	6	3	6	1
2	2	6	2	6	6	2	6	2
3	3	6	1	6	6	1	6	3

For 26 shells remaining, Mace4 found 6 models (one depicted in Fig. 11.12)

Models (26 shells)	c_1	c_2	c_3	c_4	c_5	c_6	c_7	c_8
1	1	3	6	3	3	6	3	1
2	2	3	5	3	3	5	3	2
3	3	3	4	3	3	4	3	3
4	4	3	3	3	3	3	3	4
5	5	3	2	3	3	2	3	5
6	6	3	1	3	3	1	3	6

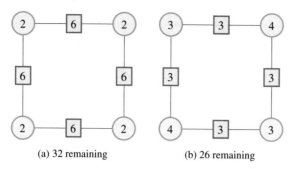

Fig. 11.12   Two models found by Mace4

Puzzle 117. The courageous garrison

A courageous garrison was defending a snow fort. The commander arranged his forces
as shown in the square frame (the inner square showing garrison's total strength of
40 boys): 11 boys defending each side of the fort. The garrison "lost" 4 boys during each of the first, second, third, and fourth assaults, and 2 during the fifth and last.
But after each charge 11 boys defended each side of the snow fort. How? (puzzle 101
from Kordemsky (1992))

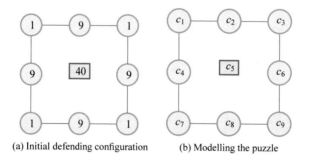

Fig. 11.13   The courageous garrison

```
1    assign(domain_size,37).
2    assign(max_models,−1).
3    set(arithmetic).
4
5    formulas(assumptions).
6      c1 + c2 + c3 = 11.        c1 + c4 + c7 = 11.
7      c7 + c8 + c9 = 11.        c3 + c6 + c9 = 11.
8
9      c1 + c2 + c3 + c4 + c6 + c7 + c8 + c9 = 36.
10   end_of_list.
```

Solution

There are eight variables c_i, with their initial sum of 40. After each assault, we have to keep 11 boys on each side given 36, 32, 28, 24, 22, 20 available. For the first assault (36 boys available), c_i is maxim 36, hence we set the domain size to this value (one can start even with a lower value, as the boys are distributed among the nodes). Given the codification in subfigure of Fig. 11.13b, we just need to write four equations, one for each side:

$$c_1 + c_2 + c_3 = c_1 + c_4 + c_7 = c_7 + c_8 + c_9 = c_3 + c_6 + c_9 = 11 \qquad (11.7)$$

For $c_5 = 36$ boys, Mace4 finds 165 models, many of them isomorphic. For $c_5 = 32$ boys, Mace4 finds 407 models. For $c_5 = 28$ defenders, Mace4 finds 329 models. For $c_5 = 24$ boys, there are 91 models, and for 22 boys only 12. Note this is the smallest number of defenders which can satisfy the initial constraint. For lower values, Mace4 fails to find a solution. A model for each case is listed below (see also Fig. 11.14):

Model	c_1	c_2	c_3	c_4	c_5	c_6	c_7	c_8	c_9
1	2	7	2	7	36	7	2	7	2
2	0	11	0	7	32	7	4	3	4
3	0	6	5	6	28	0	6	0	6
4	1	0	10	2	24	0	8	2	1
5	5	0	6	0	22	0	6	0	5

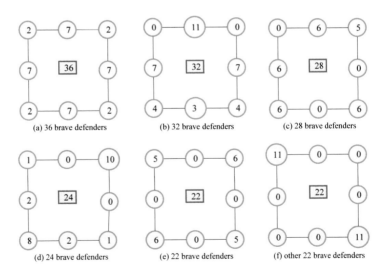

Fig. 11.14 Six solutions found by Mace4

Puzzle 118. A grouping of integers 1 through 15

See how elegantly the integers 1 through 15 can be arranged in 5 arithmetic progressions of 3 integers:

$$d_5 = 7 \begin{cases} 1 \\ 8 \\ 15 \end{cases} \quad d_4 = 5 \begin{cases} 4 \\ 9 \\ 14 \end{cases} \quad d_3 = 4 \begin{cases} 2 \\ 6 \\ 10 \end{cases} \quad d_2 = 2 \begin{cases} 3 \\ 5 \\ 7 \end{cases} \quad d_1 = 1 \begin{cases} 11 \\ 12 \\ 13 \end{cases}$$

For example, $8 - 1 = 15 - 8 = 7$, so the difference d_5 is 7 for the first triplet. Now, keeping the first triplet, make four new triplets, still with $d = 5, 4, 2,$ and 1. On your own, try arranging the integers from 1 through 15 with other values of d. (puzzle 112 from Kordemsky (1992))

Listing 11.12 Models for grouping integers in arithmetic progressions

```
1    assign(domain_size,16).
2    assign(max_models,−1).
3    set(arithmetic).
4
5    list(distinct).
6        [0,c1,c2,c3,c4,c5,c6,c7,c8,c9,c10,c11,c12,1,8,15].
7        [0,d1,d2,d3,d4,7,8,9,10,11,12,13,14,15].
8    end_of_list.
9
10   formulas(assumptions).
11       d5 = 7.    c13 = 1.    c14 = 8.    c15 = 15.    %keep the first progression.
12       d1 < d2.   d2 < d3.    d3 < d4.                 %avoid isomorphism
13
14       c2 = d1 + c1.    c3 = d1 + c2.    c5  = d2 + c4.    c6  = d2 + c5.
15       c8 = d3 + c7.    c9 = d3 + c8.    c11 = d4 + c10.   c12 = d4 + c11.
16   end_of_list.
```

Solution

Since there are 15 values to be arranged from [1..15], we use a domain size of 16. We keep the triplet $c_{13} = 1, c_{14} = 8, c_{15} = 15$. The 12 remaining values (c_i) are distinct among them and also different from 0, 1, 8, and 15.

We will try to solve a more general problem: instead of fixing $d = 5, 4, 2,$ and 1, we are searching for progressions with any four distinct ratios d_i. The maximum ratio is $15 - 8 = 7$, hence $d_1, d_2, d_3,$ and $d_4 < 7$ (line 7 in Listing 11.12). To avoid generating isomorphic models, we set an order relation between these ratios: $d_1 < d_2, d_2 < d_3, d_3 < d_4$.

We need to formalise the four arithmetic progressions:

$$c_2 = d_1 + c_1 \wedge c_3 = d_1 + c_2 \quad c_5 = d_2 + c_4 \wedge c_6 = d_2 + c_5$$
$$c_8 = d_3 + c_7 \wedge c_9 = d_3 + c_8 \quad c_{11} = d_4 + c_{10} \wedge c_{12} = d_4 + c_{11}$$

Mace4 finds two models: the initial one and the solution below:

$$d_5 = 7 \begin{cases} 1 \\ 8 \\ 15 \end{cases} \quad d_4 = 5 \begin{cases} 2 \\ 7 \\ 12 \end{cases} \quad d_3 = 4 \begin{cases} 6 \\ 10 \\ 14 \end{cases} \quad d_2 = 2 \begin{cases} 9 \\ 11 \\ 13 \end{cases} \quad d_1 = 1 \begin{cases} 3 \\ 4 \\ 5 \end{cases}$$

Puzzle 119. A star

Can you place the integers from 1 through 12 in the circles of the six-pointed star so that the sum of the numbers in each of the six rows is 26? (puzzle 324 from Kordemsky (1992))

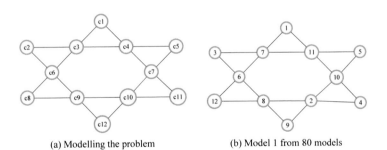

(a) Modelling the problem   (b) Model 1 from 80 models

Fig. 11.15 A star with 12 numbers

Listing 11.13 Finding models for 12 values summing up to 26

```
1   assign(domain_size,13).
2   assign(max_models,-1).
3   set(arithmetic).
4
5   list(distinct).
6      [0,c1,c2,c3,c4,c5,c6,c7,c8,c9,c10,c11,c12].
7   end_of_list.
8
9   formulas(assumptions).
10     (c1 + c3 + c6  + c8  = 26) & (c2 + c3 + c4  + c5 = 26)  &
11     (c1 + c4 + c7  + c11 = 26) & (c2 + c6 + c9  + c12 = 26) &
12     (c5 + c7 + c10 + c12 = 26) & (c8 + c9 + c10 + c11 = 26).
13
14     c1=1.   %remove some isomorphism
15  end_of_list.
```

Solution

We model the problem using the 12 functions from Fig. 11.15a. Since there are six lines, we need six equations. To remove some isomorphic models we fix a corner, i.e. $c_1 = 1$. Given the formalisation in Listing 11.13, Mace4 finds 80 models. Four such models are:

Model	c_1	c_2	c_3	c_4	c_5	c_6	c_7	c_8	c_9	c_{10}	c_{11}	c_{12}
1	1	3	7	11	5	6	10	12	8	2	4	9
2	1	5	9	10	2	12	7	4	3	11	8	6
3	1	3	12	4	7	8	11	5	9	2	10	6
4	1	7	11	5	3	4	12	10	6	2	8	9

Model number 1 is also depicted in Fig. 11.15b.

Puzzle 120. The hexagon

Enter integers from 1 through 19 in the spots of the hexagon so that each row of three (on the rim, and outward from the centre) adds to 22. Rearrange to add to 23. (puzzle 327 from Kordemsky (1992)) (Fig.11.16)

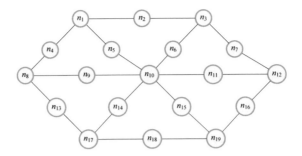

Fig. 11.16 Modelling the hexagon puzzle

Listing 11.14 Finding models for a hexagon with 19 values

```
1    assign(domain_size,20).
2    assign(max_models,−1).
3    set(arithmetic).
4
5    list(distinct).
6    [0,n1,n2,n3,n4,n5,n6,n7,n8,n9,n10,n11,n12,n13,n14,n15,n16,n17,n18,n19].
7    end_of_list.
8
9    formulas(demodulators).
10   n1  + n2  + n3  = 22.   n1  + n4  + n8  = 22.   n8  + n13 + n17 = 22.
11   n17 + n18 + n19 = 22.   n19 + n16 + n12 = 22.   n3  + n7  + n12 = 22.
12   n1  + n10 + n5  = 22.   n10 + n6  + n3  = 22.   n8  + n9  + n10 = 22.
13   n10 + n11 + n12 = 22.   n10 + n14 + n17 = 22.   n10 + n15 + n19 = 22.
14
15   n2 = 19  | n4  = 19 | n5  = 19 | n6  = 19 | n7  = 19 | n9  = 19 |
16   n11 = 19 | n13 = 19 | n14 = 19 | n15 = 19 | n16 = 19 | n18 = 19.
17   n1 != 19 & n3 != 19 & n12 != 19 & n19 != 19 & n17 != 19 & n8 != 19.
18   n10 < 10 & n1 < 16 & n3 < 16 & n8 < 16 & n12 < 16 & n17 < 16 & n19 < 16.
19   end_of_list.
```

Solution

Since integers are from 1 through 19, we need a domain size of 20. Values are distinct and not equal to zero (line 6). There are 12 lines in the hexagon, each line with its own equation to specify the sum of 22 (lines 10–13). Given this formalisation, Mace4 finds 48 models. The problem is that it takes a lot of time. To improve the search, we could add some additional constraints depending on the number of edges of the nodes. There are three types of nodes: 12 nodes with 2 edges (n_2, n_4, n_5, n_6, n_7, n_9, n_{11}, n_{13}, n_{14}, n_{15}, n_{16}, n_{18}), 6 nodes with 3 edges (n_1, n_3, n_{12}, n_{19}, n_{17}, n_8), and one node (n_{10}) with 6 edges.

First, observe that value 19 only when summing with 2 and 1 gives 22 ($19 + 1 + 2 = 22$). Thereby, only nodes with 2 edges can contain this value.

Second, consider n_{10} with 6 edges. All 6 sums should be $s = 22 - n_{10}$. The smallest value s that can be written in six different ways is 13 ($1 + 12, 2 + 11, 3 + 10, 4 + 9, 5 + 8, 6 + 7$). Hence, n_{10} has only values equal or smaller than 9 ($22-13$).

Third, the nodes with three connections can store only values ≤ 15, because we have to find a value v that can be written in three different ways. The smallest value that can be written in three different ways is $v = 7$ ($1 + 6, 2 + 5, 3 + 4$). Hence, the nodes with three connections contain values less or equal than $22 - v = 15$.

There are 48 solutions (including isomorphisms) for the sum of 22. One solution appears in Fig. 11.17b. In order to find the solutions for the sum of 23, we only need to change the sum for each pair of 3 numbers and to increase by 1 the values in the constraints used for speeding up the search. For sum 23, one model appears in Fig. 11.17a.

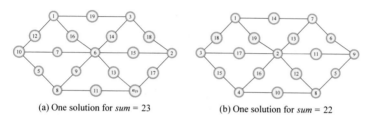

(a) One solution for *sum* = 23   (b) One solution for *sum* = 22

Fig. 11.17 Mace4 finds solutions for both values

Reference

Kordemsky, B. A. (1992). *The Moscow puzzles: 359 mathematical recreations*. Courier Corporation.

Chapter 12
Polyomino Puzzles

Abstract In which we introduce the square world of polyominoes, including tetrominoes, pentominoes Golomb (1996), or the broken chess puzzle of Dudeney (2002).
Polyomino are two-dimensional shapes made by connecting n squares of the same
size. A polyomino with $n = 2$ is a *domino*. If $n = 3$ we have a *tromino*, $n = 4$ gives
us a *tetromino*, while $n = 5$ forms a *pentomino*. Dominous have a single shape (two
squares in a line), trominoes have two shapes (three squares in a line or the L-shape),
tetrominoes have five shapes, while pentominoes have twelve distinct configurations.
A popular tetromino puzzle is the video game *Tetris*.

> $\forall x \, \forall y \, Man(x) \wedge has(x, h) \wedge hammer(h) \wedge problem(y) \rightarrow sees(x, y, nail)$
> $great(hammer, our Age) = algorithm$
>
> William Poundstone

Polyomino are two-dimensional shapes made by connecting n squares of the same
size. A polyomino with $n = 2$ is a *domino*. If $n = 3$ we have a *tromino*, $n = 4$ gives
us a *tetromino*, while $n = 5$ forms a *pentomino*. Dominous have a single shape (two
squares in a line), trominoes have two shapes (three squares in a line or the L-shape),
tetrominoes have five shapes, while pentominoes have twelve distinct configurations.
There is no known algorithm for computing how many distinct polyominoes of each
order there are Darling (2004).

A popular polyomino puzzle is the "Broken Chessboard" published in 1907 by
H. Dudeney in his The Canterbury Puzzles Dudeney (2002). The puzzle asks to
reconstruct the board from 13 pieces, 12 of which are different possible ways in which
pentominoes can be arranged (i.e. pieces of five squares) and one tetromino (i.e. a
piece of 2×2 squares). Another popular tetromino puzzle is the video game *Tetris*.

Polyominoes have been given some consideration in literature, starting from 1953
when Golomb discussed them in a talk he gave to the Harvard Mathematics Club. The
interested reader can find various polyomino puzzles in Golomb (1996) or Martin
(1991), while their practical and mathematical aspects are analysed in Guttmann
(2009).

A. Groza, *Modelling Puzzles in First Order Logic*,
https://doi.org/10.1007/978-3-030-62547-4_12

Puzzle 121. Broken chess row

This puzzle uses one monomino, two dominoes, and one tromino, for a total of eight squares (Fig. 12.1). Group the four shapes in a chess line (i.e. 8 × 1 grid). How many solutions are there?

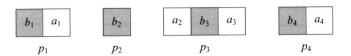

Fig. 12.1 A broken chess line into four polyomino

Listing 12.1 Finding models for a broken chess line

```
1    set ( arithmetic ).
2    assign ( domain_size , 8 ).
3    assign ( max_models , −1 ).
4
5    list ( distinct ).
6      [ a1 , a2 , a3 , a4 , b1 , b2 , b3 , b4 ].
7    end_of_list .
8
9    formulas ( utils ).
10     w( a1 ).    w( a2 ).    w( a3 ).    w( a4 ).    %white squares
11     b( b1 ).    b( b2 ).    b( b3 ).    b( b4 ).    %black squares
12     w( x ) <−> −b( x ).                            %black xor white
13     b( 0 ).                                        %first cell is black
14     left ( x , y ) <−> x + 1 = y .
15     −left ( 7 , y ).
16     left ( x , y ) & b( x ) −> w( y ).   %two adjacent squares have distinct color
17     left ( x , y ) & w( x ) −> b( y ).
18   end_of_list .
19
20   formulas ( polyomino ).
21     p1  <−>  left ( b1 , a1 ).
22     p3  <−>  left ( a2 , b3 ) & left ( b3 , a3 ).
23     p4  <−>  left ( b4 , a4 ).
24   end_of_list .
25
26   formulas ( pretty_print ).
27     c ( a1 )=1.    c ( b1 )=1.                     %p1
28     c ( b2 )=2.                                    %p2
29     c ( a2 )=3.    c ( b3 )=3.    c ( a3 )=3.       %p3
30     c ( a4 )=4.    c ( b4 )=4.                     %p4
31   end_of_list .
```

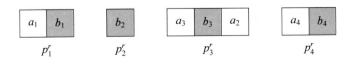

Fig. 12.2 New polyominoes by rotating the initial ones

Solution

We need eight distinct relations to represent each square: a_1, a_2, a_3, a_4, b_1, b_2, b_3, and b_4. Four of these squares are white $w(a_1)$, $w(a_2)$, $w(a_3)$, and $w(a_4)$, and four are black $b(b_1)$, $b(b_2)$, $b(b_3)$, and $b(b_4)$. A black square cannot be white, and the other way around holds: $b(x) \leftrightarrow \neg w(x)$.

To remove some isomorphic models, we assume the leftmost square is black (i.e. $b(0)$). The four shapes are modelled with the predicate $left(x, y)$:

$$left(x, y) \leftrightarrow x + 1 = y \tag{12.1}$$

The $left$ predicate does no hold on the last column: $\neg left(7, y)$. We know that two adjacent squares have distinct colours:

$$left(x, y) \wedge b(x) \rightarrow w(y) \tag{12.2}$$

$$left(x, y) \wedge w(x) \rightarrow b(y) \tag{12.3}$$

Now, we represent the four shapes. Let the first shape (denoted by p_1) consisting of the squares b_1 and a_1.

$$p_1 \leftrightarrow left(b_1, a_1) \tag{12.4}$$

Let the second shape (denoted by p_2) consisting of the square b_2. No constraints are needed for this shape. The third shape p_3 consists of a_2, b_3, and a_3:

$$p_3 \leftrightarrow left(a_2, b_3) \wedge left(b_3, a_3) \tag{12.5}$$

The last piece p_4 contains the remaining squares b_4 and a_4:

$$p_4 \leftrightarrow left(b_4, a_4) \tag{12.6}$$

Listing 12.2 The pieces cannot be rotated

```
1   formulas(no_rotation).
2     p1 & p2 & p3 & p4.          %all four pieces are needed
3   end_of_list.
```

Listing 12.3 The pieces can be rotated

```
1   formulas(rotate).
2     p1r <-> left(a1,b1).                %rotate p1
3     p3r <-> left(a3,b3) & left(b3,a2).  %rotate p2
4     p4r <-> left(a4,b4).                %rotate p3
5
6     (p1 | p1r) & p2 & (p3 | p3r) & (p4 | p4r).  %four pieces are needed
7     p1 -> -p1r .                         %p1 XOR p1 rotated
8     p3 -> -p3r.                          %p3 XOR p3 rotated
9     p4 -> -p4r.                          %p4 XOR p4 rotated
10  end_of_list.
```

Solution

We add the function $c(x)$ to collect the solution in one vector (lines 26–30 in Listing 12.1). By coding each of the four polyominoes with a value (e.g. $p_1 = 1$, $p_2 = 2$, $p_3 = 3$, $p_4 = 4$), the function $c(x)$ can be:

$$c(a_1) = 1 \wedge c(b_1) = 1 \tag{12.7}$$

$$c(b_2) = 2 \tag{12.8}$$

$$c(a_2) = 3 \wedge c(b_3) = 3 \wedge c(a_3) = 3 \tag{12.9}$$

$$c(a_4) = 4 \wedge c(b_4) = 4 \tag{12.10}$$

A solution of the form: `function(c(_), [1, 1, 2, 3, 3, 3, 4, 4])` represents the chess line. The puzzle asks to use all the four shapes:

$$p_1 \wedge p_2 \wedge p_3 \wedge p_4 \tag{12.11}$$

Given two files: Listing 12.1 and the constraint in equation 12.11 (i.e. the file `no_rotation.in`): `mace4 -f chessline.in no_rotation`, Mace4 computes six models:

Model	$c(0)$	$c(1)$	$c(2)$	$c(3)$	$c(4)$	$c(5)$	$c(6)$	$c(7)$
1	1	1	2	3	3	3	4	4
2	1	1	4	4	2	3	3	3
3	4	4	1	1	2	3	3	3
4	2	3	3	3	1	1	4	4
5	2	3	3	3	4	4	1	1
6	4	4	2	3	3	3	1	1

Now, assume that the pieces can be rotated. By rotating p_1, p_3, and p_4 (see Fig. 12.2), three new shapes are obtained p_1^r, p_3^r, and p_4^r:

$$p_1^r \leftrightarrow left(a_1, b_1) \tag{12.12}$$

$$p_3^r \leftrightarrow left(a_3, b_3) \wedge left(b_3, a_2) \tag{12.13}$$

$$p_4^r \leftrightarrow left(a_4, b_4) \tag{12.14}$$

Observe that instead of the initial piece, we can use its rotation. Hence, instead of the condition $p_1 \wedge p_2 \wedge p_3 \wedge p_4$ forcing the occurrence of all the initial pieces, we use the "exclusive-or" between a piece and its rotation:

$$\neg(p_1 \rightarrow p_2^r) \wedge \neg(p_1^r \rightarrow p_1) \tag{12.15}$$

$$\neg(p_3 \rightarrow p_3^r) \wedge \neg(p_3^r \rightarrow p_3) \tag{12.16}$$

$$\neg(p_4 \rightarrow p_4^r) \wedge \neg(p_4^r \rightarrow p_4) \tag{12.17}$$

Given two files (i.e. Listing 12.1 and Listing 12.3) `mace4 -f chessline.in rotation.in`, Mace4 computes 24 models.

Puzzle 122. A simple polyomino

This puzzle uses one monomino, one domino, and two trominoes, for a total of nine squares. Assume that you cannot rotate the shapes. Group the four shapes in a 3×3 grid.

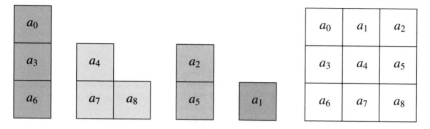

Listing 12.4 Formalising a simple polyomino

```
1   assign(domain_size ,9).
2   assign(max_models , -1).
3   set(arithmetic ).
4
5   list(distinct ).
6     [a0 ,a1 ,a2 ,a3 ,a4 ,a5 ,a6 ,a7 ,a8 ].
7   end_of_list.
8
9   formulas(assumptions ).
10    (x != 6 & x != 7 & x != 8) -> (on(x,y) <-> y = x + 3).   %x on y,
11    -(x != 6 & x != 7 & x != 8) -> -on(x,y).                  %3x3 grid
12
13    (x != 2 & x != 5 & x != 8) -> (left(x,y) <-> y = x + 1). %x left of y
14    -(x != 2 & x != 5 & x != 8) -> -left(x,y).               %3x3 grid
15
16    on(a0 ,a3 ) & on(a3 ,a6 ).                                %green shape
17    on(a2 ,a5 ).                                              %red shape
18    on(a4 ,a7 ) & left(a7 ,a8 ).                              %yellow shape
19  end_of_list.
```

Solution

We will use the functions a_i, i in [0..8] for each position in the 3×3 grid. These values should be distinct in order to cover the grid. Hence, we have to fill nine distinct positions (from 0 to 8).

Each shape is modelled with two predicates: $on(x, y)$ is true when square x is on square y and $left(x, y)$ is true when x is left to y. Since the grid is 3×3, there is a distance of 3 between the two squares satisfying the on relation: $on(x, y) \leftrightarrow y = x + 3$. We need to pay attention to the bottom line. A square situated anywhere on the bottom line ($x = 6$, or $x = 7$, or $x = 9$) cannot be above another one. Hence, we define the predicate $on(x, y)$ as follows:

$$(x \neq 6 \wedge x \neq 7 \wedge x \neq 8) \rightarrow (on(x, y) \leftrightarrow y = x + 3) \qquad (12.18)$$
$$\neg(x \neq 6 \wedge x \neq 7 \wedge x \neq 8) \rightarrow \neg on(x, y) \qquad (12.19)$$

Solution

The same modelling approach is applied to the *left* predicate. If a square x is to the left of y, its value is less by one from y's value. A square x situated anywhere on the last column ($x = 2$, or $x = 5$, or $x = 8$) cannot be left to another square:

$$(x \neq 2 \wedge x \neq 5 \wedge x \neq 8) \rightarrow (on(x, y) \leftrightarrow y = x + 1) \tag{12.20}$$

$$\neg(x \neq 2 \wedge x \neq 5 \wedge x \neq 8) \rightarrow \neg left(x, y) \tag{12.21}$$

Then we need to model the four shapes. The green tromino is formalised with the predicate *on*. Let the three squares be a_0, a_3, and a_6:

$$on(a_0, a_3) \wedge on(a_3, a_6) \tag{12.22}$$

The yellow tromino needs also the *left* predicate. For this tromino, we allocate the functions a_4, a_7, and a_8.

$$on(a_4, a_7) \wedge left(a_7, a_8) \tag{12.23}$$

The red domino requires a single relation, for which we allocate the functions a_2 and a_5.

$$on(a_2, a_5) \tag{12.24}$$

The blue monomino is modelled with the single remaining value a_1. Since a monomino can be anywhere on the grid, we do not have to specify any constraint for it.
Given the implementation in Listing 12.4, Mace4 finds two models:

Models	a_0	a_1	a_2	a_3	a_4	a_5	a_6	a_7	a_8
1	0	1	2	3	4	5	6	7	8
2	2	0	1	5	3	4	8	6	7

These two models correspond to the solutions depicted in Fig. 12.3. For instance, in the second model, a_0 is on position 2, a_1 on position 0, or a_2 is on position 1.

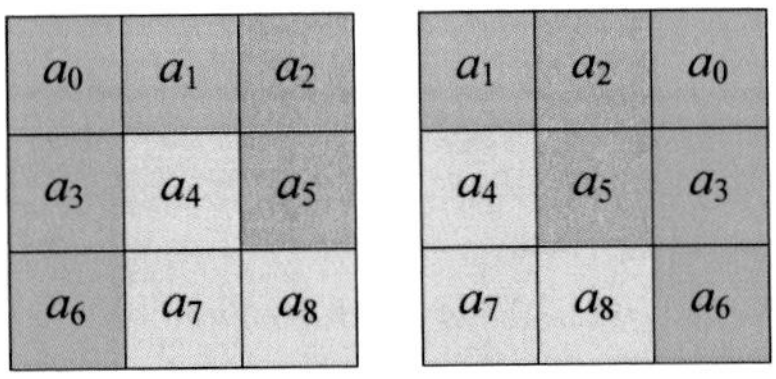

Fig. 12.3 Two solutions found by Mace4 for the simple polyomino puzzle

Puzzle 123. Rotating polyomino

This puzzle uses one monomino, one domino, and two trominoes, for a total of nine squares. Now, assume that you can rotate the shapes. Group the four shapes in a 3×3 grid. How many solutions are there?

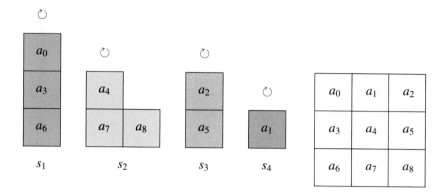

Allowing rotations will generate five new shapes (see Fig. 12.4). It suffices to use only the rotation to the right $\circlearrowright$. If we apply any number of rotations on the green and red polyminoes, we obtain only one extra shape for each of them. If we apply successive rotations on the yellow shape, we obtain three additional shapes. By rotating the blue monomino, we obtain no extra shapes.

We will extend the model from the previous puzzle. We will use the functions a_i for each position in the 3×3 grid. These values should be distinct in order to cover the grid.

The shapes are modelled with the predicates $on(x, y)$ and $left(x, y)$. Their definition considers the grid size and the borders:

$$(x \neq 6 \wedge x \neq 7 \wedge x \neq 8) \rightarrow (on(x, y) \leftrightarrow y = x + 3) \tag{12.25}$$

$$\neg(x \neq 6 \wedge x \neq 7 \wedge x \neq 8) \rightarrow \neg on(x, y) \tag{12.26}$$

$$(x \neq 2 \wedge x \neq 5 \wedge x \neq 8) \rightarrow (on(x, y) \leftrightarrow y = x + 1) \tag{12.27}$$

$$\neg(x \neq 2 \wedge x \neq 5 \wedge x \neq 8) \rightarrow \neg left(x, y) \tag{12.28}$$

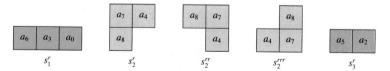

Fig. 12.4 Additional five shapes generated through rotations

Listing 12.5 Finding models for rotating polyominoes

```
1    assign(domain_size ,9).
2    assign(max_models , -1).
3    set(arithmetic).
4
5    list(distinct).
6      [a0 , a1 , a2 , a3 , a4 , a5 , a6 , a7 , a8 ].
7    end_of_list.
8
9    formulas(utils).
10     (x != 6 & x != 7 & x != 8) -> (on(x,y) <-> y = x + 3).    %x on y,
11     -(x != 6 & x != 7 & x != 8) -> -on(x,y).                  %3x3 grid
12     (x != 2 & x != 5 & x != 8) -> (left(x,y) <-> y = x + 1).  %x left of y
13     -(x != 2 & x != 5 & x != 8) -> -left(x,y).                %3x3 grid
14   end_of_list.
15
16   formulas(polyominoes).
17     s1   <-> on(a0,a3)     & on(a3,a6).          %green shape
18     s1r <-> left(a0,a3) & left(a3,a6).           %s1 rotated right
19     -(s1  -> s1r) | -(s1r -> s1).                %xor
20
21     s2   <-> on(a2,a5).                          %red shape
22     s2r <-> left(a5,a2).                         %s2 rotated right
23     -(s2  -> s2r) | -(s2r -> s2).                %xor
24
25     s3   <-> on(a4,a7) & left(a7,a8).            %yellow shape
26     s3r <-> on(a7,a8) & left(a7,a4).             %s2 rotated right once
27     s3rr <-> on(a7,a4) & left(a8,a7).            %s2 rotated right twice
28     s3rrr <-> on(a8,a7) & left(a4,a7).           %s2 rotated right
29     s3 | s3r | s3rr | s3rrr.                     %at least one
30     s3    -> -s3r & -s3rr & -s3rrr.              %at most one
31     s3r   -> -s3  & -s3rr & -s3rrr.
32     s3rr  -> -s3  & -s3r  & -s3rrr.
33     s3rrr -> -s3  & -s3r  & -s3rr.
34   end_of_list.
35
36   formulas(pretty_print).
37     c(a0) = 0. c(a3) = 0. c(a6) = 0.   %pretty print (green=0)
38     c(a4) = 1. c(a7) = 1. c(a8) = 1.   %pretty print (yellow=1)
39     c(a2) = 2. c(a5) = 2.              %pretty print (red=2)
40     c(a1) = 3.                         %pretty print (blue=3)
41   end_of_list.
```

Solution

Let the green shape (denoted by s_1) consist of the squares a_0, a_3, and a_6: $s_1 \leftrightarrow on(a_0, a_3) \wedge on(a_3, a_6)$. By rotating s_1, the new shape s_1^r contains the same squares with the following relations among them:

$$s_1^r \leftrightarrow left(a_0, a_3) \wedge left(a_3, a_6) \tag{12.29}$$

Either s_1 or s_1^r should appear on the grid, but not both of them. This is formalised with the exclusive-or (recall the formula $A \; xor \; B \equiv \neg(A \rightarrow B) \vee \neg(B \rightarrow A)$):

$$\neg(s_1 \rightarrow s_1^r) \vee \neg(s_1^r \rightarrow s_1) \tag{12.30}$$

Solution

We apply the same modelling pattern for the red shape s_2: first we rotate s_2 to obtain s_2^r, and then we state the xor relation between s_2 and its rotation s_2^r:

$$s_2 \leftrightarrow on(a_2, a_5) \tag{12.31}$$

$$s_2^r \leftrightarrow left(a_5, a_2) \tag{12.32}$$

$$\neg(s_2 \rightarrow s_2^r) \vee \neg(s_2^r \rightarrow s_2) \tag{12.33}$$

Let the yellow shape s_3 contains the squares a_7 and a_8: $s_3 \leftrightarrow on(a_4, a_7) \wedge left(a_7, a_8)$. Each rotation to the right of s_3 generates a new shape:

$$s_3^r \leftrightarrow on(a_7, a_8) \wedge left(a_7, a_4) \tag{12.34}$$

$$s_3^{rr} \leftrightarrow on(a_7, a_4) \wedge left(a_8, a_7) \tag{12.35}$$

$$s_3^{rrr} \leftrightarrow on(a_8, a_7) \wedge left(a_4, a_7) \tag{12.36}$$

One of these four shapes $(s_3, s_3^r, s_3^{rr}, s_3^{rrr})$ should appear in the solution:

$$s_3 \vee s_3^r \vee s_3^{rr} \vee s_3^{rrr} \tag{12.37}$$

$$s_3 \rightarrow \neg s_3^r \wedge \neg s_3^{rr} \wedge \neg s_3^{rrr} \tag{12.38}$$

$$s_3^r \rightarrow \neg s_3 \wedge \neg s_3^{rr} \wedge \neg s_3^{rrr} \tag{12.39}$$

$$s_3^{rr} \rightarrow \neg s_3 \wedge \neg s_3^r \wedge \neg s_3^{rrr} \tag{12.40}$$

$$s_3^{rrr} \rightarrow \neg s_3 \wedge \neg s_3^r \wedge \neg s_3^{rr} \tag{12.41}$$

The blue monomino s_4 is modelled with the single remaining value a_1. Since monominoes can be anywhere on the grid, we do not have to specify any constraint for it.
One can add the colour function $c(x)$ to collect the solution in one vector. By coding each of the four colours with a value (e.g. $green = 0, yellow = 1, red = 2, blue = 3$), the function $c(x)$ can be:

$$c(a_0) = 0 \wedge c(a_3) = 0 \wedge c(a_6) = 0 \tag{12.42}$$

$$c(a_4) = 1 \wedge c(a_7) = 1 \wedge c(a_8) = 1 \tag{12.43}$$

$$c(a_2) = 2 \wedge c(a_5) = 2 \tag{12.44}$$

$$c(a_1) = 3 \tag{12.45}$$

A solution of the form: `function(c(_), [ 1, 2, 2, 1, 1, 3, 0, 0, 0 ])` represents the grid:

1	2	2
1	1	3
0	0	0

Knowing the colours for each value, one can easily see polyominoes within the 3×3 solution.

Solution

Given the implementation in Listing 12.5, Mace4 finds 48 models. Three such models based on the c function are:

Models	0	1	2	3	4	5	6	7	8
47	2	1	1	2	1	3	0	0	0
36	1	1	0	2	1	0	2	3	0
26	3	1	0	1	1	0	2	2	0

These three models correspond to the following grids:

2	1	1
2	1	3
0	0	0

1	1	0
2	1	0
2	3	0

3	1	0
1	1	0
2	2	0

For each model, the values of each relation s_i signals whether the original shapes have been rotated or not. For instance, for the model 47 above (i.e. the grid on the left), Mace4 displays $s_1 = 0$, $s_1^r = 1$, $s_2 = 1$, $s_2^r = 0$, $s_3 = 0$, $s_3^r = 1$, $s_3^{rr} = 0$, and $s_2^{rrr} = 0$. That is, for each model only the polyominoes with value 1 occur in the solution: $s_1^r = 1$, $s_2 = 1$, $s_3^r = 1$ (see Fig. 12.5). Recall that $s_1^r = 1$ is the original green shape rotated once to the right, s_2^r is the original yellow shape rotated once to the right, and s_3 is the original red shape.

Fig. 12.5 Model 47 from 48 solutions found by Mace4

Puzzle 124. Ten-Yen

Ten-yen puzzle consists of ten polyominoes as depicted below (Fig. 12.6). For now, we assume that: shapes cannot be rotated and there are no constraints on the colours of each shape. The task is to assemble the shapes in a 6×6 grid. (taken from http://www.gamepuzzles.com/polycub3.htm and published in 1950 by Multiple Products Corporation and used here for educational purpose)

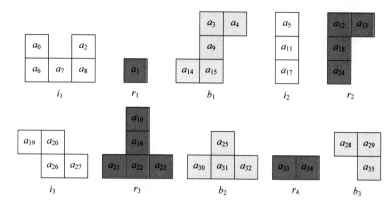

Fig. 12.6 Ten polyominoes in the ten-yen puzzle

Listing 12.6 Formalising the domain for the Ten-Yen puzzle

```
1    assign(domain_size,36).          %6x6 grid
2    assign(max_models,-1).
3    set(arithmetic).
4
5    list(distinct).
6      [a0,   a1,   a2,   a3,   a4,   a5,
7       a6,   a7,   a8,   a9,   a10,  a11,
8       a12,  a13,  a14,  a15,  a16,  a17,
9       a18,  a19,  a20,  a21,  a22,  a23,
10      a24,  a25,  a26,  a27,  a28,  a29,
11      a30,  a31,  a32,  a33,  a34,  a35].
12   end_of_list.
13
14   formulas(utils).
15     x < 30 -> (on(x,y) <-> y = x + 6).
16     x > 29 -> -on(x,y).
17     (x != 5 & x != 11 & x != 17 & x != 23 & x != 29 & x != 35)
18                -> (left(x,y) <-> y = x + 1).
19     -(x != 5 & x != 11 & x != 17 & x != 23 & x != 29 & x != 35)
20                -> -left(x,y).
21   end_of_list.
```

Listing 12.7 Formalising the polyominoes in the Ten-Yen puzzle

```
1    formulas(ten_yen).
2      i1 <-> left(a6,a7)   & on(a0,a6)    & left(a7,a8) & on(a2,a8).
3      b1 <-> left(a14,a15) & on(a9,a15)   & on(a3,a9)   & left(a3,a4).
4      i2 <-> on(a5,a11)    & on(a11,a17).
5      r2 <-> left(a12,a13) & on(a12,a18)  & on(a18,a24).
6      i3 <-> left(a19,a20) & on(a20,a26)  & left(a26,a27).
7      r3 <-> left(a21,a22) & left(a22,a23) & on(a16,a22) & on(a10,a16).
8      b2 <-> left(a30,a31) & on(a25,a31)  & left(a31,a32).
9      r4 <-> left(a33,a34).
10     b3 <-> left(a28,a29) & on(a29,a35).
11
12     i1 & i2 & i3 & r1 & r2 & r3 & r4 & b1 & b2 & b3.
13   end_of_list.
```

```
14   formulas ( pretty_print ).                    %0 = white , 1 = black , 2 = red
15     c ( a0 )  = 0.  c ( a6 )  = 0.  c ( a7 )  = 0.  c ( a8 )  = 0.  c ( a2 )  = 0.     %i1
16     c ( a1 )  = 2.                                                                     %r1
17     c ( a14 ) = 1.  c ( a15 ) = 1.  c ( a9 )  = 1.  c ( a3 )  = 1.  c ( a4 )  = 1.     %b1
18     c ( a22 ) = 2.  c ( a21 ) = 2.  c ( a23 ) = 2.  c ( a16 ) = 2.  c ( a10 ) = 2.     %r2
19     c ( a5 )  = 0.  c ( a11 ) = 0.  c ( a17 ) = 0.                                     %i2
20     c ( a12 ) = 2.  c ( a13 ) = 2.  c ( a18 ) = 2.  c ( a24 ) = 2.                     %r3
21     c ( a19 ) = 0.  c ( a20 ) = 0.  c ( a26 ) = 0.  c ( a27 ) = 0.                     %i3
22     c ( a30 ) = 1.  c ( a31 ) = 1.  c ( a25 ) = 1.  c ( a32 ) = 1.                     %b2
23     c ( a33 ) = 2.  c ( a34 ) = 2.                                                     %r4
24     c ( a28 ) = 1.  c ( a29 ) = 1.  c ( a35 ) = 1.                                     %b3
25   end_of_list .
```

Solution

Since the grid has 6×6 we need a domain of 35 elements (Fig. 12.7). The variables a_i, $i \in [0..35]$ represent distinct locations in the grid (lines 5–12 in Listing 12.6). In lines 14–21, we define the predicates $on(x, y)$ and $left(x, y)$. We make sure that the grid is 6×6 and that the *on* predicate does not hold on the last line (locations < 30), while *left* does not hold on the last column (i.e. locations 5, 11, 17, 23, 29, and 35). We define each shape in Listing 12.7, where we used i for white shapes, b for black, and r for red. For instance, the first white polyomino i_1 (see Fig. 12.8) is formalised with:

$$i_1 \leftrightarrow left(a_6, a_7) \wedge on(a_0, a_6) \wedge left(a_7, a_8) \wedge on(a_2, a_8) \tag{12.46}$$

Line 12 states that all shapes should occur in the grid. The function $c(a_i)$ prints the colours in each solution. All variables occurring in the white shapes i_1, i_2, i_3 have colour 0. Variables occurring in the black shapes b_1, b_2, b_3 have colour 1. Variables defining the red polyominoes r_1, r_2, r_3, r_4 have colour 2. Given the implementation in Listing 12.7, Mace4 finds two models. Mace4 computes a position in $[0..35]$ for each variable a_i and also the colour code:

$$
\begin{array}{cccccc}
0 & 2 & 0 & 1 & 1 & 0 \\
0 & 0 & 0 & 1 & 2 & 0 \\
2 & 2 & 1 & 1 & 2 & 0 \\
2 & 2 & 2 & 2 & 2 & 2 \\
2 & 1 & 0 & 0 & 1 & 1 \\
1 & 1 & 1 & 0 & 0 & 1
\end{array}
\qquad
\begin{array}{cccccc}
0 & 2 & 0 & 1 & 1 & 0 \\
0 & 0 & 0 & 1 & 2 & 0 \\
2 & 2 & 1 & 1 & 2 & 0 \\
2 & 0 & 0 & 2 & 2 & 2 \\
2 & 1 & 0 & 0 & 1 & 1 \\
1 & 1 & 1 & 2 & 2 & 1
\end{array}
$$

These two models correspond to the solutions depicted in Fig. 12.7. The only difference between these models is between the shapes r_4 and i_3.

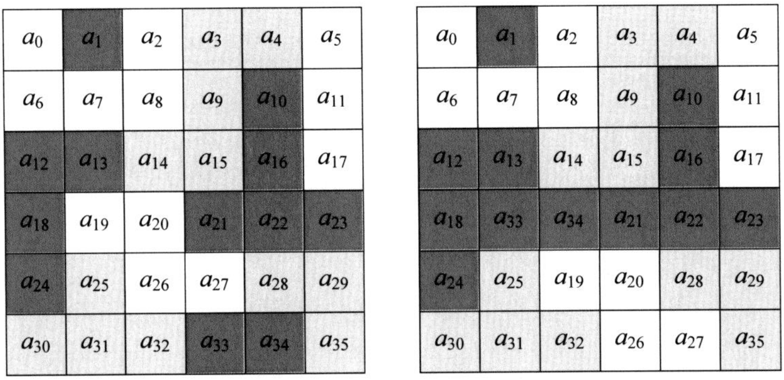

Fig. 12.7 Two models found by Mace4 for Ten-Yen puzzle

> **Puzzle 125. Rotating Ten-Yen**
>
> Let us now rotate some of the polyominoes from the Ten-Yen puzzle. Namely, we can rotate i_2, r_4, and b_3. Which would be a new solution in this case? (adapted from the Ten-Yen puzzle published in 1950 by Multiple Products Corporation and used here for educational purpose)

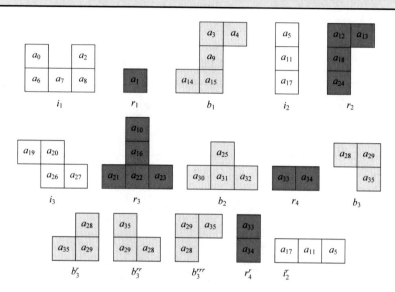

Fig. 12.8 Ten polyominoes in the Ten-Yen puzzle

Listing 12.8 The shapes i_2, r_4 and b_3 can be rotated

```
1   formulas(rotating_r4).
2      i2r    <-> left(a5,a11) & left(a11,a17).
3      r4r    <-> on(a33,a34).
4      b3r    <-> on(a28,a29) & left(a35,a29).
5      b3rr   <-> left(a29,a28) & on(a35,a29).
6      b3rrr  <-> on(a29,a28) & left(a29,a35).
7   end_of_list.
8
9   formulas(ten_yen).
10     i1 & (i2 | i2r) & i3 & r1 & r2 & r3 & (r4 | r4r) &
11     b1 & b2 & (b3 | b3r | b3rr | b3rrr).
12     r4      -> -r4r.
13     i2      -> -i2r.
14     b3      -> -b3r   & -b3rr & -b3rrr.
15     b3r     -> -b3    & -b3rr & -b3rrr.
16     b3rr    -> -b3    & -b3r  & -b3rrr.
17     b3rrr   -> -b3    & -b3r  & -b3rr.
18  end_of_list.
```

Solution

Now we can use some additional shapes. By rotating i_2, we obtain:

$$i_2^r \leftrightarrow left(a_5, a_{11}) \wedge left(a_{11}, a_{17}) \tag{12.47}$$

By rotating r_4, we obtain:

$$r_4^r \leftrightarrow on(a_{33}, a_{34}) \tag{12.48}$$

By rotating b_3, we obtain three new shapes:

$$b_3^r \leftrightarrow on(a28, a29) \wedge left(a35, a29) \tag{12.49}$$
$$b_3^{rr} \leftrightarrow left(a29, a28) \wedge on(a35, a29) \tag{12.50}$$
$$b_3^{rrr} \leftrightarrow on(a29, a28) \wedge left(a29, a35) \tag{12.51}$$

Since we can use either the initial shapes or the rotated one, the goal is relaxed:

$$i_1 \wedge (i_2 \vee i_2^r) \wedge i_3 \wedge r_1 \wedge r_2 \wedge r_3 \wedge (r_4 | r_4^r) \wedge b_1 \wedge b_2 \wedge (b_3 \vee b_3^r \vee b_3^{rr} \vee b_3^{rrr}) \tag{12.52}$$

We have to pay attention that a shape and its rotation cannot appear both in the solution:

$$r_4 \rightarrow \neg r_4^r \tag{12.53}$$
$$i_2 \rightarrow \neg i_2^r \tag{12.54}$$
$$b_3 \rightarrow \neg b_3^r \wedge \neg b_3^{rr} \wedge \neg b_3^{rrr} \tag{12.55}$$
$$b_3^r \rightarrow \neg b_3 \wedge \neg b_3^{rr} \wedge \neg b_3^{rrr} \tag{12.56}$$
$$b_3^{rr} \rightarrow \neg b_3 \wedge \neg b_3^r \wedge \neg b_3^{rrr} \tag{12.57}$$
$$b_3^{rrr} \rightarrow \neg b_3 \wedge \neg b_3^r \wedge \neg b_3^{rr} \tag{12.58}$$

Given the implementation in Listing 12.8, Mace4 finds four models. Two models using rotating shapes are depicted in Fig. 12.9. The model on the left uses $r_4^r(a_{33}, a_{34})$, while the model on the right uses two rotated shapes: $r_4^r(a_{33}, a_{34})$ and $b_3^r(a_{28}, a_{29}, a_{35})$. Note that Mace4 signals which shapes are used in each model by allocating the value 1 to those shapes.

---

a_{33}	a_{12}	a_{13}	a_{25}	a_{28}	a_{29}
a_{34}	a_{18}	a_{30}	a_{31}	a_{32}	a_{35}
a_0	a_{24}	a_2	a_3	a_4	a_5
a_6	a_7	a_8	a_9	a_{10}	a_{11}
a_{19}	a_{20}	a_{14}	a_{15}	a_{16}	a_{17}
a_1	a_{26}	a_{27}	a_{21}	a_{22}	a_{23}

a_0	a_1	a_2	a_3	a_4	a_5
a_6	a_7	a_8	a_9	a_{10}	a_{11}
a_{12}	a_{13}	a_{14}	a_{15}	a_{16}	a_{17}
a_{18}	a_{19}	a_{20}	a_{21}	a_{22}	a_{23}
a_{24}	a_{25}	a_{26}	a_{27}	a_{28}	a_{33}
a_{30}	a_{31}	a_{32}	a_{35}	a_{29}	a_{34}

Fig. 12.9 Two new models use one rotation r_4^r (left), and two rotations b_3^r and r_4^r (right)

Puzzle 126. A 4 × 5 rectangle

Use the following four pieces to fill a 4 × 5 grid. Each shape can be rotated (taken from MCRuffy Pentomino Puzzle Book)

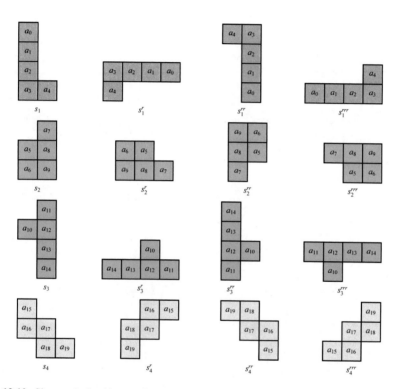

Fig. 12.10 Shapes obtained by rotating the initial four pentominoes

Listing 12.9 Finding models for the 4x5 grid

```
1   assign(domain_size ,20).    %grid 4x5
2   assign(max_models , -1).
3   set( arithmetic ).
4
5   list( distinct ).
6     [a0 ,a1 ,a2 ,a3 ,a4 ,
7      a5 ,a6 ,a7 ,a8 ,a9 ,
8      a10 ,a11 ,a12 ,a13 ,a14 ,
9      a15 ,a16 ,a17 ,a18 ,a19 ].
10  end_of_list .
11
12  formulas( utils ).
13    x < 15 -> (on(x,y) <-> y = x + 5).    %x on y,
```

```
14   x > 14 -> -on(x,y).                        %4x5 grid
15
16   (x != 4 & x != 9 & x != 14 & x != 19) -> (left(x,y) <-> y = x + 1).
17   -(x != 4 & x != 9 & x != 14 & x != 19) -> -left(x,y).
18   end_of_list.
19
20   formulas(shapes).
21   s1    <-> on(a0,a1) & on(a1,a2)  & on(a2,a3) & left(a3,a4).
22   s1r   <-> left(a1,a0) & left(a2,a1) & left(a3,a2) & on(a3,a4).
23   s1rr  <-> on(a1,a0) & on(a2,a1) & on(a3,a2) & left(a4,a3).
24   s1rrr <-> left(a0,a1) & left(a1,a2) & left(a2,a3) & on(a4,a3).
25   s1 | s1r | s1rr | s1rrr.                %at least one
26   s1    -> -s1r & -s1rr & -s1rrr.         %at most one
27   s1r   -> -s1  & -s1rr & -s1rrr.
28   s1rr  -> -s1  & -s1r  & -s1rrr.
29   s1rrr -> -s1  & -s1r  & -s1rr.
30
31   s2    <-> on(a5,a6) & on(a7,a8) & on(a8,a9) & left(a5,a8) & left(a6,a9).
32   s2r   <-> left(a6,a5) & left(a8,a7) & left(a9,a8) & on(a5,a8) & on(a6,a9).
33   s2rr  <-> on(a6,a5) & on(a8,a7) & on(a9,a8) & left(a8,a5) & left(a9,a6).
34   s2rrr <-> left(a5,a6) & left(a7,a8) & left(a8,a9) & on(a8,a5) & on(a9,a6).
35   s2 | s2r | s2rr | s2rrr.
36   s2    -> -s2r & -s2rr & -s2rrr.
37   s2r   -> -s2  & -s2rr & -s2rrr.
38   s2rr  -> -s2  & -s2r  & -s2rrr.
39   s2rrr -> -s2  & -s2r  & -s2rr.
40
41   s3    <-> on(a11,a12) & on(a12,a13) & on(a13,a14) & left(a10,a12).
42   s3r   <-> left(a12,a11) & left(a13,a12) & left(a14,a13) & on(a10,a12).
43   s3rr  <-> on(a12,a11) & on(a13,a12) & on(a14,a13) & left(a12,a10).
44   s3rrr <-> left(a11,a12) & left(a12,a13) & left(a13,a14) & on(a12,a10).
45   s3 | s3r | s3rr | s3rrr.
46   s3    -> -s3r & -s3rr & -s3rrr.
47   s3r   -> -s3  & -s3rr & -s3rrr.
48   s3rr  -> -s3  & -s3r  & -s3rrr.
49   s3rrr -> -s3  & -s3r  & -s3rr.
50
51   s4    <-> on(a15,a16) & left(a16,a17) & on(a17,a18) & left(a18,a19).
52   s4r   <-> left(a16,a15) & on(a16,a17) & left(a18,a17) & on(a18,a19).
53   s4rr  <-> on(a16,a15) & left(a17,a16) & on(a18,a17) & left(a19,a18).
54   s4rrr <-> left(a15,a16) & on(a17,a16) & left(a17,a18) & on(a19,a18).
55   s4 | s4r | s4rr | s4rrr.
56   s4    -> -s4r & -s4rr & -s4rrr.
57   s4r   -> -s4  & -s4rr & -s4rrr.
58   s4rr  -> -s4  & -s4r  & -s4rrr.
59   s4rrr -> -s4  & -s4r  & -s4rr.
60   end_of_list.
```

Solution

Since the grid is 4×5 we need a domain of 20 elements. Each element a_i represents a position in the grid from 0 to 19. Hence, each position a_i should be distinct.
We start by defining the predicates used to represent the shapes. Since there are 5 columns, there is a distance of 5 between two squares for which the relation *on* holds: $on(x, y) \leftrightarrow y = x + 5$. Squares on the bottom line cannot be above other squares:

$$x \leq 14 \rightarrow (on(x, y) \leftrightarrow y = x + 5) \qquad x > 14 \rightarrow \neg on(x, y)$$

For the *left* predicate, if a square x is to the left of y, its value is less by one from y's value. A square x situated anywhere on the last column ($x = 4$, $x = 9$, $x = 14$, or $x = 19$) cannot be to the left of other square:

$$(x \neq 4 \wedge x \neq 9 \wedge x \neq 14 \wedge x \neq 19) \rightarrow (on(x, y) \leftrightarrow y = x + 1) \quad (12.59)$$
$$\neg(x \neq 4 \wedge x \neq 9 \wedge x \neq 14 \wedge x \neq 19) \rightarrow \neg left(x, y) \quad (12.60)$$

Next we need to formalise each of the four shapes and their rotations. Let, for instance, the red shape s_1 and its consecutive rotations to the right (see Fig. 12.10):

$$s_1 \leftrightarrow on(a_0, a_1) \wedge on(a_1, a_2) \wedge on(a_2, a_3) \wedge left(a_3, a_4) \quad (12.61)$$
$$s_1^r \leftrightarrow left(a_1, a_0) \wedge left(a_2, a_1) \wedge left(a_3, a_2) \wedge on(a_3, a_4) \quad (12.62)$$
$$s_1^{rr} \leftrightarrow on(a_1, a_0) \wedge on(a_2, a_1) \wedge on(a_3, a_2) \wedge left(a_4, a_3) \quad (12.63)$$
$$s_1^{rrr} \leftrightarrow left(a_0, a_1) \wedge left(a_1, a_2) \wedge left(a_2, a_3) \wedge on(a_4, a_3) \quad (12.64)$$

Recall that exactly one of the above shapes can be used:

$$s_1 \vee s_1^r \vee s_1^{rr} \vee s_1^{rrr}$$
$$(s_1 \rightarrow \neg s_1^r \neg s_1^{rr} \neg s_1^{rrr}) \wedge (s_1^r \rightarrow \neg s_1 \neg s_1^{rr} \neg s_1^{rrr}) \wedge (s_1^{rr} \rightarrow \neg s_1 \neg s_1^r \neg s_1^{rrr}) \wedge (s_1^{rrr} \rightarrow \neg s_1 \neg s_1^r \neg s_1^{rr})$$

Given the implementation in Listing 12.9, Mace4 finds two models. The shapes used in each model are:

Model	s_1	s_1^r	s_1^{rr}	s_1^{rrr}	s_2	s_2^r	s_2^{rr}	s_2^{rrr}	s_3	s_3^r	s_3^{rr}	s_3^{rrr}	s_4	s_4^r	s_4^{rr}	s_4^{rrr}
1	0	0	0	1	0	0	0	1	0	0	1	0	0	1	0	0
2	0	1	0	0	0	1	0	0	1	0	0	0	0	0	1	0

These two models correspond to the solutions depicted in Fig. 12.11. Note that the second solution can be obtained by rotating the first model twice.

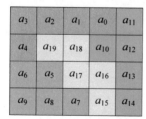

Fig. 12.11 Two models found by Mace4 for the 4×5 puzzle

Puzzle 127. The 12 pentominoes

Show that the 12 pentominoes form a 3×20 rectangle. (taken from Golomb (1996))

0	1	2	3	4	5	6	7	8	9	10	11	12	12	14	15	16	17	18	19
20	21	22	23	24	25	26	27	28	29	30	31	32	33	34	35	36	37	38	39
40	41	42	43	44	45	46	47	48	49	50	51	52	53	54	55	56	57	58	59

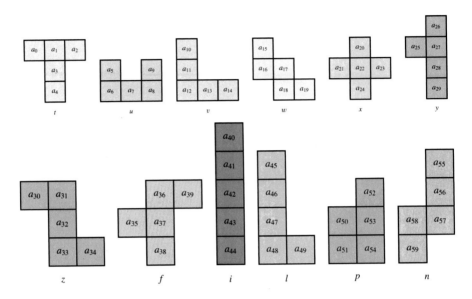

Listing 12.10 Formalising a 3x20 grid

```
1   assign(domain_size,60).    %grid 3x20
2   assign(max_models,1).
3   assign(max_megs,  -1).
4   set(arithmetic).
5
6   list(distinct).
7    [a0,a1,a2,a3,a4,a5,a6,a7,a8,a9,a10,a11,a12,a13,a14,a15,a16,a17,a18,a19,
8     a20,a21,a22,a23,a24,a25,a26,a27,a28,a29,a30,a31,a32,a33,a34,a35,a36,
9                                                             a37,a38,a39,
10    a40,a41,a42,a43,a44,a45,a46,a47,a48,a49,a50,a51,a52,a53,a54,a55,a56,
11                                                             a57,a58,a59].
12   end_of_list.
13
14   formulas(utils).
15     x < 40 -> (on(x,y) <-> y = x + 20).   %x on y,
16     x > 39 -> -on(x,y).                    %3x20 grid
17
18     (x != 19 & x != 39 & x != 59) -> (left(x,y) <-> y = x + 1).
19     -(x != 19 & x != 39 & x != 59) -> -left(x,y).
20   end_of_list.
```

Listing 12.11 The T-pentomino

```
1   formulas ( t_shape ).
2    t    <-> left (a0,a1) & left (a1,a2) &    on (a1,a3) &    on (a3,a4).
3    tr   <->   on (a0,a1) &   on (a1,a2) & left (a3,a1) & left (a4,a3).
4    trr  <-> left (a1,a0) & left (a2,a1) &    on (a3,a1) &    on (a4,a3).
5    trrr <->   on (a1,a0) & on (a2,a1) & left (a1,a3) & left (a3,a4).
6    t I tr I trr I trrr.
7    t    -> -tr & -trr & -trrr.
8    tr   -> -t  & -trr & -trrr.
9    trr  -> -t  & -tr  & -trrr.
10   trrr -> -t  & -tr  & -trr.
11  end_of_list.
```

Listing 12.12 The U-pentonymo

```
1   formulas ( u_shape ).
2    U    <->   on (a5,a6) & left (a6,a7) & left (a7,a8) &    on (a9,a8).
3    Ur   <-> left (a6,a5) &   on (a6,a7) &   on (a7,a8) & left (a8,a9).
4    Urr  <->   on (a6,a5) & left (a7,a6) & left (a8,a7) &    on (a8,a9).
5    Urrr <-> left (a5,a6) &   on (a7,a6) &   on (a8,a7) & left (a9,a8).
6    U I Ur I Urr I Urrr.
7    U    -> -Ur & -Urr & -Urrr.
8    Ur   -> -U  & -Urr & -Urrr.
9    Urr  -> -U  & -Ur  & -Urrr.
10   Urrr -> -U  & -Ur  & -Urr.
11  end_of_list.
```

Listing 12.13 The V-pentonymo

```
1   formulas ( v_shape ).
2    V    <->   on (a10,a11) &   on (a11,a12) & left (a12,a13) & left (a13,a14).
3    Vr   <-> left (a11,a10) & left (a12,a11) &   on (a12,a13) &   on (a13,a14).
4    Vrr  <->   on (a11,a10) &   on (a12,a11) & left (a13,a12) & left (a14,a13).
5    Vrrr <-> left (a10,a11) & left (a11,a12) &   on (a13,a12) &   on (a14,a13).
6    V I Vr I Vrr I Vrrr.
7    V    -> -Vr & -Vrr & -Vrrr.
8    Vr   -> -V  & -Vrr & -Vrrr.
9    Vrr  -> -V  & -Vr  & -Vrrr.
10   Vrrr -> -V  & -Vr  & -Vrr.
11  end_of_list.
```

Solution

Let us start by defining the 3×20 grid (see Listing 12.10). We need 60 distinct elements a_i, $i \in [0..59]$. That is, each variable a_i occupies a position from 0 to 59 in the grid. Since there are 20 columns, the predicate $on(x, y)$ assumes a difference of 20 between y and x. This predicate does not hold for the positions in the last line:

$$x < 40 \to (on(x, y) \leftrightarrow y = x + 20) \tag{12.65}$$

$$x \geq 40 \to \neg on(x, y) \tag{12.66}$$

The predicate $left(x, y)$ assumes a difference of 1 between y and x, and it does not hold on the elements of the last column:

$$x \neq 19 \wedge x \neq 39 \wedge x \neq 59 \to (left(x, y) \leftrightarrow y = x + 1) \tag{12.67}$$

$$\neg(x \neq 19 \wedge x \neq 39 \wedge x \neq 59) \to \neg left(x, y) \tag{12.68}$$

We continue by defining each of the 12 pentominoes and all their rotations. We assume that polyominoes can be rotated (turned 90, 180, or 270 degrees) or reflected (flipped over).

Let the T-shape (see Listing 12.11) for which we allocate the variables a_0, a_1, a_2, a_3, and a_4:

$$T \leftrightarrow left(a_0, a_1) \wedge left(a_1, a_2) \wedge on(a_1, a_3) \wedge on(a3, a4) \tag{12.69}$$

By rotating to the right with 90, 180, and 270 degree, we obtain three new shapes:

$$T^r \leftrightarrow on(a_0, a_1) \wedge on(a_1, a_2) \wedge left(a_3, a_1) \wedge left(a_4, a_3) \tag{12.70}$$

$$T^{rr} \leftrightarrow left(a_1, a_0) \wedge left(a_2, a_1) \wedge on(a_3, a_1) \wedge on(a_4, a_3) \tag{12.71}$$

$$T^{rrr} \leftrightarrow on(a_1, a_0) \wedge on(a_2, a_1) \wedge left(a_1, a_3) \wedge left(a_3, a_4) \tag{12.72}$$

Note that by reflecting these shapes we do not obtain a new one. Hence, the T-pentomino has only four shapes. Don't forget to add that only one of its shapes can be used:

$$T \vee T^r \vee T^{rr} \vee T^{rrr} \tag{12.73}$$

$$T \rightarrow \neg T^r \wedge \neg T^{rr} \wedge \neg T^{rrr} \tag{12.74}$$

$$T^r \rightarrow \neg T \wedge \neg T^{rr} \wedge \neg T^{rrr} \tag{12.75}$$

$$T^{rr} \rightarrow \neg T \wedge \neg T^r \wedge \neg T^{rrr} \tag{12.76}$$

$$T^{rrr} \rightarrow \neg T \wedge \neg T^r \wedge \neg T^{rr} \tag{12.77}$$

Solution

Assume now the U-pentomino has the variables a_5, a_6, a_7, a_8, and a_9. With three successive rotations to the right, we obtain:

$$U \leftrightarrow on(a_5, a_6) \wedge left(a_6, a_7) \wedge left(a_7, a_8) \wedge on(a_9, a_8) \tag{12.78}$$

$$U^r \leftrightarrow left(a_6, a_5) \wedge on(a_6, a_7) \wedge on(a_7, a_8) \wedge left(a_8, a_9) \tag{12.79}$$

$$U^{rr} \leftrightarrow on(a_6, a_5) \wedge left(a_7, a_6) \wedge left(a_8, a_7) \wedge on(a_8, a_9) \tag{12.80}$$

$$U^{rrr} \leftrightarrow left(a_5, a_6) \wedge on(a_7, a_6) \wedge on(a_8, a_7) \wedge left(a_9, a_8) \tag{12.81}$$

By reflecting these shapes, we do not obtain a new one. Hence, the U-pentomino has only four shapes. Similar to the T-pentomino, don't forget to add the conditions that exactly one of the U, U^r, U^{rr}, U^{rrr} can be used to fill the rectangle (lines 6–10 in Listing 12.12).

The V-pentomino and W-pentomino have only four shapes each. Let the next five variables a_{10} to a_{14} for the V-shape:

$$V \leftrightarrow on(a_{10}, a_{11}) \wedge on(a_{11}, a_{12}) \wedge left(a_{12}, a_{13}) \wedge left(a_{13}, a_{14}) \tag{12.82}$$

$$V^r \leftrightarrow left(a_{11}, a_{10}) \wedge left(a_{12}, a_{11}) \wedge on(a_{12}, a_{13}) \wedge on(a_{13}, a_{14}) \tag{12.83}$$

$$V^{rr} \leftrightarrow on(a_{11}, a_{10}) \wedge on(a_{12}, a_{11}) \wedge left(a_{13}, a_{12}) \wedge left(a_{14}, a_{13}) \quad (12.84)$$
$$V^{rrr} \leftrightarrow left(a_{10}, a_{11}) \wedge left(a_{11}, a_{12}) \wedge on(a_{13}, a_{12}) \wedge on(a_{14}, a_{13}) \quad (12.85)$$

The W-pentomino is similarly formalised with the variables a_{15} to a_{19}.
The X-pentomino has only one shape, no matter the rotation or reflection. Given the variables a_{20} to a_{24}, the X-pentonymo is formalised with:

$$X \leftrightarrow on(a_{20}, a_{22}) \wedge on(a_{22}, a_{24}) \wedge left(a_{21}, a_{22}) \wedge left(a_{22}, a_{23}) \quad (12.86)$$

Since X-pentomino should appear in the solution, we set its value to true (i.e. X).

Solution

The Y-pentomino has eight shapes, from which three are obtained from rotations and four by reflection. Given the variables a_{25} to a_{29}, the initial shape Y and its rotations with 90 (Y^r), 180 (Y^{rr}), and 270 (Y^{rrr}) degrees are:

$$Y \leftrightarrow on(a_{26}, a_{27}) \wedge on(a_{27}, a_{28}) \wedge on(a_{28}, a_{29}) \wedge left(a_{25}, a_{27}) \quad (12.87)$$
$$Y^r \leftrightarrow left(a_{27}, a_{26}) \wedge left(a_{28}, a_{27}) \wedge left(a_{29}, a_{28}) \wedge on(a_{25}, a_{27}) \quad (12.88)$$
$$Y^{rr} \leftrightarrow on(a_{27}, a_{26}) \wedge on(a_{28}, a_{27}) \wedge on(a_{29}, a_{28}) \wedge left(a_{27}, a_{25}) \quad (12.89)$$
$$Y^{rrr} \leftrightarrow left(a_{26}, a_{27}) \wedge left(a_{27}, a_{28}) \wedge left(a_{28}, a_{29}) \wedge on(a_{27}, a_{25}) \quad (12.90)$$

By reflecting each of the above shapes, we obtain four new ones:

$$Y_u \leftrightarrow on(a_{26}, a_{27}) \wedge on(a_{27}, a_{28}) \wedge on(a_{28}, a_{29}) \wedge left(a_{27}, a_{25}) \quad (12.91)$$
$$Y_u^r \leftrightarrow left(a_{27}, a_{26}) \wedge left(a_{28}, a_{27}) \wedge left(a_{29}, a_{28}) \wedge on(a_{27}, a_{25}) \quad (12.92)$$
$$Y_u^{rr} \leftrightarrow on(a_{27}, a_{26}) \wedge on(a_{28}, a_{27}) \wedge on(a_{29}, a_{28}) \wedge left(a_{25}, a_{27}) \quad (12.93)$$
$$Y_u^{rrr} \leftrightarrow left(a_{26}, a_{27}) \wedge left(a_{27}, a_{28}) \wedge left(a_{28}, a_{29}) \wedge on(a_{25}, a_{27}) \quad (12.94)$$

From these eight Y-pentomino forms, exactly one can be used in the solution:

$$Y \vee Y^r \vee Y^{rr} \vee Y^{rrr} \vee Y_u \vee Y_u^r \vee Y_u^{rr} \vee Y_u^{rrr} \quad (12.95)$$
$$Y \rightarrow \neg Y^r \wedge \neg Y^{rr} \wedge \neg Y^{rrr} \wedge \neg Y_u \wedge \neg Y_u^r \wedge \neg Y_u^{rr} \wedge \neg Y_u^{rrr} \quad (12.96)$$
$$Y^r \rightarrow \neg Y \wedge \neg Y^{rr} \wedge \neg Y^{rrr} \wedge \neg Y_u \wedge \neg Y_u^r \wedge \neg Y_u^{rr} \wedge \neg Y_u^{rrr} \quad (12.97)$$
$$Y^{rr} \rightarrow \neg Y \wedge \neg Y^r \wedge \neg Y^{rrr} \wedge \neg Y_u \wedge \neg Y_u^r \wedge \neg Y_u^{rr} \wedge \neg Y_u^{rrr} \quad (12.98)$$
$$Y^{rrr} \rightarrow \neg Y \wedge \neg Y^r \wedge \neg Y^{rr} \wedge \neg Y_u \wedge \neg Y_u^r \wedge \neg Y_u^{rr} \wedge \neg Y_u^{rrr} \quad (12.99)$$
$$Y_u \rightarrow \neg Y \wedge \neg Y^r \wedge \neg Y^{rr} \wedge \neg Y^{rrr} \wedge \neg Y_u^r \wedge \neg Y_u^{rr} \wedge \neg Y_u^{rrr} \quad (12.100)$$
$$Y_u^r \rightarrow \neg Y \wedge \neg Y^r \wedge \neg Y^{rr} \wedge \neg Y^{rrr} \wedge \neg Y_u \wedge \neg Y_u^{rr} \wedge \neg Y_u^{rrr} \quad (12.101)$$
$$Y_u^{rr} \rightarrow \neg Y \wedge \neg Y^r \wedge \neg Y^{rr} \wedge \neg Y^{rrr} \wedge \neg Y_u \wedge \neg Y_u^r \wedge \neg Y_u^{rrr} \quad (12.102)$$
$$Y_u^{rrr} \rightarrow \neg Y \wedge \neg Y^r \wedge \neg Y^{rr} \wedge \neg Y^{rrr} \wedge \neg Y_u \wedge \neg Y_u^r \wedge \neg Y_u^{rr} \quad (12.103)$$

The Z-pentomino has three more shapes (excepting the initial one): one shape obtained by rotation and two obtained by reflection:

$$Z \leftrightarrow left(a_{30}, a_{31}) \wedge on(a_{31}, a_{32}) \wedge on(a_{32}, a_{33}) \wedge left(a_{33}, a_{34}) \qquad (12.104)$$

$$Z^r \leftrightarrow on(a_{30}, a_{31}) \wedge left(a_{32}, a_{31}) \wedge left(a_{33}, a_{32}) \wedge on(a_{33}, a_{34}) \qquad (12.105)$$

$$Z_u \leftrightarrow left(a_{31}, a_{30}) \wedge on(a_{31}, a_{32}) \wedge on(a_{32}, a_{33}) \wedge left(a_{34}, a_{33}) \qquad (12.106)$$

$$Z_u^r \leftrightarrow on(a_{31}, a_{30}) \wedge left(a_{32}, a_{31}) \wedge left(a_{33}, a_{32}) \wedge on(a_{34}, a_{33}) \qquad (12.107)$$

Note that the I-pentomino has only two shapes. The number of shapes for each pentomino is:

Pentomino	T	U	V	W	X	Y	Z	F	I	L	P	N	All
Shapes	4	4	4	4	1	8	4	8	2	8	8	8	63

There are then 63 distinct shapes if rotation and reflection are allowed. For each pentomino, we use a distinct file. Hence, we can import only the shapes required by a given puzzle. Since the current puzzle uses all 12 pentominoes, we call Mace4 with the 3×20 rectangle and all the pentominoes files:

```
mace4 -f 3x20.in t.in u.in v.in w.in x.in y.in z.in f.in
                 i.in l.in p.in n.in
```

One model computed by Mace4 is depicted in Fig. 12.12. The model uses the following shapes: U^r, V^{rrr}, W^{rr}, Y^r, Z^r, F_u^{rrr}, I_r, L_u^{rrr}, N_u^{rrr}, P_u^{rrr}, and T.

a_6	a_5	a_{20}	a_{44}	a_{43}	a_{42}	a_{41}	a_{40}	a_{30}	a_{19}	a_{18}	a_0	a_1	a_2	a_{35}	a_{45}	a_{46}	a_{47}	a_{48}	a_{14}
a_7	a_{21}	a_{22}	a_{23}	a_{50}	a_{51}	a_{33}	a_{32}	a_{31}	a_{25}	a_{17}	a_{16}	a_3	a_{36}	a_{37}	a_{38}	a_{58}	a_{59}	a_{49}	a_{13}
a_8	a_9	a_{24}	a_{52}	a_{53}	a_{54}	a_{34}	a_{29}	a_{28}	a_{27}	a_{26}	a_{15}	a_4	a_{39}	a_{55}	a_{56}	a_{57}	a_{10}	a_{11}	a_{12}

Fig. 12.12 One model found by Mace4 for the 3×20 rectangle

Find two solutions to fill a 5×6 rectangle with the following five pentonimoes: T, W, Y, Z, I, and L. You can rotate and flip over each pentonimo. (adapted from Golomb (1996))

0	1	2	3	4	5
6	7	8	9	10	11
12	12	14	15	16	17
18	19	20	21	22	23
24	25	26	27	28	29

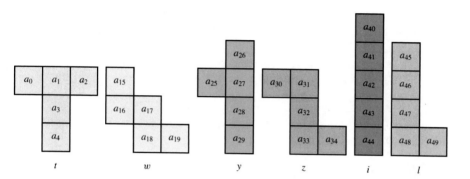

Listing 12.14 Formalising a 5x6 grid

```
1   assign(domain_size ,30).   %grid 5x6
2   assign(max_models ,2).
3   set(arithmetic).
4
5   formulas(utils).
6     x < 24 -> (on(x,y) <-> y = x + 6).    %x on y,
7     x > 23 -> -on(x,y).                    %4x5 grid
8
9     (x != 5 & x != 11 & x != 17 & x != 23 & x != 29) ->
10                                (left(x,y) <-> y = x + 1).
11    -(x != 5 & x != 11 & x != 17 & x != 23 & x != 29) -> -left(x,y).
12  end_of_list.
```

Listing 12.15 Matching variables with the active pentonimoes

```
1   list(distinct).
2     [a0 ,a1 ,a2 ,a3 ,a4 ,            %t [0..4]
3      a15 ,a16 ,a17 ,a18 ,a19 ,       %w [15..19]
4      a25 ,a26 ,a27 ,a28 ,a29 ,       %y [25..29],
5      a30 ,a31 ,a32 ,a33 ,a34 ,       %z [30..34]
6      a40 ,a41 ,a42 ,a43 ,a44 ,       %i [40-44]
7      a45 ,a46 ,a47 ,a48 ,a49 ].      %l [45..49]
8   end_of_list.
```

Solution

We start by defining the 5×6 rectangle (see Listing 12.14). We need 30 distinct elements a_i, $i \in [0..29]$. That is, each variable a_i occupies a position from 0 to 29 in the grid. Since there are six columns, the predicate $on(x, y)$ assumes a difference of six between y and x. This predicate does not hold for the positions in the last line:

$$x < 24 \rightarrow (on(x, y) \leftrightarrow y = x + 20) \tag{12.108}$$

$$x \geq 24 \rightarrow \neg on(x, y) \tag{12.109}$$

The predicate $left(x, y)$ assumes a difference of 1 between y and x, and it does not hold on the elements of the last column:

$$x \neq 5 \wedge x \neq 11 \wedge x \neq 17 \wedge x \neq 23 \wedge x \neq 29 \rightarrow (left(x, y) \leftrightarrow y = x + 1) \tag{12.110}$$

$$\neg(x \neq 5 \wedge x \neq 11 \wedge x \neq 17 \wedge x \neq 23 \wedge x \neq 29) \rightarrow -left(x, y) \tag{12.111}$$

The active pentominoes are: T, W, Y, Z, I, L. We will introduce only the 30 variables allocated for these pentominoes. These variables are a_0 to a_4 for T, a_{15} to a_{19} for W, a_{25} to a_{29} for Y, a_{30} to a_{34} for Z, a_{40} to a_{44} for I, and a_{45} to a_{49} for L (see Listing 12.15).

Next we need to import the six pentonimoes T, W, Y, Z, I, and L. We exemplify here only the L-pentomino (see Fig. 12.13 and Listing12.16). We start by defining the L-shape (line 2):

$$L \leftrightarrow on(a_{45}, a_{46}) \wedge on(a_{46}, a_{47}) \wedge on(a_{47}, a_{48}) \wedge left(a_{48}, a_{49}) \tag{12.112}$$

Then, we have its three successive rotations (lines 3–5) and four reflections (lines 7–10). The exclusive-or between these eight shapes is formalised in lines 12–20.

Listing 12.16 Formalising the L-pentomino and its rotations and reflexions

```
1   formulas(l_shape).
2   l     <->    on(a45,a46) &    on(a46,a47) &    on(a47,a48) & left(a48,a49).
3   lr    <->  left(a46,a45) & left(a47,a46) & left(a48,a47) &    on(a48,a49).
4   lrr   <->    on(a46,a45) &    on(a47,a46) &    on(a48,a47) & left(a49,a48).
5   lrrr  <->  left(a45,a46) & left(a46,a47) & left(a47,a48) &    on(a49,a48).
6
7   lu    <->    on(a45,a46) &    on(a46,a47) &    on(a47,a48) & left(a49,a48).
8   lur   <->  left(a46,a45) & left(a47,a46) & left(a48,a47) &    on(a49,a48).
9   lurr  <->    on(a46,a45) &    on(a47,a46) &    on(a48,a47) & left(a48,a49).
10  lurrr <->  left(a45,a46) & left(a46,a47) & left(a47,a48) &    on(a49,a48).
11
12  l | lr | lrr | lrrr | lu | lur | lurr | lurrr.
13  l     -> -lr  & -lrr & -lrrr & -lu   & -lur  & -lurr & -lurrr.
14  lr    -> -l   & -lrr & -lrrr & -lu   & -lur  & -lurr & -lurrr.
15  lrr   -> -l   & -lr  & -lrr  & -lu   & -lur  & -lurr & -lurrr.
16  lrrr  -> -l   & -lr  & -lrr  & -lu   & -lur  & -lurr & -lurrr.
17  lu    -> -l   & -lr  & -lrr  & -lrrr & -lur  & -lurr & -lurrr.
18  lur   -> -l   & -lr  & -lrr  & -lrrr & -lu   & -lurr & -lurrr.
19  lurr  -> -l   & -lr  & -lrr  & -lrrr & -lu   & -lur  & -lurrr.
20  lurrr -> -l   & -lr  & -lrr  & -lrrr & -lu   & -lur  & -lurr.
21  end_of_list.
```

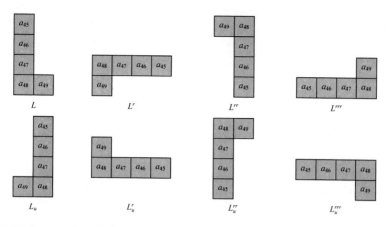

Fig. 12.13 L-pentomino plus its rotations (L^r, L^{rr}, L^{rrr}) and reflections (L_u, L_u^r, L_u^{rr}, L_u^{rrr})

Solution

To search for a model, we just need to import the 5×6 grid, the list of distinct variables, and the six pentominoes:

```
mace4 -f 5x6.in vars_t_w_y_z_i_l t.in w.in y.in z.in i.in
                              l.in
```

Two models computed by Mace4 are depicted in Fig. 12.14. The first model uses the following shapes: W^{rr}, Y_u, Z, I^r, L_u^r, and T^r. The second model uses W^r, Y, Z^r, I^r, L^{rrr}, and T^{rrr}.

a_{44}	a_{43}	a_{42}	a_{41}	a_{40}	a_0
a_{26}	a_{30}	a_{31}	a_4	a_3	a_1
a_{27}	a_{25}	a_{32}	a_{19}	a_{18}	a_2
a_{28}	a_{49}	a_{33}	a_{34}	a_{17}	a_{16}
a_{29}	a_{48}	a_{47}	a_{46}	a_{45}	a_{15}

a_2	a_{44}	a_{43}	a_{42}	a_{41}	a_{40}
a_1	a_3	a_4	a_{31}	a_{30}	a_{26}
a_0	a_{16}	a_{15}	a_{32}	a_{25}	a_{27}
a_{18}	a_{17}	a_{34}	a_{33}	a_{49}	a_{28}
a_{19}	a_{45}	a_{46}	a_{47}	a_{48}	a_{29}

Fig. 12.14 The first two models found by Mace4 for the 5×6 rectangle using T, W, Y, Z, I, and L

Puzzle 129. Importing other six pentominoes

Find two solutions to fill a 5 × 6 rectangle with the following five pentominoes: U, V, X, F, P, and N. You can rotate and flip over each pentomino. (adapted from Golomb (1996))

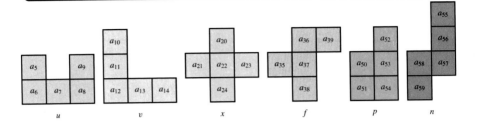

Listing 12.17 Matching variables with the active pentonimoes

```
1    list ( distinct ) .
2      [ a5 , a6 , a7 , a8 , a9 ,      %u [ 5 . . 9 ]
3        a10 , a11 , a12 , a13 , a14 ,   %v [ 10 . . 14 ]
4        a20 , a21 , a22 , a23 , a24 ,   %x [ 20 . . 24 ] ,
5        a35 , a36 , a37 , a38 , a39 ,   %f [ 35 . . 39 ]
6        a50 , a51 , a52 , a53 , a54 ,   %p [ 50 . . 54 ]
7        a55 , a56 , a57 , a58 , a59 ] .  %n [ 55 . . 59 ]
8    end_of_list .
```

Solution

We use the same 5 × 6 rectangle formalised in Listing 12.14. The active pentominoes are: U, V, X, F, P, and N. We will introduce only the 30 variables appearing in these pentominoes. These variables are a_5 to a_9 for U, a_{10} to a_{14} for V, a_{20} to a_{24} for X, a_{35} to a_{39} for F, a_{50} to a_{54} for P, and a_{55} to a_{59} for N (see Listing 12.17).
Next we need to import the six pentominoes U, V, X, F, P, and N.
We exemplify here only the P-pentomino (see Fig. 12.15 and Listing 12.18). We start by defining the P-shape (line 2):

$$P \leftrightarrow on(a_{50}, a_{51}) \wedge on(a_{52}, a_{53}) \wedge on(a_{53}, a_{54}) \wedge left(a_{51}, a_{54}) \qquad (12.113)$$

Then, we have its three successive rotations (lines 3–5) and four reflections (lines 7–10). The "exclusive-or" between these eight shapes is formalised in lines 12–20.
To search for a model, we just need to import the 5 × 6 grid, the list of distinct variables, and the six pentominoes:

```
mace4 -f 5x6.in vars_u_v_x_f_p_n u.in v.in x.in f.in p.in
                                n.in
```

Two models computed by Mace4 are depicted in Fig. 12.16. The first model uses the following shapes: U^{rrr}, V^r, X, F^{rr}, and N_u^{rrr}. The second model uses U^{rrr}, V^r, X, F^{rr}, and N.

Listing 12.18 Formalising the P-pentonimo and its rotations and reflexions

```
1    formulas ( p_shape ).
2    p     <->  on ( a50 , a51 )  &    on ( a52 , a53 )  &    on ( a53 , a54 )  &  left ( a51 , a54 ).
3    pr    <->  left ( a51 , a50 )  &  left ( a53 , a52 )  &  left ( a54 , a53 )  &    on ( a51 , a54 ).
4    prr   <->  on ( a51 , a50 )  &    on ( a53 , a52 )  &    on ( a54 , a53 )  &  left ( a54 , a51 ).
5    prrr  <->  left ( a50 , a51 )  &  left ( a52 , a53 )  &  left ( a53 , a54 )  &    on ( a54 , a51 ).
6
7    pu    <->  on ( a50 , a51 )  &    on ( a52 , a53 )  &    on ( a53 , a54 )  &  left ( a54 , a51 ).
8    pur   <->  left ( a51 , a50 )  &  left ( a53 , a52 )  &  left ( a54 , a53 )  &    on ( a54 , a51 ).
9    purr  <->  on ( a51 , a50 )  &    on ( a53 , a52 )  &    on ( a54 , a53 )  &  left ( a51 , a54 ).
10   purrr <->  left ( a50 , a51 )  &  left ( a52 , a53 )  &  left ( a53 , a54 )  &    on ( a51 , a54 ).
11
12   p | pr | prr | prrr | pu | pur | purr | purrr .
13   p     ->  -pr  & -prr  & -prrr  & -pu   & -pur  & -purr  & -purrr .
14   pr    ->  -p   & -prr  & -prrr  & -pu   & -pur  & -purr  & -purrr .
15   prr   ->  -p   & -pr   & -prrr  & -pu   & -pur  & -purr  & -purrr .
16   prrr  ->  -p   & -pr   & -prr   & -pu   & -pur  & -purr  & -purrr .
17   pu    ->  -p   & -pr   & -prr   & -prrr & -pur  & -purr  & -purrr .
18   pur   ->  -p   & -pr   & -prr   & -prrr & -pu   & -purr  & -purrr .
19   purr  ->  -p   & -pr   & -prr   & -prrr & -pu   & -pur   & -purrr .
20   purrr ->  -p   & -pr   & -prr   & -prrr & -pu   & -pur   & -purr .
21   end_of_list .
```

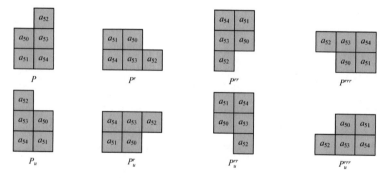

Fig. 12.15 P-pentomino plus its rotations (P^r, P^{rr}, P^{rrr}) and reflections (P_u, P_u^r, P_u^{rr}, P_u^{rrr})

a_{12}	a_{11}	a_{10}	a_{20}	a_9	a_8
a_{13}	a_{38}	a_{21}	a_{22}	a_{23}	a_7
a_{14}	a_{37}	a_{35}	a_{24}	a_5	a_6
a_{39}	a_{36}	a_{58}	a_{59}	a_{50}	a_{51}
a_{55}	a_{56}	a_{57}	a_{52}	a_{53}	a_{54}

a_{12}	a_{11}	a_{10}	a_{20}	a_9	a_8
a_{13}	a_{55}	a_{21}	a_{22}	a_{23}	a_7
a_{14}	a_{56}	a_{38}	a_{24}	a_5	a_6
a_{58}	a_{57}	a_{37}	a_{35}	a_{50}	a_{51}
a_{59}	a_{35}	a_{36}	a_{52}	a_{53}	a_{54}

Fig. 12.16 The first two models found by Mace4 for the 5×6 rectangle using U, V, X, F, P, and N

Puzzle 130. Twelve pentominoes on a chessboard

Using each of the twelve pentominoes, find a solution to the 8×8 grid with the four-square hole in the middle. (adapted from Golomb (1996))

Listing 12.19 Twelve pentominoes on a 8×8 grid

```
1    assign(domain_size,64).   %grid 8x8
2    assign(max_models,1).
3    assign(max_megs,-1).
4    set(arithmetic).
5
6    list(distinct).
7      [a0,a1,a2,a3,a4,a5,a6,a7,
8       a8,a9,a10,a11,a12,a13,a14,a15,
9       a16,a17,a18,a19,a20,a21,a22,a23,
10      a24,a25,a26,a27,a28,a29,a30,a31,
11      a32,a33,a34,a35,a36,a37,a38,a39,
12      a40,a41,a42,a43,a44,a45,a46,a47,
13      a48,a49,a50,a51,a52,a53,a54,a55,
14      a56,a57,a58,a59,27,28,35,36].  %the center is free
15   end_of_list.
16
17   formulas(utils).
18     x < 56 -> (on(x,y) <-> y = x + 8).   %x on y,
19     x > 55 -> -on(x,y).                  %8x8 grid
20
21     (x != 7 & x != 15 & x != 23 & x != 31 &
22      x != 39 & x != 47 & x != 55 & x != 63) -> (left(x,y) <-> y = x + 1).
23     -(x != 7 & x != 15 & x != 23 & x != 31 &
24      x != 39 & x != 47 & x != 55 & x != 63) -> -left(x,y).
25   end_of_list.
```

Solution

The *on* and *left* predicates are defined as usual for the 8×8 grid. The $on(x, y)$ predicate requires a difference of 8 between the position of x and the position of y. The predicate does not hold on the last line:

$$x < 56 \rightarrow (on(x, y) \leftrightarrow y = x + 8) \tag{12.114}$$

$$x \geq 56 \rightarrow \neg on(x, y) \tag{12.115}$$

The predicate $left(x, y)$ assumes a difference of 1 between y and x, and it does not hold on the elements of the last column: $x \bmod 8 = 7 \rightarrow \neg left(x, y)$. Explicitly, these elements are 7, 15, 23, 31, 39, 47, 55, and 59 (lines 21–24 in Listing 12.19).

To cover the chessboard, we need a domain of 64 elements. In the current formalisation, the hole in the middle covers the positions 27, 28, 35, and 36 (see Fig. 12.17). The trick here is just to assure that each variable a_i is distinct for these empty positions. We model this with the distinct list in Listing 12.19.

To search for a model, we just need to import the 8×8 grid, the lists of distinct variables, and all 12 pentominoes:

```
mace4 -f 8x8.in all12.in
```

The first model computed by Mace4 is depicted in Fig. 12.17. Note that Mace4 requires some computational effort to find this model. The model uses the following shapes: U^{rr}, V^{rr}, W, X, Y, Z_u^r, F_u, I, L_u^r, N^r, and P^{rr}.

a_{40}	a_8	a_7	a_6	a_{34}	a_{14}	a_{13}	a_{12}
a_{41}	a_9	a_{20}	a_5	a_{33}	a_{32}	a_{31}	a_{11}
a_{42}	a_{21}	a_{22}	a_{23}	a_{39}	a_{36}	a_{30}	a_{10}
a_{43}	a_{15}	a_{24}	27	28	a_{37}	a_{35}	a_{26}
a_{44}	a_{16}	a_{17}	35	36	a_{38}	a_{25}	a_{27}
a_{59}	a_{58}	a_{18}	a_{19}	a_{54}	a_{51}	a_4	a_{28}
a_{49}	a_{57}	a_{56}	a_{55}	a_{50}	a_{53}	a_3	a_{29}
a_{48}	a_{47}	a_{46}	a_{45}	a_{52}	a_2	a_1	a_0

Fig. 12.17 Putting all pentominoes on a chessboard

Puzzle 131. Five tetrominoes on a strange shape

Fill the following shape with five distinct tetrominoes. You can rotate and flip over each piece. (adapted from www.geogebra.org)

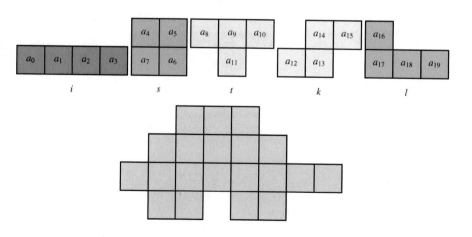

Listing 12.20 Formalising a 4x8 grid

```
1    assign(domain_size ,32).    %grid 4x8
2    assign(max_models ,3).
3    set(arithmetic).
4
5    list(distinct).
6      [a0,   a1 ,  a2 ,  a3 ,              %i [0..3]
7       a4 ,  a5 ,  a6 ,  a7 ,              %s [4..7]
8       a8 ,  a9 ,  a10 ,a11 ,              %t [8..11]
9       a12 ,  a13 ,a14 ,  a15 ,            %k [12..15]
10      a16 ,a17 ,a18 ,a19 ,                %l [16..19]
11      0 ,1 ,5 ,6 ,7 ,8 ,14 ,15 ,24 ,27 ,30 ,31]. %the given shape
12   end_of_list.
13
14   formulas(utils).
15     left(x,y) <-> x mod 8 != 7 & y = x + 1.
16     on(x,y)   <-> x < 24     & y = x + 8.
17   end_of_list.
```

0	1	2	3	4	5	6	7
8	9	10	11	12	13	14	15
16	17	18	19	20	21	22	23
24	25	26	27	28	29	30	31

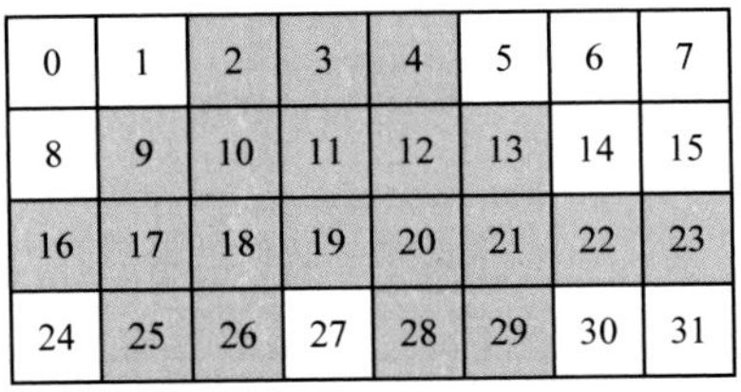

Fig. 12.18 Encapsulating the given shape in a 4×8 rectangular

Listing 12.21 Formalising the five tetrominoes

```
1    formulas(straight).   %i [0..3]
2    i   <-> left(a0,a1) & left(a1,a2) & left(a2,a3).
3    ir <-> on(a0,a1) & on(a1,a2) & on(a2,a3).
4    i | ir.
5    i -> -ir.
6    end_of_list.
7
8    formulas(square).     %s [4..7]
9    s   <-> left(a4,a5) & on(a5,a6) & left(a7,a6).
10   s.
11   end_of_list.
12
13   formulas(t).          %t [8..11]
14   t    <-> left(a8,a9) & left(a9,a10) & on(a9,a11).
15   tr <-> on(a8,a9) & on(a9,a10) & left(a11,a9).
16   trr <-> left(a9,a8) & left(a10,a9) & on(a11,a9).
17   trrr <-> on(a9,a8) & on(a10,a9) & left(a9,a11).
18   t | tr | trr | trrr.
19   t     -> -tr & -trr & -trrr.
20   tr    -> -t  & -trr & -trrr.
21   trr   -> -t  & -tr  & -trrr.
22   trrr -> -t  & -tr  & -trr.
23   end_of_list.
```

```
24
25   formulas ( skew ).          %k [ 1 2 .. 1 5 ].
26    k       <-> left(a12,a13) & on(a14,a13) & left(a14,a15).
27    kr      <-> on(a12,a13) & left(a13,a14) & on(a14,a15).
28    krr     <-> left(a13,a12) & on(a13,a14) & left(a15,a14).
29    krrr    <-> on(a13,a12) & left(a14,a13) & on(a15,a14).
30    ku      <-> left(a13,a11) & on(a14,a13) & left(a15,a14).
31    kur     <-> on(a13,a12) & left(a13,a14) & on(a15,a14).
32    kurr    <-> left(a12,a13) & on(a13,a14) & left(a14,a15).
33    kurrr   <-> on(a12,a13) & left(a14,a13) & on(a14,a15).
34    k | kr | krr | krrr | ku | kur | kurr | kurrr.
35    k       -> -kr  & -krr  & -krrr  & -ku   & -kur & -kurr & -kurrr.
36    kr      -> -k   & -krr  & -krrr  & -ku   & -kur & -kurr & -kurrr.
37    krr     -> -k   & -kr   & -krrr  & -ku   & -kur & -kurr & -kurrr.
38    krrr    -> -k   & -kr   & -krr   & -ku   & -kur & -kurr & -kurrr.
39    ku      -> -k   & -kr   & -krr   & -krrr & -kur & -kurr & -kurrr.
40    kur     -> -k   & -kr   & -krr   & -krrr & -ku  & -kurr & -kurrr.
41    kurr    -> -k   & -kr   & -krr   & -krrr & -ku  & -kur  & -kurrr.
42    kurrr   -> -k   & -kr   & -krr   & -krrr & -ku  & -kur  & -kurr.
43   end_of_list.
44
45   formulas ( l ).              %l [ 1 6 .. 1 9 ]
46    l       <-> on(a16,a17) & left(a17,a18) & left(a18,a19).
47    lr      <-> left(a17,a16) & on(a17,a18) & on(a18,a19).
48    lrr     <-> on(a17,a16) & left(a18,a17) & left(a18,a19).
49    lrrr    <-> left(a16,a17) & on(a18,a17) & on(a19,a18).
50    lu      <-> on(a16,a17) & left(a18,a17) & left(a19,a18).
51    lur     <-> left(a17,a16) & on(a18,a17) & on(a19,a18).
52    lurr    <-> on(a17,a16) & left(a17,a18) & left(a18,a19).
53    lurrr   <-> left(a16,a17) & on(a17,a18) & on(a18,a19).
54    l | lr | lrr | lrrr | lu | lur | lurr | lurrr.
55    l       -> -lr  & -lrr  & -lrrr  & -lu   & -lur & -lurr & -lurrr.
56    lr      -> -l   & -lrr  & -lrrr  & -lu   & -lur & -lurr & -lurrr.
57    lrr     -> -l   & -lr   & -lrrr  & -lu   & -lur & -lurr & -lurrr.
58    lrrr    -> -l   & -lr   & -lrr   & -lu   & -lur & -lurr & -lurrr.
59    lu      -> -l   & -lr   & -lrr   & -lrrr & -lur & -lurr & -lurrr.
60    lur     -> -l   & -lr   & -lrr   & -lrrr & -lu  & -lurr & -lurrr.
61    lurr    -> -l   & -lr   & -lrr   & -lrrr & -lu  & -lur  & -lurrr.
62    lurrr   -> -l   & -lr   & -lrr   & -lrrr & -lu  & -lur  & -lurr.
63   end_of_list.
```

Solution

We handle this strange shape by encapsulating it in a 4×8 rectangular (see Fig. 12.18). In this rectangular, there are 32 locations (from 0 to 31). Some of these locations cannot be occupied by a tetromino. These locations are depicted with white: 0, 1, 5, 6, 7, 8, 14, 15, 24, 27, 30, and 31. This is formalising in Listing 12.20 by forcing these locations (line 11) to be distinct of the locations a_i, $i \in [0..19]$ that can be used by tetrominoes. Indeed, five tetrominoes each with four squares require 20 locations.

By encapsulating the irregular shape with a rectangular, we can define the *on* and *left* predicates like usual. In the 4×8 grid, the $on(x, y)$ predicate does not hold on the last line (locations ≥ 24)) (line 16). The predicate left does not hold on the last column, given by $x \bmod 8 \neq 7$ (line 15).

Next, we formalise the tetrominoes, their rotations and reflections (Listing 12.21). For the straight tetromino (i.e. i), we allocate functions a_0, a_1, a_2, and a_3. From i we can obtain only one additional piece i^r through rotation:

$$i \leftrightarrow left(a_0, a_1) \wedge left(a_1, a_2) \wedge left(a_2, a_3) \tag{12.116}$$

$$i^r \leftrightarrow on(a_0, a_1) \wedge on(a_1, a_2) \wedge on(a_2, a_3) \tag{12.117}$$

The square s is formalised with a_4 to a_7 and has only one form:

$$s \leftrightarrow left(a_4, a_5) \wedge on(a_5, a_6) \wedge left(a_2, a_3) \tag{12.118}$$

The t-tetromino (a_8 to a_{11}) has four shapes obtained through rotation (lines 14–17). Lines 18–22 specify that exactly one shape from t, t^r, t^{rr}, or t^{rrr} can be used. The skew-tetromino (a_{12} to a_{15}) has three additional forms (k^r, k^{rr}, or k^{rrr}) obtained through successive rotations, and four forms (k_u, k_u^r, k_u^{rr}, or k_u^{rrr}) obtained by reflection. The exclusive-or among these eight shapes is modelled in lines 34–42. The l-tetromino (a_{16} to a_{19}) has eight versions, too (lines 46–53). There are then 23 shapes if rotation and reflection are allowed:

Tetromino	i	s	t	k	l	All
Shapes	2	1	4	8	8	23

Since the current puzzle uses all five tetrominoes, we call Mace4 with the 4×8 rectangle and all tetrominoes:

```
mace4 -f 4x8.in 5tetros.in
```

Mace4 finds a single model (see Fig. 12.19), with the following shapes: i^r, k_u^{rr}, l, s, and t^{rr}.

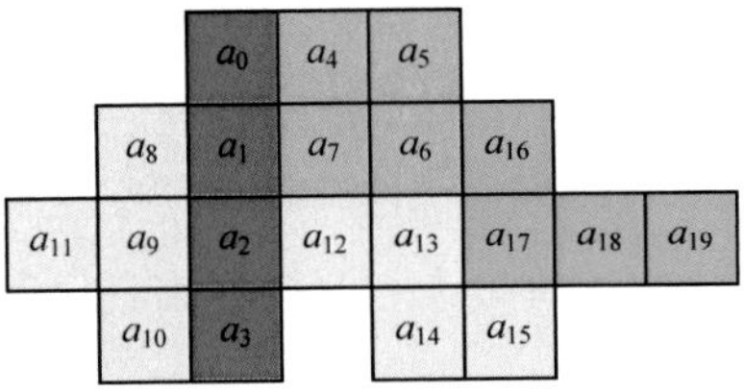

Fig. 12.19 The single model found by Mace4 for the tetrominoes puzzle

Puzzle 132. A cut-up chessboard

A merry chess player cuts his chessboard into 14 parts, as shown. Friends who wanted
to play chess with him first had to put the parts back together again. (In this version,
you cannot rotate the shapes) (adapted from Kordemsky (1992))

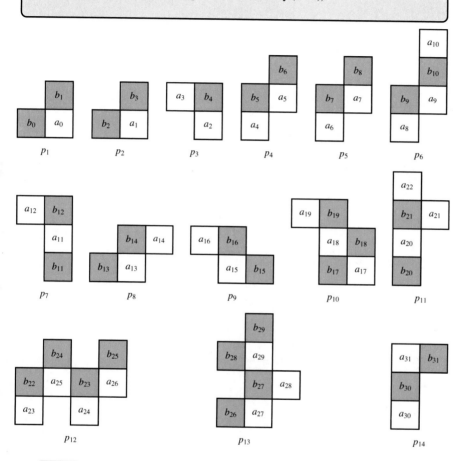

Solution

We need 64 elements: 32 are white (denoted with a_i, $i \in [0..31]$) and 32 are black
(denoted with b_i, $i \in [0..31]$).

First, the on and $left$ predicates are defined as usual for the 8×8 grid. The $on(x, y)$
predicate requires a difference of 8 between the position of x and the position of y. The
predicate does not hold on the last line:

$$x < 56 \rightarrow (on(x, y) \leftrightarrow y = x + 8) \tag{12.119}$$

$$x \geq 56 \rightarrow \neg on(x, y) \tag{12.120}$$

The predicate $left(x, y)$ assumes a difference of 1 between y and x, and it does not
hold on the elements of the last column: 7, 15, 23, 31, 39, 47, 55, and 59.

Listing 12.22 A cut-up chessboard

```
1   assign(domain_size,64).
2   assign(max_models,1).
3   set(arithmetic).
4
5   list(distinct).
6    [a0,  a1,   a2,   a3,   a4,   a5,   a6,   a7,   a8,   a9,
7     a10,  a11,  a12,  a13,  a14,  a15,  a16,  a17,  a18,  a19,
8     a20,  a21,  a22,  a23,  a24,  a25,  a26,  a27,  a28,  a29,  a30,  a31,
9     b0,   b1,   b2,   b3,   b4,   b5,   b6,   b7,   b8,   b9,
10    b10,  b11,  b12,  b13,  b14,  b15,  b16,  b17,  b18,  b19,
11    b20,  b21,  b22,  b23,  b24,  b25,  b26,  b27,  b28,  b29,  b30,  b31].
12  end_of_list.
13
14  formulas(utils).
15   left(x,y) <-> x mod 8 !=7 & y = x + 1.
16   on(x,y)   <-> x < 56 & y = x + 8.
17  end_of_list.
18
19  formulas(black_white).
20   left(x,y) & w(x) -> b(y).  %adjacent squares have different colors
21   left(x,y) & b(x) -> w(y).
22   on(x,y) & b(x) -> w(y).
23   on(x,y) & w(x) -> b(y).
24
25   w(x) <-> x=0  | x=2  | x=4  | x=6  | x=9  | x=11 | x=13 | x=15 |
26          x=16 | x=18 | x=20 | x=22 | x=25 | x=27 | x=29 | x=31 |
27          x=32 | x=34 | x=36 | x=38 | x=41 | x=43 | x=45 | x=47 |
28          x=48 | x=50 | x=52 | x=54 | x=57 | x=59 | x=61 | x=63.
29   w(x) <-> -b(x).  % if white not black
30
31   w(a0).   w(a1).   w(a2).   w(a3).   w(a4).   w(a5).   w(a6).    w(a7).
32   w(a8).   w(a9).   w(a10).  w(a11).  w(a12).  w(a13).  w(a14).   w(a15).
33   w(a16).  w(a17).  w(a18).  w(a19).  w(a20).  w(a21).  w(a22).   w(a23).
34   w(a24).  w(a25).  w(a26).  w(a27).  w(a28).  w(a29).  w(a30).   w(a31).
35
36   b(b0).   b(b1).   b(b2).   b(b3).   b(b4).   b(b5).   b(b6).    b(b7).
37   b(b8).   b(b9).   b(b10).  b(b11).  b(b12).  b(b13).  b(b14).   b(b15).
38   b(b16).  b(b17).  b(b18).  b(b19).  b(b20).  b(b21).  b(b22).   b(b23).
39   b(b24).  b(b25).  b(b26).  b(b27).  b(b28).  b(b29).  b(b30).   b(b31).
40  end_of_list.
41
42  formulas(shapes).
43   p1  -> left(b0,a0)   & on(b1,a0).
44   p2  -> left(b2,a1)   & on(b3,a1).
45   p3  -> left(a3,b4)   & on(b4,a2).
46   p4  -> left(b5,a5)   & on(b5,a4)   & on(b6,a5).
47   p5  -> left(b7,a7)   & on(b7,a6)   & on(b8,a7).
48   p6  -> left(b9,a9)   & on(b9,a8)   & on(b10,a9)   & on(a10,b10).
49   p7  -> left(a12,b12) & on(b12,a11) & on(a11,b11).
50   p8  -> left(b13,a13) & on(b14,a13) & left(b14,a14).
51   p9  -> left(a16,b16) & on(b16,a15) & left(a15,b15).
52   p10 -> left(a19,b19) & on(b19,a18) & on(a18,b17)   & on(b18,a17) &
53          left(a18,b18).
54   p11 -> left(b21,a21) & on(a22,b21) & on(b21,a20)   & on(a20,b20).
55   p12 -> left(b22,a25) & on(b22,a23) & on(b24,a25)   & left(a25,b23) &
56          left(b23,a26) & on(b23,a24) & on(b25,a26).
57   p13 -> left(b28,a29) & on(b29,a29) & on(a29,b27)   & left(b27,a28) &
58          left(b26,a27) & on(b27,a27).
59   p14 -> left(a31,b31) & on(a31,b30) & on(b30,a30).
60
61   p1. p2. p3. p4. p5. p6. p7. p8. p9. p10. p11. p12. p13. p14.
62  end_of_list.
```

Solution

Second, we focus on the constraints related to black and white positions (lines 19–40 in Listing 12.22). We use the predicate $w(x)$ for x is white and $b(x)$ for x is black. Adjacent squares have different colours:

$$left(x, y) \wedge w(x) \rightarrow b(y) \qquad\qquad left(x, y) \wedge b(x) \rightarrow w(y)$$
$$on(x, y) \wedge b(x) \rightarrow w(y) \qquad\qquad on(x, y) \wedge w(x) \rightarrow b(y)$$

This is also explicitly stated in lines 25–28. Of course, the colours are disjoint: $w(x) \leftrightarrow \neg b(x)$. We also state which variables are white and which are black (lines 31–39).

Third, we define the shapes. For instance, $p_1 \rightarrow left(b_0, a_0) \wedge on(b_1, a_0)$. Don't forget to specify that all the shapes are required (line 61).

Fourth, one could try to reduce the search space, by formalising additional constraints. These constraints are obtained by analysing the available shapes. The following optional observations can reduce the search space (see Listing 12.23):

1. Position i of shape p_{12} can be occupied only by a_2, a_4, a_6, a_8, or a_{30} (see Fig. 12.20): $on(a_2, b_{23}) \vee on(a_4, b_{23}) \vee on(a_6, b_{23}) \vee on(a_8, b_{23}) \vee on(a_{30}, b_{23})$.
2. Position j of shape p_{12} can be occupied only by b_3, b_6, b_8, or b_{29}: $on(a_{25}, b_3) \vee on(a_{25}, b_6) \vee on(a_{25}, b_8) \vee on(a_{25}, b_{29})$.
3. Position k of shape p_{13} can be occupied only by a_{14} or a_{21}: $on(b_{28}, a_{14}) \vee on(b_{28}, a_{21})$.
4. Top-left corner can be occupied only by p_{14}: $a_{31} = 0 \wedge b_{31} = 1 \wedge b_{30} = 8 \wedge a_{30} = 16$.
5. Bottom-right corner can be occupied only by p_1 or p_2. Since p_1 and p_2 are identical, we do not care which of them is in the corner. Let p_1 in the bottom-right corner: $a_0 = 63 \wedge b_0 = 62 \wedge b_1 = 55$.
6. Shapes p_3, p_7, p_9, and p_{10} cannot be on column 0, since there is no other piece to match them: $a_3 \bmod 8 \neq 0 \wedge a_{12} \bmod 8 \neq 0 \wedge a_{16} \bmod 8 \neq 0 \wedge a_{19} \bmod 8 \neq 0$.

The first model computed by Mace4 is depicted in Fig. 12.21.

Listing 12.23 Helping Mace4

```
1   formulas(help).
2     on(a2,b23)  |  on(a4,b23)  |  on(a6,b23)  |  on(a8,b23)  |  on(a30,b23).  %pos i
3     on(a25,b3)  |  on(a25,b6)  |  on(a25,b8)  |  on(a25,b29).                 %pos j
4     on(b28,a14) |  on(b28,a21).                                              %pos k
5     a31 =  0.   b31 =  1.   b30 =  8.   a30 = 16.         %top-left corner
6     a0  = 63.   b0  = 62.   b1  = 55.                     %bottom-right corner
7     a3  mod 8 != 0.     a12 mod 8 != 0.     %p3 and p7 cannot be on column 0
8     a16 mod 8 != 0.     a19 mod 8 != 0.     %p9 and p10 cannot be on column 0
9   end_of_list.
```

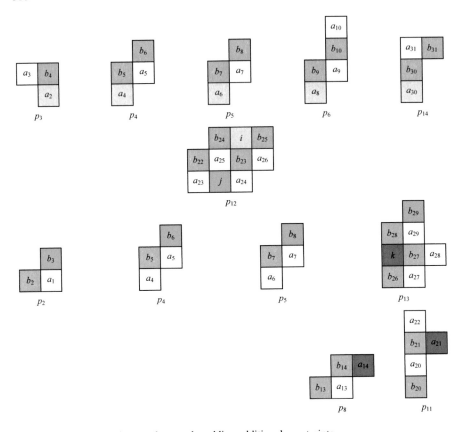

Fig. 12.20 Reducing the search space by adding additional constraints

a_{31}	b_{31}	a_{16}	b_{16}	a_{19}	b_{19}	a_{12}	b_{12}
b_{30}	a_3	b_4	a_{15}	b_{15}	a_{18}	b_{18}	a_{11}
a_{30}	b_{24}	a_{22}	b_{25}	a_{22}	b_{17}	a_{17}	b_{11}
b_{22}	a_{25}	b_{23}	a_{26}	b_{21}	a_{21}	b_6	a_{10}
a_{23}	b_8	a_{24}	b_{29}	a_{20}	b_5	a_5	b_{10}
b_7	a_7	b_{28}	a_{29}	b_{20}	a_4	b_9	a_9
a_6	b_{14}	a_{14}	b_{27}	a_{28}	b_3	a_8	b_1
b_{13}	a_{13}	b_{26}	a_{27}	b_2	a_1	b_0	a_0

Fig. 12.21 Repairing the chessboard

References

Darling, D. (2004). *The Universal Book of Mathematics from Abracadabra to Zeno's Paradoxes.* John Wiley & Sons Inc.

Dudeney, H. E. (2002). *The Canterbury Puzzles.* Courier Corporation.

Golomb, S. W. (1996). *Polyominoes: puzzles, patterns, problems, and packings,* (Vol. 16). Princeton University Press.

Guttmann, A. J., editor (2009). *Polygons, Polyominoes and Polycubes,* volume 775 of *Springer Lecture Notes in Physics.* Springer Nederlands.

Kordemsky, B. A. (1992). *The Moscow puzzles: 359 mathematical recreations.* Courier Corporation.

Martin, G. (1991). *Polyominoes: A guide to puzzles and problems in tiling.* Cambridge University Press.

Chapter 13
Self-reference and Other Puzzles

Abstract In which we formalise some self-referencing brain teasers. Well-known examples of such puzzles are the liar paradox (i.e. "I am lying"), or fill in the blanks sentences (e.g. "There are _____ e's in this sentence" which has solution "eight". Solving such puzzles requires a disciplined mind, since most of them demand a sort of recursive reasoning. Hence the joke: *To understand recursion, you must first understand recursion.* Some of these puzzles are not easily modelled in FOL, too.

$$\forall x \, \forall y \, ((link(x) \wedge chain(y) \wedge contains(y, x) \wedge weakest(x) \rightarrow canBreak(x, y)) \rightarrow strongest(x))$$

Stanislaw J. Lec

I formalised here some self-referencing brain teasers. Well-known examples of such puzzles are the liar paradox (i.e. "I am lying"), or fill in the blanks sentences (e.g. "There are _____ e's in this sentence" which has solution "eight". Solving such puzzles requires a disciplined mind, since most of them demand a sort of recursive reasoning. Hence the joke: *To understand recursion, you must first understand recursion.* Some of these puzzles are not easily modelled in FOL, too.

The chapter starts with some puzzles on tricky messages that argue one against the other which is true or false. This can be seen as a generalisation of the puzzles from the chapter "Island of truth". The current chapter also includes three "fill in the blank" self-referencing sentences. For instance, in a list of ten statements the nth says there are *exactly n* false items in the list. The task is to figure out, which of the statements, if any, are false. A variant of this puzzle assumes a list of ten statements, where the nth statements says there are *at least n* false items in the list. In both cases, five logical equations suffice for modelling them. The last puzzle, "self-counting sentence" did pose technical difficulties when expressed in FOL. The puzzle asks to insert numbers in the blanks to make the following sentence true: *In this sentence, the number of occurrences of 0 is ___, of 1 is ___, of 2 is ___, of 3 is ___, of 4 is ___, of 5 is ___, of 6 is ___, of 7 is ___, of 8 is ___, and of 9 is ___.* Mace4 found two solutions. Can you find them?

Puzzle 133. Tricky messages

Which of the following statements are true?

1. The next two statements are false.
2. The last statement is false.
3. The next statement is false.
4. The first statement was false. (puzzle from Walicki (2016))

Listing 13.1 Modelling tricky messages in propositional logic

```
1  set(arithmetic).
2  assign(max_models, -1).
3  assign(domain_size ,2).
4  formulas(assumptions).
5    m1 <-> -m2 & -m3.
6    m2 <-> -m4.
7    m3 <-> -m4.
8    m4 <-> -m1.
9  end_of_list.
```

Solution

We use propositional logic to model the tricky messages. Let m_1, m_2, m_3, and m_4 denote each message (see Listing 13.1). Since the messages are true or false, we use a domain size of two. In line 5, message m_1 states that the next two statements (i.e. m_2 and m_3) are false: $m_1 \leftrightarrow \neg m_2 \wedge \neg m_3$. With no solution computed by Mace4, the discourse is inconsistent. Prover9 proves the inconsistency using the steps in Fig. 13.1.

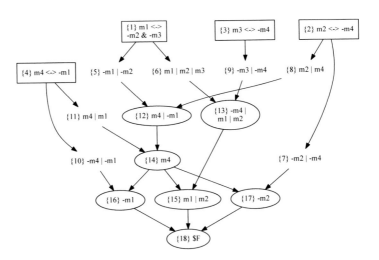

Fig. 13.1 Proving the inconsistency of the given messages

Puzzle 134. Not so tricky messages

Which of the following statements are true?

1. The next statement is false.
2. The next two statements are false.
3. The next statement is false.
4. The first statement was false.

(taken from Walicki (2016))

Listing 13.2 Modelling tricky messages that are consistent

```
1   set(arithmetic).
2   assign(max_models, -1).
3   assign(domain_size,2).
4   formulas(assumptions).
5     m1 <-> -m2.
6     m2 <-> -m3 & -m4.
7     m3 <-> -m4.
8     m4 <-> -m1.
9   end_of_list.
```

Solution

We use the propositional variables m_i to model the statement i. The four messages are consistent, as Mace4 finds a model: $m_1, \neg m_2, m_3, \neg m_4$. For instance, one can prove that m_1 is true (see Fig. 13.2).

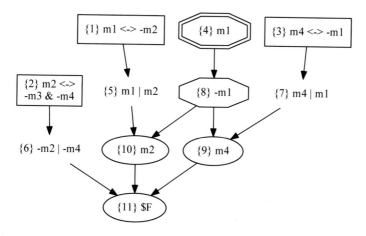

Fig. 13.2 Proving that the first sentence is true

Puzzle 135. Self-referring sentences

Which answer in the list is the correct answer to this question?

1. All of the below.

2. None of the below.

3. All of the above.

4. One of the above.

5. None of the above.

6. None of the above.

Listing 13.3 Modelling six self-referring statements in propositional logic

```
1   set ( arithmetic ).
2   assign ( max_models , − 1 ).
3   assign ( domain_size , 2 ).
4   formulas ( assumptions ).
5     m1 <−> m2 & m3 & m4 & m5 & m6.
6     m2 <−> −m3 & −m4 & −m5 & −m6.
7     m3 <−> m1 & m2.
8     m4 <−> (m1 & −m2 & −m3) | (−m1 & m2 & −m3) | (−m1 & −m2 & m3).
9     m5 <−> −m1 & −m2 & −m3 & −m4.
10    m6 <−> −m1 & −m2 & −m3 & −m4 & −m5.
11  end_of_list .
```

Listing 13.4 Modelling six self-referring statements in first order logic

```
1   set ( arithmetic ).
2   assign ( max_models , − 1 ).
3   assign ( domain_size , 6 ).
4
5   formulas ( six_sentences ).
6     m(0) <−> ( all  x  (x > 0 −>  m( x ))).
7     m(1) <−> ( all  x  (x > 1 −> −m( x ))).
8     m(2) <−> ( all  x  (x < 2 −>  m( x ))).
9     m(3) <−> ( all  x  all  y  ((x < 3 & x < 3 & m(x) & m(y)) −> x = y) &
10               ( exists  x  (x < 3 & m( x )))).
11    m(4) <−> ( all  x  (x < 4 −> −m( x ))).
12    m(5) <−> ( all  x  (x < 5 −> −m( x ))).
13  end_of_list .
14
15  formulas ( test ). %test  that  in  all  models  the  formula  is  true
16    −m(0) & −m(1) & −m(2) & −m(3) & m(4) & −m(5).
17  end_of_list .
```

We use the propositional variables m_i to model the statement i (see Listing 13.3). Mace4 finds a single model in which only m_5 is true.

One can also use first-order logic to model the puzzle (see Listing 13.4). Here, we use values 0 to 5 to formalise the six messages. The first message $m(0)$ states that all future messages $m(x)$, $x > 0$ are true:

$$m(0) \leftrightarrow (\forall x \ (x > 0 \rightarrow m(x))) \tag{13.1}$$

The second message $m(1)$ states that all future messages $m(x)$, $x > 1$ are false:

$$m(1) \leftrightarrow (\forall x \ (x > 1 \rightarrow \neg m(x))) \tag{13.2}$$

The third message $m(2)$ states that all previous messages $m(x)$, $x < 2$ are true:

$$m(2) \leftrightarrow (\forall x \ (x < 2 \rightarrow m(x))) \tag{13.3}$$

The fourth message $m(3)$ states that exactly one of the previous messages $m(x)$, $x < 3$ is true. We need the "at-most" constraint:

$$(\forall x \ \forall y \ ((x < 3 \land x < 3 \land m(x) \land m(y)) \rightarrow x = y) \tag{13.4}$$

and the "at-least" constraint $\exists x \ (x < 3 \land m(x))$.

The fifth message $m(4)$ states that all previous messages $m(x)$, $x < 4$ are false:

$$m(4) \leftrightarrow (\forall x \ (x < 4 \rightarrow \neg m(x))) \tag{13.5}$$

The sixth message $m(5)$ states that all previous messages $m(x)$, $x < 5$ are also false:

$$m(5) \leftrightarrow (\forall x \ (x < 5 \rightarrow \neg m(x))) \tag{13.6}$$

Based on the formalisation in FOL in Listing 13.4, Mace4 finds 10,368 isomorphic models. In all of them, only the message $m(5)$ is true. This can be verified by adding the formula:

$$\neg m(0) \land \neg m(1) \land \neg m(2) \land \neg m(3) \land m(4) \land \neg m(5) \tag{13.7}$$

Since Mace4 returns the same number of models (10,368), it means that the formula is true in all initial 10,368 models.

Puzzle 136. Which hand?

Give a friend an "even" coin (say, a dime—ten is an even number) and an "odd" coin (say, a nickel). Ask him to hold one coin in his right hand and the other in his left. Tell him to triple the value of the coin in his right hand and double the value of the coin in his left, then add the two. If the sum is even, the dime is in his right hand; if odd, in his left. Explain, and think up some variations. (puzzle 44 from Kordemsky (1992)).

Listing 13.5 Finding models for "Which hand?" puzzle

```
1   set ( arithmetic ) .
2   assign ( domain_size , 100 ) .
3   assign ( max_models , -1 ) .
4
5   formulas ( assumptions ) .
6     even ( x )  <-> x mod 2 = 0 .        odd ( x )   <-> x mod 2 = 1 .
7     even ( r )  -> odd ( l ) .           odd ( r )   -> even ( l ) .
8     even ( r )  | odd ( r ) .            even ( l )  | odd ( l ) .
9
10    sum = 3*r + 2*l .
11
12    -( even ( sum )  -> even ( r )) .
13    %-( odd ( sum )  -> even ( l )) .
14  end_of_list .
```

Solution

First we define the *odd* and *even* predicates. The right hand (r) can contain an even or an odd coin: $even(r) \lor odd(r)$. The same for the left hand: $even(l) \lor odd(l)$. Of course, if right is even, then left is odd: $even(r) \rightarrow odd(l)$. We add the equation $sum = 3 * r + 2 * l$ (line 10 in Listing 13.5).

A human agent may reason as follows. If $odd(l)$ then $even(2 * l)$. Also r will be even and also $even(3 * r)$. That is, the sum is even. Otherwise, if $even(l)$ then r will be odd and therefore $odd(3 * r)$. Hence, the sum is odd.

For the software agent, since Prover9 is not aware of the *mod* operator, Listing 13.5 cannot be used to prove this. Instead, we can use Mace4 to validate the given statement. To do this, we just add to the knowledge base the negation of the sentence we want to prove: $even(sum) \rightarrow even(r)$. As Mace4 is not able to find any model, this negated sentence is proved (line 16). Similarly, to check the validity of $odd(sum) \rightarrow even(l)$, we add its negation to the knowledge base (line 13). Failing to find a model proves the sentence.

Puzzle 137. An ornament for a window

A store selling semiprecious stones used a five-pointed star made of circular spots held together by a wire. The 15 spots hold from 1 through 15 stones (each number used once). Each of the 5 circles holds 40 stones, and at the 5 ends of the star, there are 40 stones. (puzzle 326 from Kordemsky (1992))

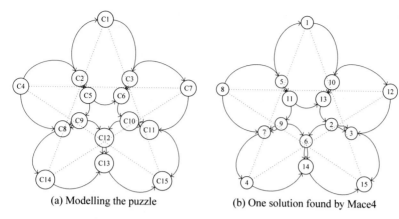

(a) Modelling the puzzle   (b) One solution found by Mace4

Fig. 13.3 An ornament for a window

Listing 13.6 Finding ornaments for a window

```
1   assign(domain_size,16).
2   assign(max_models,-1).
3   set(arithmetic).
4
5   list(distinct).
6     [0,c1,c2,c3,c4,c5,c6,c7,c8,c9,c10,c11,c12,c13,c14,c15].
7   end_of_list.
8
9   formulas(assumptions).
10    c1  + c2  + c3  + c5  + c6  = 40.        %circles
11    c2  + c5  + c9  + c8  + c4  = 40.
12    c8  + c9  + c12 + c13 + c14 = 40.
13    c10 + c11 + c12 + c13 + c15 = 40.
14    c6  + c7  + c3  + c10 + c11 = 40.
15    c1  + c7  + c15 + c14 + c4  = 40.        %values from the ends of the star
16  end_of_list.
```

Solution

Since the 15 distinct values are from the interval [1..15], we set a domain size of 16, and we eliminate value 0 (line 6 in Listing 13.6). We need to write one equation for each circle (lines 10–14), and one equation for the values at the end of the star (line 15). One model is shown in Fig. 13.3b.

Puzzle 138. At the brook

A man goes to the brook with two measures of 15 pints and 16 pints. How is he to measure exactly 8 pints of water, in the fewest possible transactions? Filling or emptying a vessel or pouring any quantity from one vessel to another counts as a transaction. (puzzle 403 from Dudeney (2016))

Listing 13.7 Measuring the value 8 with two vessels of 15 and 16 pints. Eight actions can be used to reach the goal state $\exists x \, \exists t \, (J(8, x, t) \lor J(x, 8, t))$ from the init state $J(15, 16, 0)$

```
1    set(production).
2
3    formulas(usable).
4      all x all y all t (J(x,y,t) -> J(15,y,t+1)).   %fill 1
5      all x all y all t (J(x,y,t) -> J(0,y,t+1)).    %empty 1
6
7      all x all y all t (J(x,y,t) -> J(x,16,t+1)).   %fill 2
8      all x all y all t (J(x,y,t) -> J(x,0,t+1)).    %empty 2
9
10     all x all y all t (J(x,y,t) & x+y<=16 -> J(0,y+x,t+1)).   %empty(1,2)
11     all x all y all t (J(x,y,t) & x+y> 16 -> J(x+ -(16+ -y),16,t+1)).%p(1,2)
12     all x all y all t (J(x,y,t) & x+y<=15 -> J(x+y,0,t+1)).   %empty(2,1)
13     all x all y all t (J(x,y,t) & x+y> 15 -> J(15,y+ -(15+ -x),t+1)).%p(2,1)
14   end_of_list.
15
16   formulas(assumptions).
17     J(15,16,0).       %init state
18   end_of_list.
19
20   formulas(goals).
21     exists x exists t (J(8,x,t) | J(x,8,t)).
22   end_of_list.
```

Solution

We model this problem as a planning task. Let the state $J(x, y, t)$, where x is the content of the first vessel, y the content of the second vessel, and t the number of transactions. The init state is $J(15, 16, 0)$. The goal state is to measure 8 pints of water. Since we do not care if it is measured in the first or in the second vessel, we need to reach one of the following two states:

$$\exists x \, \exists t \, (J(8, x, t) \lor J(x, 8, t)) \tag{13.8}$$

There are eight possible actions (see Listing 13.7):

1. Fill 1: $\forall x \, \forall y \, \forall t \, (J(x, y, t) \to J(15, y, t + 1))$
2. Empty 1: $\forall x \, \forall y \, \forall t \, (J(x, y, t) \to J(0, y, t + 1))$
3. Fill 2: $\forall x \, \forall y \, \forall t \, (J(x, y, t) \to J(x, 16, t + 1))$
4. Empty 2: $\forall x \, \forall y \, \forall t \, (J(x, y, t) \to J(x, 0, t + 1))$
5. Empty 1 into 2: $\forall x \, \forall y \, \forall t \, (J(x, y, t) \land x + y \leq 16 \to J(0, y + x, t + 1))$
6. Pour 1 into 2: $\forall x \, \forall y \, \forall t \, (J(x, y, t) \land x + y > 16 \to J(x - (16 - y), 16, t + 1))$
7. Empty 2 into 1: $\forall x \, \forall y \, \forall t \, (J(x, y, t) \land x + y \leq 15 \to J(x + y, 0, t + 1))$
8. Pour 2 into 1: $\forall x \, \forall y \, \forall t \, (J(x, y, t) \land x + y > 15 \to J(15, y - (15 - x), t + 1))$

Note that with every transaction, t is incremented. Based on the implementation in Listing 13.7, Prover9 finds a solution in 28 steps, as indicated by the time variable in the state:

$$
\begin{aligned}
&J(15, 16, 0) \;\to\; J(15, 0, 1) \;\;\to\; J(0, 15, 2) \;\;\to\; J(15, 15, 3) \;\to\; J(14, 16, 4) \\
&J(14, 0, 5) \;\;\to\; J(0, 14, 6) \;\;\to\; J(15, 14, 7) \;\to\; J(13, 16, 8) \;\to\; J(13, 0, 9) \\
&J(0, 13, 10) \;\to\; J(15, 13, 11) \to J(12, 16, 12) \to J(12, 0, 13) \;\to\; J(0, 12, 14) \\
&J(15, 12, 15) \to J(11, 16, 16) \to J(11, 0, 17) \;\to\; J(0, 11, 18) \;\to\; J(15, 11, 19) \\
&J(10, 16, 20) \to J(10, 0, 21) \;\to\; J(0, 10, 22) \to J(15, 10, 23) \to J(9, 16, 24) \\
&J(9, 0, 25) \;\;\to\; J(0, 9, 26) \;\;\to\; J(15, 9, 27) \;\to\; J(8, 16, 28)
\end{aligned}
$$

Note that the requested quantity of 8 is measured in the first vessel.

Puzzle 139. A car tour

A man started in a car from town *A*, and wished to make a complete tour of these roads, going along every one of them once, and once only. How many different routes are there from which he can select? It is puzzling unless you can devise some ingenious method. Every route must end at town *A*, from which you start, and you must go straight from town to town never turning off at crossroads. (puzzle 425 from Dudeney (2016))

Listing 13.8 How many Eulerian tours exist in the given graph?

```
1   set(arithmetic).     assign(domain_size,20).        assign(max_models,-1).
2
3   formulas(demodulators).
4     x<19 -> s(x) = x+1.              s(19)=0.
5     c(0)=0  |  c(0)=1  |  c(0)=2  |  c(0)=3.      %start with  AB, AC, AD, AE
6     c(9)=10 |  c(9)=11 |  c(9)=12 |  c(9)=13.     %end with    BA, CA, DA, EA
7     x>9 -> c(x)=0.                                %there are only 10 edges
8   end_of_list.
9
10  formulas(assumptions).
11    all x all y ((x != y & x <= 9 & y <= 9) -> (c(x) != c(y))).
12    all x all y ((x = y  & x <= 9 & y <= 9) -> (c(x) = c(y))).
13    %if car uses XY then it cannot use YX (XY + 10 = YX)
14    all x all y all z ((z < 10 & c(x) = z & x != y) -> (c(y) != z + 10)).
15
16    %continutiy between edges
17    all x ((x<9 & c(x)=0)  -> ((c(s(x))=4)  | (c(s(x))=5)  | (c(s(x))=6))).
18    all x ((x<9 & c(x)=1)  -> ((c(s(x))=7)  | (c(s(x))=8)  | (c(s(x))=14))).
19    all x ((x<9 & c(x)=2)  -> ((c(s(x))=9)  | (c(s(x))=15) | (c(s(x))=17))).
20    all x ((x<9 & c(x)=3)  -> ((c(s(x))=16) | (c(s(x))=18) | (c(s(x))=19))).
21    all x ((x<9 & c(x)=4)  -> ((c(s(x))=7)  | (c(s(x))=8)  | (c(s(x))=11))).
22    all x ((x<9 & c(x)=5)  -> ((c(s(x))=9)  | (c(s(x))=12) | (c(s(x))=17))).
23    all x ((x<9 & c(x)=6)  -> ((c(s(x))=13) | (c(s(x))=18) | (c(s(x))=19))).
24    all x ((x<9 & c(x)=7)  -> ((c(s(x))=9)  | (c(s(x))=12) | (c(s(x))=15))).
25    all x ((x<9 & c(x)=8)  -> ((c(s(x))=13) | (c(s(x))=16) | (c(s(x))=19))).
26    all x ((x<9 & c(x)=9)  -> ((c(s(x))=13) | (c(s(x))=16) | (c(s(x))=18))).
27    all x ((x<9 & c(x)=10) -> ((c(s(x))=1)  | (c(s(x))=2)  | (c(s(x))=3))).
28    all x ((x<9 & c(x)=11) -> ((c(s(x))=0)  | (c(s(x))=2)  | (c(s(x))=3))).
29    all x ((x<9 & c(x)=12) -> ((c(s(x))=0)  | (c(s(x))=1)  | (c(s(x))=3))).
30    all x ((x<9 & c(x)=13) -> ((c(s(x))=0)  | (c(s(x))=1)  | (c(s(x))=2))).
31    all x ((x<9 & c(x)=14) -> ((c(s(x))=5)  | (c(s(x))=6)  | (c(s(x))=10))).
32    all x ((x<9 & c(x)=15) -> ((c(s(x))=4)  | (c(s(x))=6)  | (c(s(x))=10))).
33    all x ((x<9 & c(x)=16) -> ((c(s(x))=4)  | (c(s(x))=5)  | (c(s(x))=10))).
34    all x ((x<9 & c(x)=17) -> ((c(s(x))=8)  | (c(s(x))=11) | (c(s(x))=14))).
35    all x ((x<9 & c(x)=18) -> ((c(s(x))=7)  | (c(s(x))=11) | (c(s(x))=14))).
36    all x ((x<9 & c(x)=19) -> ((c(s(x))=12) | (c(s(x))=15) | (c(s(x))=17))).
37  end_of_list.
```

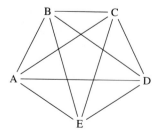

Since there are 20 edges, we set the domain size at this value. For each edge, let the following codification with values from the domain: $AB = 0$, $BA = 10$, $AC = 1$, $CA = 11$, $AD = 2$, $DA = 12$, $AE = 3$, $EA = 13$, $BC = 4$, $CB = 14$, $BD = 5$, $DB = 15$, $BE = 6$, $EB = 16$, $CD = 7$, $DC = 17$, $CE = 8$, $EC = 18$, $DE = 9$, and $ED = 19$. The solution will be identified by the function $c(x)$ that returns the edge used at the step x. There are four ways to start at step 0 (line 5):

$$c(0) = AB \vee c(0) = AC \vee c(0) = AD \vee c(0) = AE \qquad (13.9)$$

A complete route requires 10 steps (since each route XY is also marked by YX). Hence, at the last step (i.e. 9) we need to get to the node A. There are four edges for reaching A (line 6):

$$c(9) = BA \vee c(9) = CA \vee c(9) = DA \vee c(9) = EA \qquad (13.10)$$

For the values larger than 9, we fix in line 7 the value of c: $x > 9 \rightarrow c(x) = 0$. Next, we model the following three constraints: First, if the car uses the route XY, it cannot use its pair YX (line 14). Note that we codified routes XY with values from $[0..9]$ and YX from $[10..19]$. Thus, for any edge XY from $[0..9]$, $XY + 10 = YX$. Our formalisation says that, for all steps x and y, for any edge $z < 10$ (of type XY) if at the current step x the car uses the edge z (i.e. $c(x) = z$), for all other steps y the car cannot use the corresponding edge $z + 10$ (i.e. its pair YX):

$$\forall x \, \forall y \, \forall z \, ((z < 10 \wedge c(x) = z \wedge x! = y) \rightarrow (c(y) \neq z + 10)) \qquad (13.11)$$

Second, the car uses different edges at different moments $x \, y$ of time (line 11):

$$\forall x \, \forall y \, ((x \neq y \wedge x \leq 9 \wedge y \leq 9) \rightarrow (c(x) \neq c(y))) \qquad (13.12)$$

Third, at the same time instances, the car uses only one edge (line 12):

$$\forall x \, \forall y \, ((x = y \wedge x \leq 9 \wedge y \leq 9) \rightarrow (c(x) = c(y))) \qquad (13.13)$$

The last step is to define the transition model (lines 16–36). For instance, for all time steps x in $[0..9]$ after using the edge 0 (AB), only edges 4 (BC), 5 (BD), or 6 (BE) can be used the next time $s(x)$:

$$\forall x((x < 9 \wedge c(x) = 0) \rightarrow ((c(s(x)) = 4) \vee (c(s(x)) = 5) \vee (c(s(x)) = 6))) \quad (13.14)$$

We provide similar sentences for all the 19 remaining edges. Mace4 finds 264 solutions. Three solutions are:

$c(x)$	0	1	2	3	4	5	6	7	8	9
1	0	4	7	9	13	1	8	16	5	12
2	2	17	8	13	0	6	19	15	4	11
3	3	19	17	14	6	18	11	0	5	12

The first model here corresponds to the following route:

$$AB \rightarrow BC \rightarrow DC \rightarrow DE \rightarrow EA \rightarrow AC \rightarrow CE \rightarrow EB \rightarrow BD \rightarrow DA$$

Puzzle 140. A diamond ring

We have a ring with diamond stone whose "atoms" are joined in 10 rows of 3 atoms each. Select 13 integers, of which 12 are different, and place them in the "atoms" so that each row totals 20. The smallest number needed is 1, the largest is 15. There are six solutions, with one of them illustrated below. Can you find the other five? (adapted from puzzle 325 from Kordemsky (1992))

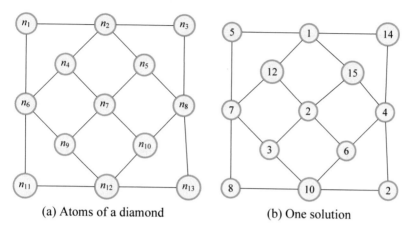

(a) Atoms of a diamond   (b) One solution

Fig. 13.4 A crystal with three atoms in each row

Listing 13.9 Finding models in a diamond with 12 distinct values

```
1   assign(domain_size,16).
2   assign(max_models,-1).
3   set(arithmetic).
4
5   list(distinct).
6     [0,n1,n2,n3,n4,n5,n6,n7,n8,n9,n10,n11,n12].
7     [0,n13].
8   end_of_list.
9
10  formulas(assumptions).
11    n1  + n2  + n3  = 20.     n1 + n6 + n11 = 20.    n11 + n12 + n13 = 20.
12    n13 + n8  + n3  = 20.     n2 + n4 + n6  = 20.    n5  + n7  + n9  = 20.
13    n8  + n10 + n12 = 20.     n2 + n5 + n8  = 20.    n4  + n7  + n10 = 20.
14    n6  + n9  + n12 = 20.
15
16    n2 < 15 & n6 < 15 & n7 < 15 & n8 < 15 & n12 < 15.
17    n4 < 17 & n5 < 17 & n9 < 17 & n10 < 17.
18  end_of_list.
```

Solution

Since the values are from 1 to 15 we need a domain size of 16. All values but one (n_{13}) are distinct and not equal to zero (line 6 in Listing 13.9). There are 10 lines in the figure, each line with its own equation to specify the sum of 20 (lines 11–14). Given this formalisation, Mace4 finds six models, one of them being illustrated. The problem is that it takes a lot of time. To improve the search, we could add some additional constraints depending on the number of edges of the nodes. There are three types of nodes: two edges (n_1, n_3, n_{11}, n_{13}), three edges (n_4, n_5, n_9, n_{10}), and four edges (n_2, n_8, n_{12}, n_7, n_7). One can include the following two observations:

First, a node with four connections has to be equal or smaller than 14 (i.e. 20-6), since 6 is the smallest number that can be written in three different ways, so that only one number is not different (i.e. 1+5, 2+4, 3+3).

Second, a node with three connections has to be equal or smaller than 16 (i.e. 20-4), since 4 is the smallest number that can be written in two different ways, so that only one number is not different (i.e. 1+2, 2+2).

Mace4 finds six solutions, one being illustrated also in Fig. 13.4b. Note that in all models, $n_7 = 2$.

Model	n_1	n_2	n_3	n_4	n_5	n_6	n_7	n_8	n_9	n_{10}	n_{11}	n_{12}	n_{13}
1	5	1	14	12	15	7	2	4	3	6	8	10	2
2	5	7	8	12	3	1	2	10	15	6	14	4	2
3	8	1	11	12	15	7	2	4	3	6	5	10	5
4	8	7	5	12	3	1	2	10	15	6	11	4	5
5	11	1	8	15	12	4	2	7	6	3	5	10	5
6	11	4	5	15	6	1	2	10	12	3	8	7	5

Puzzle 141. Drinker paradox

For any pub, there is always a customer in the pub so that, if he is drinking, every customer in the pub is drinking. (taken from Smullyan (2011)).

Listing 13.10 Drinker paradox as a provable theorem

```
1   formulas ( goals ).
2     exists  x ( drinks ( x )  -> all  y  drinks ( y )).
3   end_of_list.
```

Listing 13.11 Drinker paradox as a not-provable theorem

```
1   formulas ( goals ).
2     exists  x  drinks (x)  ->  all  p  ( inPub (p)  ->  drinks (p)).
3   end_of_list .
```

Solution

This paradox can be stated as a provable theorem (see Listing 13.10):

$$\exists x \ (drinks(x) \rightarrow \forall p(inPub(p) \rightarrow drinks(p))) \tag{13.15}$$

Prover9 outputs a three-step proof. By negating the goal, it infers both $\neg drinks(f1(x))$ (here $f1(x)$ is a Skolem function) and $drinks(x)$ that contradict each other
The drinker paradox is an instance of the more general statement: *There is some x, where, if p applies to x then p applies to every y:* $\exists x \ (p(x) \rightarrow \forall y \ p(y))$.
By implication elimination, we get: $T : \exists x \ (\neg p(x) \lor \forall y \ p(y))$. Sentence $\forall y \ p(y)$ can be true or false. If it is true, the entire disjunction T is true. If it is false, then exists x for each D does not apply (i.e. $\exists x \ \neg D(x)$). This also makes the statement T true. What we did not prove is:

$$\exists x \ (drinks(x) \rightarrow \forall p(inPub(p) - > drinks(p))) \tag{13.16}$$

Prover9 is not able to prove the goal in Listing 13.11).

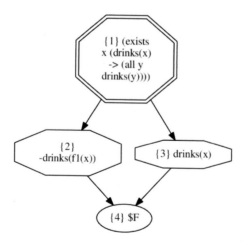

Puzzle 142. Ten sentences

In a list of 10 statements, the nth statement says there are exactly n false items in the list. Which of the statements, if any, are false?

1. In this list, exactly 1 statement is false.
2. In this list, exactly 2 statements are false.
3. In this list, exactly 3 statements are false.
4. In this list, exactly 4 statements are false.
5. In this list, exactly 5 statements are false.
6. In this list, exactly 6 statements are false.
7. In this list, exactly 7 statements are false.
8. In this list, exactly 8 statements are false.
9. In this list, exactly 9 statements are false.
10. In this list, exactly 10 statements are false.

Listing 13.12 Modelling ten tricky sentences

```
1    set(arithmetic).
2    assign(max_models,-1).
3    assign(domain_size,11).
4
5    formulas(assumptions).
6      m(x) = 0 | m(x) = 1.
7      NoSentences = 10.
8      NoTrue = NoSentences + -NoFalse.
9
10     m(x) = 1 <-> NoFalse = x.
11     m(1)+m(2)+m(3)+m(4)+m(5)+m(6)+m(7)+m(8)+m(9)+m(10) = NoTrue.
12   end_of_list.
```

Solution

Let the function $m(x)$ stating if the xth message is true (i.e. $m(x) = 1$) or false (i.e. $m(x) = 0$). Each xth statement says there are exactly x false items in the list:

$$m(x) = 1 \leftrightarrow NoFalse = x \qquad (13.17)$$

Let the number of true sentences be the difference between the number of given sentences (i.e. 10) and the number of false statements: $NoTrue = NoSentences - NoFalse$. The number of true sentences should equal the variable $NoTrue$:

$$m(1) + m(2) + m(3) + m(4) + m(5) + m(6) + m(7) + m(8) + m(9) + m(10) = NoTrue \qquad (13.18)$$

Mace4 finds a single model:

```
function(NoFalse,     [9]),
function(NoTrue,      [1]),
function(NoSentences, [10]),
function(m(_),        [0, 0, 0, 0, 0, 0, 0, 0, 0, 1, 0 ])
```

There are 9 false sentences. The single true sentence is the 9th one, given by the function $m(9) = 1$. Note that we set the domain size to 11 and there is no message $m(0)$.

In a list of 10 statements, the nth statement says there are at least n false items in the list. Which of the statements, if any, are false?

Listing 13.13 Modelling ten tricky sentences

```
1    set ( arithmetic ).
2    assign ( max_models , -1).
3    assign ( domain_size ,11).
4
5    formulas ( assumptions ).
6      m( x ) = 0 | m( x ) = 1.
7      NoSentences = 10.
8      NoTrue = NoSentences + -NoFalse .
9
10     m( x ) = 1 <-> NoFalse >= x .
11     m(1)+m(2)+m(3)+m(4)+m(5)+m(6)+m(7)+m(8)+m(9)+m(10) = NoTrue .
12   end_of_list .
```

Solution

Let the function $m(x)$ be stating if the xth message is true (i.e. $m(x) = 1$) or false (i.e. $m(x) = 0$). Each xth statement says there are at least x false items in the list:

$$m(x) = 1 \leftrightarrow NoFalse \geq x \tag{13.19}$$

Let the number of true sentences be the difference between the number of given sentences (i.e. 10) and number of false statements: $NoTrue = NoSentences - NoFalse$. The number of true sentences should equal the variable $NoTrue$:

$$m(1) + m(2) + m(3) + m(4) + m(5) + m(6) + m(7) + m(8) + m(9) + m(10) = NoTrue \tag{13.20}$$

Mace4 finds a single model:

```
function(NoFalse,      [5]),
function(NoTrue,       [5]),
function(NoSentences, ()),
function(m(_),         [X, 1, 1, 1, 1, 1, 0, 0, 0, 0, 0 ])
```

Except sentence $m(0)$ which does not appear in the list, there are five false and five true sentences. The first five sentences are true, while the last five are false.

Puzzle 144. Self-counting sentence

Insert numbers in the blanks to make this sentence true:
*In this sentence, the number of occurrences of 0 is ___, of 1 is ___, of 2 is ___, of 3 is ___, of
4 is ___, of 5 is ___, of 6 is ___, of 7 is ___, of 8 is ___, and of 9 is _.*
There are exactly two solutions.

Listing 13.14 Self counting sentence

```
1    assign(domain_size ,12).
2    assign(max_models ,2).
3    set(arithmetic).
4
5    formulas(utils).
6      all x all y (less(x,y) < 2 & greater(x,y) < 2 & equal(x,y) < 2).
7      less(x,y) = 1 -> (x <= y).
8      less(x,y) = 0 -> (x > y).
9      greater(x,y) = 1 -> (x >= y).
10     greater(x,y) = 0 -> (x < y).
11     equal(x,y) = 1 -> (x = 1 & y = 1).
12     equal(x,y) = 0 -> (x != 1 | y != 1).
13     f(x,y) = equal(less(c(x),y),greater(c(x),y)).
14     h(x)   = f(0,x) + f(1,x) + f(2,x) + f(3,x) + f(4,x) +
15               f(5,x) + f(6,x) + f(7,x) + f(8,x) + f(9,x).
16   end_of_list.
17
18   formulas(self_counting_sentence).
19     c(0) = 1 | c(0) = 2 | c(0)=3.
20     c(1) != 1.
21     all x ((x > 1 & x <10)  -> (c(x) = 1 | c(x) = 2 |c(x) =3)).
22
23     c(0) = 1 + h(10).                    %value 1 appears once in number 10
24
25     %value 1 appears once in numbers 1 or 10, and twice in 11
26     c(1) = 1 + 2 * h(11) + h(1) + h(10).
27
28     (x > 1 & x <10) -> c(x) =  1 + h(x). %value x appears once in number x
29
30     all x  (x > 9 -> c(x) = 11).         %avoid isomorphic models
31   end_of_list.
```

Solution

Let function $c(x)$ counting the number of occurrences of the input x: $c(x) : [0..9] \rightarrow$
$[0..12]$ Since the domain of $c(x)$ is from 0 to 9, we need to fix the values for $c(10)$ and
$c(11)$ as we are not interested in these two values. Hence, $\forall x \ (x > 9 \rightarrow c(x) = 11)$,
which avoids generating isomorphic models.
To help Mace4 during search, we start by restricting the domain. For instance, since
any digit is already within the sentence, then $\forall x \ c(x) \neq 0$. Similarly, assume the value
1 appears only once. Then we have to fill the sentence with "1 is 1". This is a contradiction, as there are two occurrences of 1. Hence, 1 appears more than once: $c(1) \neq 1$.

continued

Moreover, assume that a value appears four times. It means that four positions are occupied by that value. It also means that the digits where that value occurs should also appear several times. That is, there are not enough positions for values larger than 3. Hence, we restrict the search to the three values 1, 2, or 3:

$$\forall x \, ((x > 1 \land x < 10) \to (c(x) = 1 \lor c(x) = 2 \lor c(x) = 3)) \tag{13.21}$$

Note that this reasoning line does not apply to the value 1, for which the number of occurrences can be larger.

We define the functions "less" (i.e. $l(x, y)$), "greater" (i.e. $g(x, y)$), and "equal" (i.e. $e(x, y)$) in lines 3–9 (Listing 13.14), since we need their output in computations:

$$
\begin{aligned}
(x > 1 \land x < 10) &\to c(x) = 1 \\
+e(l(c(0), x), g(c(0), x)) &+ e(l(c(1), x), g(c(1), x)) \\
+e(l(c(2), x), g(c(2), x)) &+ e(l(c(3), 2), g(c(3), x)) \\
+e(l(c(4), x), g(c(4), x)) &+ e(l(c(5), 2), g(c(5), x)) \\
+e(l(c(6), x), g(c(6), x)) &+ e(l(c(7), 2), g(c(7), x)) \\
+e(l(c(8), x), g(c(8), x)) &+ e(l(c(9), 2), g(c(9), x))
\end{aligned}
\tag{13.22}
$$

For the value 1, there is a supplementary condition: in number 11, there are two occurrences (lines 24–33). Note that the digit 0 can appear also in number 10 (lines 18–23). Mace4 finds two solutions: $\langle 1, 11, 2, 1, 1, 1, 1, 1, 1, 1 \rangle$ and $\langle 1, 7, 3, 2, 1, 1, 1, 2, 1, 1 \rangle$, corresponding to the sentences:

1. *In this sentence, the number of occurrences of 0 is 1, of 1 is 11, of 2 is 2, of 3 is 1, of 4 is 1, of 5 is 1, of 6 is 1, of 7 is 1, of 8 is 1, and of 9 is 1.*
2. *In this sentence, the number of occurrences of 0 is 1, of 1 is 7, of 2 is 3, of 3 is 2, of 4 is 1, of 5 is 1, of 6 is 1, of 7 is 2, of 8 is 1, and of 9 is 1.*

It would be interesting to use Mace4 to signal if there are (or aren't) solutions to more general self-counting sentences (e.g. up to values 12).

References

Dudeney, H. E. (2016). 536 Puzzles and curious problems. Courier Dover Publications.

Kordemsky, B. A. (1992). *The Moscow puzzles: 359 mathematical recreations*. Courier Corporation.

Smullyan, R. M. (2011). What is the name of this book? The riddle of Dracula and other logical puzzles. Dover Publications.

Walicki, M. (2016). *Introduction to Mathematical Logic: Extended Edition*. World Scientific Publishing Company.

Appendix A
The Epigraph Puzzle

$dubito(I) \rightarrow cogito(I), cogito(I) \rightarrow sum(I)$
Dubito, ergo cogito, ergo sum
I doubt, therefore I think, therefore I am (René Descartes)

$\forall x\ refuses(x, do(arithmetic)) \rightarrow doomed(x, talk(nonsense))$
He who refuses to do arithmetic is doomed to talk nonsense (John McCarthy)

$\exists x\ \exists y\ (x \neq y \wedge (\forall z\ (z = x \vee z = y)) \wedge Master(x) \wedge Apprentice(y))$
Always two there are: a Master and an Apprentice (Yoda)

$\forall x\ \forall y\ can(x, do(y)) \rightarrow does(x, y)$
$\forall x\ \forall y\ \neg can(x, do(y)) \rightarrow teaches(x, y)$
He who can, does. He who cannot, teaches. (George Bernard Shaw)

$\exists x\ (interrupt(I, x) \wedge (\forall y\ \neg interrupt(y, I)))$
Don't interrupt me, while I'm interrupting (Winston S. Churchill)

$\exists x\ \exists y\ ((x \neq y \wedge (\forall z\ wayToLive(z) \leftrightarrow (z = x \vee z = y))) \wedge$
$(wayToLive(x) \rightarrow (\neg \exists u\ miracle(u))) \wedge$
$(wayToLive(y) \rightarrow (\forall u\ miracle(u))$
There are only two ways to live your life. One is as though nothing is a miracle. The
other is as though everything is a miracle (Albert Einstein)

$(\forall x\quad learns(x) \wedge \neg thinks(x) \rightarrow lost(x)) \wedge (\forall x\quad thinks(x) \wedge \neg learns(x) \rightarrow$
$greatDanger(x))$
He who learns but does not think, is lost. He who thinks but does not learn is in great
danger (Confucius)

© The Editor(s) (if applicable) and The Author(s), under exclusive license
to Springer Nature Switzerland AG 2021
A. Groza, *Modelling Puzzles in First Order Logic*,
https://doi.org/10.1007/978-3-030-62547-4

$\forall x \ (man(x) \wedge wife(x) = c \rightarrow (\forall z \ \neg attractive(x, z) \leftrightarrow z \neq c))$
No married man is ever attractive except to his wife (Oscar Wilde)

$\forall x \ (place(x) \wedge worth(x) \rightarrow \neg \exists y \ shortcut(y, x)$
There are no shortcuts to any place worth going (Beverly Sills)

$Mind(undisciplined) \wedge \forall x \ (Mind(x) \wedge x \neq undisciplined \rightarrow more Disobe -$
$dient(undisciplined, x))$
$Mind(disciplined) \wedge \forall x \quad\quad (Mind(x) \wedge x \neq disciplined \rightarrow more Obedient$
$(disciplined, x))$
$more Obedient(x, y) \leftrightarrow more Disobedient(y, x).$
There is nothing so disobedient as an undisciplined mind, and there is nothing so
obedient as a disciplined mind (Gautama Buddha)

$\forall x \ \forall y \ (old(x) \rightarrow belive(x, y))$
$\forall x \ \forall y \ (middleAged(x) \rightarrow suspect(x, y))$
$\forall x \ \forall y \ (young(x) \rightarrow know(x, y))$
The old believe everything, the middle-aged suspect everything, and the young know
everything (Oscar Wilde)

$\forall x \ \forall y \ ((link(x) \wedge chain(y) \wedge contains(y, x) \wedge weakest(x) \rightarrow canBreak(x, y))$
$\rightarrow strongest(x))$
The weakest link in a chain is the strongest because it can break it (Stanislaw J. Lec)

$\forall x \ \forall y \ Man(x) \wedge has(x, h) \wedge hammer(h) \wedge problem(y) \rightarrow sees(x, y, nail)$
$great(hammer, our Age) = algorithm$
The man with a hammer sees every problem as a nail. Our age's great hammer is the
algorithm (William Poundstone)

Printed in the United States
by Baker & Taylor Publisher Services